FIGHTING AT SEA:
NAVAL BATTLES FROM THE AGES
OF SAIL AND STEAM

Two single ships fighting to the finish occurred often enough in the age of sail. The deadly nature of these actions is displayed in this print published in 1846 by Sarony and Major, which shows the devastated HMS *Java* lying helpless and defeated under the guns of USS *Constitution*. Isolated combats such as this seldom affected the balance of naval power, but in matters of prestige and honour the results remained very important. (Library of Congress)

FIGHTING AT SEA

Naval Battles from the Ages of Sail and Steam

....

Edited by Douglas M. McLean

Contributors
Donald E. Graves
William S. Dudley
Andrew Lambert
Douglas M. McLean
Michael Whitby
Malcolm Llewellyn-Jones

Maps by Christopher Johnson

ROBIN BRASS STUDIO

Published 2008 by Robin Brass Studio Inc.
www.rbstudiobooks.com

ISBN-13: 978-1-896941-56-1
ISBN-10: 1-896941-56-7

Printed and bound in Canada by Marquis Imprimeur Inc., Cap-Saint-Ignace, Quebec

Library and Archives Canada Cataloguing in Publication

    Fighting at sea : naval battles from the ages of sail and steam / edited by Douglas
M. McLean ; contributors: Donald E. Graves ... [et al.] ; maps by Christopher
Johnson.

Includes bibliographical references and index.

ISBN 978-1-896941-56-1

    1. Naval battles – History. 2. Naval history. I. McLean, D. M. (Douglas
Malcolm), 1956– . II. Graves, Donald E. (Donald Edward), 1949– . III. Johnson,
Christopher

D27.F53 2008   359.4'09   C2008-905598-5

Contents

Introduction

"A willing foe and sea room"
TRADITIONAL NAVAL TOAST OF THE DAY

SHIPS OF WAR FILL MANY ROLES; ultimately, however, warships are built to
fight. The study of what makes a warship fight effectively at sea is of significant interest. History provides a lens to examine aspects that contribute to,
or alternatively undermine, fighting effectiveness. This book provides six historical situations in which the actions of warships and their crews, faced with
formidable challenges, can be examined. The book spans a period of almost two
centuries, and advances in technology over this period ensure that there will be
differences between events set in the age of sail and those in the age of steam. Yet
many aspects remain important across the entire period, which allows enduring
insights to be gained.[1]

The human element is perhaps the most enduring constant in naval warfare.
Ships are often characterized as having personalities, but the most important consideration in any warship's effectiveness is her captain and crew. The critical importance of command ensures that the captain's character is a key component of
fighting effectiveness in any ship. Each of the six studies provides insights into
the challenge of commanding fighting ships. In the engagements commanders
on both sides are seen to be skilled professionals. In some cases they meet their
match in an even more skilled opponent. In other cases the impact of other vari-
ables, such as inferior technology, numbers or intelligence, contribute to defeat.
Occasionally events that can only be characterized as chance help determine the
outcome.

Chance is an inherent part of combat and is another of the enduring con-
stants in fighting at sea. Signals missed or misinterpreted, commands misheard,
unexpected changes in weather, all these and more happen in battle. Whether
labelled chaos or friction, unforeseen events are common to the point of pre-
dictability in combat. Maintaining effectiveness in spite of the unpredictable is
often the difference between success and failure. There is no certain recipe for

overcoming the unexpected, but well trained crews commanded by skilful and adaptable commanders generally have a greater chance of success.

An aspect of fighting at sea that is common if not necessarily enduring is that small actions often have significant consequences. The loss of even a single ship can have far-reaching impacts. The ease with which small actions can have strategic impacts makes using the tactical, operational and strategic level classifications common in land warfare more difficult to employ at sea. Naval activity is almost inherently operational in nature. The actual act of fighting may be labelled tactical, but the results of combat are so frequently strategic that it is difficult to consider just tactical aspects. The ease with which naval forces can move across great distances is certainly one of the reasons for this.

A further enduring aspect of fighting at sea is that the ultimate end of the battle is not ownership of the sea, but control so as to enable its use for friendly purposes, or prevent its use by an enemy. In short the sea is but a means to an end, and allowing or preventing its use is the usual reason for battle. The sea provides easy transportation for trade or troops. Battles between warships are often about allowing or preventing this transportation.

There are also a number of aspects of fighting at sea that have changed over the years. Technology is the biggest factor and is responsible for a number of attendant changes. The shift from sail to steam liberated ships from the wind, though even today weather remains a critically important factor in fighting at sea. The changes in weaponry associated with new technology have expanded the scope of naval warfare not only in range, as effective gunnery range increased from point-blank to the horizon, but also in spatial dimensions as weapons were developed to attack underwater and airborne opponents. New technology facilitated platforms that operated in all three dimensions, and aircraft and submarines have tremendously complicated the art of fighting at sea. Visual signals – lanterns and flags – were the latest in high technology in the age of sail, but the advent of radios introduced new ways of coordinating ships at sea, as well as being an attractive means of gaining intelligence on an opponent.

The increasing sophistication of fighting at sea, much of it resulting from accelerating technology, also promoted the development of new shore-based organizations designed to enhance fighting efficiency. Navies have always been one of the most complex of state instruments. In the age of sail warships were not only the most technologically complicated machines of their day, but they also represented an enormous investment in facilities such as dockyards, arsenals and foundries. Sending a ship to sea required teams of skilled shipwrights and sailmakers, as well as a vast organization to find the men and stores for the ship. No nation could expect its naval ships to fight effectively without an efficient shore organization. However, once ships or a fleet was dispatched, there were significant limits on the interaction between the home authorities and the deployed ship. This is not to say that home authorities had no role, but once a ship slipped her moorings, she was autonomous for long periods. Coordinating intelligence ashore, such as the way the British Admiralty often acted as the central clearing house for information in the age of sail, did occur, but the long delays in getting word back to England and then once again out to deployed ships meant that independent efforts and judgment were required on the part of commanders.

Radio communication changed the situation in many ways, although the judgment of those at the sharp end remained of paramount importance. The capability to rapidly move information from sea to shore and back again, however, provided significant opportunities for doing things differently. Ships and fleets could be coordinated from ashore, and determining how much command and coordination should be exercised from ashore became an important factor in naval strategy. Different navies developed different answers to this question, and variations depending on the operational mission were also apparent. The *Kriegsmarine* used radios to centrally control the search for convoys by wolfpacks, but actual attacks were left much more to the independent decisions of U-boat commanders. The growing volumes of radio traffic offered intelligence opportunities for the enemy and presented a security challenge for the originator.

In some navies shore organizations grew as the need for command, intelligence and security ashore, as well as for more sophisticated training, maintenance of increasingly complex weaponry and research that would allow improvements in all of these facets became more apparent. In other navies staffs were sometimes reduced in size in an effort to avoid the compromise of operational information. The resultant strain on the small number of staffs proved significant indeed, and this approach to improving security in fact provided no benefit in that regard. Different approaches taken to cope with the growing complexity provides insight into what might work and what might not in the future.

The relationship between naval shore establishments and ships at sea therefore changed in significant ways as technology improved. The challenge of finding an optimal arrangement for how shore organizations should interact with forces at sea became more acute as complexity increased. Commanders at sea remained responsible for decisions in the face of the enemy, but how best to prepare or position forces for finding and defeating the enemy required coordination. Developing effective coordination between organizations that often grew rapidly in the face of new threats or opportunities proved very challenging historically, and remain a significant challenge today.

One of the mechanisms that successful navies have always used to improve the fighting effectiveness of their warships has been doctrine. The concept of doctrine is comparatively simple, but precisely describing it can be endlessly complex. Today NATO uses a definition for doctrine that might be described as commonly accepted:

> Fundamental principles by which military forces guide their actions in support of objectives. It is authoritative but requires judgment in application.[2]

This definition is useful, but determining what doctrine actually is can be very complex. Modern doctrine is generally written down explicitly, as suggested in the definition above, but this has not always been the case in the past. In the age of sail one of the most successful fighting admirals in history, Horatio Nelson, developed his ideas regarding naval combat into what is perhaps best described as a set of 'shared assumptions' amongst his captains, a group he referred to as his 'band of brothers'. The impact of shared assumptions, be they culturally derived or the result of a strong leader imparting his views can have an effect so similar to doctrine that it is difficult to distinguish the difference. The reference to judgment in the modern definition also suggests another complication: doctrine does not require blind obedience in all situations. The combined impact of doctrine and shared assumptions can be quite significant, but also difficult to determine with precision, as identifying what actions resulted from them or from individual initiative is not always possible.

Doctrine and shared assumption are important because effective navies are generally ones that have a tradition of fighting well. Tradition matters, and part of the reason it matters is that it contributes to the determination and resolve that the commanders and crews will exhibit in action. They expect that they will succeed, and believe that their nation expects them to succeed, and this mind-set influences their actions. However, doctrine and tradition are not always enough to secure victory. In the battles examined here most combatants demonstrate both good skills and determination. Where doctrine and shared assumptions can be clearly traced, they can be seen to have some influence, but not usually enough to entirely account for the final result. In some instances doctrine can be more of an obstacle than an aid, as the book answer does not always work in the cold light of reality. Something more than a good understanding of the tools of war and how to use them according to the book is usually needed.

Training is an important aspect of warfare, and one that transcends the periods studied here. In that sense training is a constant, but it has also evolved as time passed. In the age of sail training often varied from ship to ship, generally driven by the will and intention of the captain involved, or at most, the admiral of a fleet. Twentieth century warfare saw the institution of massive training establishments that honed not only individual skills, but those of entire ships and groups of ships. All major navies succeeded in developing effective training establishments, but circumstances often made it difficult or impossible to train for all eventualities, or even to train as regularly as doctrine prescribed.

Commanders on the scene often had to prioritize what would be the focus of training, and even what aspect would be practised the most. For example, in gunnery the focus could be on speed, or on accuracy. Ideally both are best, but if time is short choices have to be made. When different types of weapons or ammunition are available, then choices may have to be made as to what deserves priority. In the age of sail there was roundshot as well as a variety of munitions designed to ruin rigging or incapacitate enemy crews. The basics of using different types of ammunition were similar, but how the guns were laid differed somewhat. The focus in gunnery practice presaged the type of action anticipated or most likely to occur.

By the Second World War there were completely different classes of weapons to use against opposing warships. The best method of manoeuvring and firing torpedoes differed substantially from that required for gunnery, and therefore choices had to be made as to where the training focus would be. Commanders had to make this choice with a host of considerations in mind, and differences of views could make these debates difficult. As warfare grew in complexity the issue of training for new weapons also became important. Training with new weapons should be developed and instituted before, or at least at the same time as, the weapon is deployed operationally, but this ideal was seldom achieved in the stress of wartime. All too often men needed to learn how to use new weapons with only modest exposure to training on them. As a result the initial effectiveness of new weapons seldom achieved the level that theory suggested.

In many ways, the most important quality or aspect that contributed to success or failure might be described as adaptability. The ability of the commander of a warship or group of warships to adjust his plans effectively in the face of unexpected realities – and realities are seldom exactly as expected in combat – has often proved critical to success. In some instances this adaptability can be made easier by plans that anticipate the likely nature of an opponents actions, while in other circumstances it is demonstrated by a more rapid adjustment to new initiatives on the part of an enemy than that enemy can counter.

With the growing complexity of naval warfare, shore organizations can contribute to adaptability. The design and evolution of shore staffs proved an interesting aspect of naval warfare in the world wars. Traditionally naval staffs were quite small, but as new weapons, sensors and tactics were employed, the need for more experts grew. The tension between traditional naval culture and increasing complexity provides a fertile field for study, whether it be the effort to develop a General Staff equivalent in the Admiralty immediately before the Great War or the different approaches to harnessing scientific methods in various navies during the Second World War.

The most effective staffs were clearly those associated with the Royal Navy and United States Navy's anti-submarine warfare efforts. Perhaps as a result of the magnitude and urgency of the threat, barriers to adopting new techniques proved more easily overcome in this area of warfare than others. New staff functions were put in place specializing not only in traditional areas such as tracking enemy initiatives (intelligence) but in the application of the scientific method to study the effectiveness of tactical methods (operations research) and the development of entirely new tactics (Western Approaches Tactical Unit). These new staff arms provided an important advantage in closely matched situations. In the Battle of the Atlantic, for example, German U-boats were crewed by generally well-trained and motivated sailors, known for their skill and courage in battle. Combatting such difficult opponents took more than courage and skill on the part of escort crews, and the advantage the Allies gained through their superior shore staffs in this contest proved important to final victory in that prolonged campaign.

The wide range of factors involved in determining success or failure in modern naval battles makes it difficult or impossible to identify single aspects as decisive on their own. Even adaptability is inadequate, as in some battles it seems clear that both sides fought innovatively as well as courageously. The many considerations involved in each specific event make broad generalizations dangerous, for as always the devil is in the details. The unique aspects of each incident presented in the chapters of this book ensure that they are all worthy of individual study, even though there are many parallels in the types of situations examined.

A focus on combat and the small number of incidents ensures this. The opening chapter studies the challenges of propelling a landing force ashore. The next two chapters revolve around commerce raiding and the efforts required to defeat or neutralize such attacks in the age of sail. Chapter 4 examines how the Allies countered the coordinated commerce attacks made on convoys by groups of U-boats. The penultimate study reviews a deliberate Allied effort to neutralize the threat from German surface forces near the Normandy coast shortly after D-Day. The final chapter analyzes the remarkable efforts required by the Allies to locate and destroy a single U-boat in the complex inshore operations of the last year of the war, to prevent that boat attacking shipping.

Stripped of the specifics of their historical periods, the texts show the timelessness of naval warfare: the first chapter seeks to examine how control of the sea can be used to advantage by putting forces ashore, while the subsequent chapters revolve around attaining local sea control, either to allow attacks on commerce or prevent them, or to allow the use of the sea for projecting land forces across a shore or to prevent such operations.

The impact and analysis of military intelligence is a common theme in both the age of sail and of steam. However, the time delays involved in intelligence in the first chapters are measured in days, weeks or months, while in the later chapters the delays are usually shorter, ranging from weeks to days to hours and sometimes even minutes. The increasing pace of intelligence operations placed a greater premium on interpretation, and therefore heightened the importance of organizations devoted to assessing and disseminating timely intelligence.

Adapting to technological change is a process that can be found in both the early and late period, but again the pace of change is higher in the later chapters. The human dimension is common to all chapters, but the diversity of people ensures that differences in this aspect will abound. In short, for all the similarities, the importance of studying each chapter for its unique qualities remains high. Since each chapter also tells an interesting story in its own right, it is now time to turn to them directly.

Douglas M. McLean
Chilliwack, British Columbia, 2008

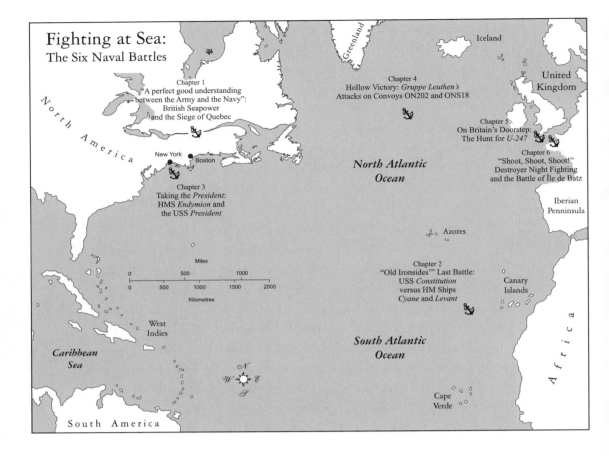

Fighting at Sea:
The Six Naval Battles

Greenland

Iceland

United Kingdom

North America

Chapter 1
"A perfect good understanding
between the Army and the Navy":
British Seapower
and the Siege of Quebec

Chapter 4
Hollow Victory: *Gruppe Leuthen's*
Attacks on Convoys ON202 and ONS18

Chapter 5
On Britain's Doorstep:
The Hunt for *U-247*

New York   Boston

Chapter 6
"Shoot, Shoot, Shoot!"
Destroyer Night Fighting
and the Battle of Île de Batz

North Atlantic
Ocean

Chapter 3
Taking the *President*:
HMS *Endymion* and
the USS *President*

Iberian
Penninsula

Azores

Miles

0   500   1000

0   500   1000   1500   2000

Kilometres

Chapter 2
"Old Ironsides'" Last Battle:
USS *Constitution*
versus HM Ships
Cyane and *Levant*

Canary
Islands

West
Indies

South Atlantic
Ocean

Africa

Caribbean
Sea

N
W   E
S

Cape
Verde

South America

PRELUDE

The Royal Navy
at Quebec, 1759

THE FIRST THREE CHAPTERS in this book all examine naval events in the age
of sail. The first chapter, by Donald E. Graves, considers one of the most
complex, and successful, amphibious operations of the 18th century. The two
subsequent chapters are more traditional naval battle studies, focused on engagements between small groups of warships in the early part of the 19th century.
The different focus of these studies provides a good example of the complexity
of considerations involved in naval warfare, as navies need to be able either to
accomplish both kinds of mission, or at the very least prevent their accomplish-
ment, to be considered effective.

The Seven Years' War was one of many 18th-century conflicts between France
and England. The label for this war only properly applies to Europe, where hos-
tilities went from 1756 to 1763. Better known as the French and Indian War in
North America, fighting there started in 1754, when a certain Lieutenant-Colonel
George Washington engaged the French in the Ohio Valley.[1] Fighting also ended
early, when New France was decisively defeated in 1760.

The North American part of this global war clearly resulted in dramatic
changes for both the French and British Empires. The British decision to conduct a holding action in Europe while focusing enormous resources overseas contributed to one of the most successful wars in British history. The Royal Navy
played a pivotal role in this effort. Strategies and tactics that had been developing
throughout the earlier wars of the 18th century came to fruition, resulting in an
impressive string of naval and amphibious victories. The opening battles had
hardly been glorious, with an inconclusive action in the Mediterranean resulting
in the loss of Minorca by the British and the subsequent court-martial and execution of the Royal Navy's Admiral Byng – *pour encourager les autres*, as Voltaire
wryly described it. Despite this inglorious start, rapid expansion of the Royal
Navy and initiation of a much more effective blockade of French ports gradually
reduced the French navy to little more than a spectator of the war. While the
new art of supporting blockading forces for prolonged periods was perfected by

British squadrons, other strong squadrons were dispatched overseas to support sizeable contingents of British soldiers. As the blockades became increasingly effective, French colonies became increasingly isolated. Nowhere did this French isolation and focus of British strength on a target have more conclusive a result than in North America. France succeeded in reinforcing their colony of New France for several years, but ultimately British naval strength proved too much and in 1760 no succour arrived to stave off final defeat.

Supporting roles are often less glamorous than the star turns taken by those that triumphantly succeed in battle. The first study demonstrates this truth while highlighting the key importance of the supporting actions of the Royal Navy. The Battle of the Plains of Abraham, fought on land in front of the walls of Quebec City on 13 September 1759 is generally remembered as a critical victory in the British conquest. Less well remembered is the pivotal role played by the Royal Navy. Yet the battle itself was perhaps one of the less risky parts of the ambitious British plan, which required an immense British fleet to travel far up one of the great tidal rivers of the world, and then land a force in the face of well prepared opponents. This plan had many possible points of failure, from the navigational hazards in the St Lawrence through the absolute necessity of close cooperation between the land and naval commanders to the hazardous night landing in fast running waters.

Today the name of General James Wolfe remains reasonably well remembered, yet few recall the leader of British naval forces at Quebec. Vice-Admiral Charles Saunders proved both skilful and cooperative, ably ensuring his fleet's arrival before Quebec and then providing flexible and timely support throughout the long summer of 1759. The drama of Wolfe's death at the moment of victory certainly contributed to his enshrinement as hero, but Saunders's contributions were at least as important as Wolfe's in achieving success, and arguably were more vital.

Many factors contributed to British success, not least the experience gained in a number of costly amphibious failures in the first half of the 18th century. The many wars of those years assisted in the development of an effective British doctrine for combined or, as they were known at the time, conjunct operations. This doctrine was formally written down only in 1759, and so practical experience and common knowledge appear more important, and this experience was indeed common in the British forces dispatched to North America.

There were also tangible expressions of this experience, in the form of specialized landing craft developed and specifically included in the expedition. These craft clearly represented the state of the art for the period and played an important part in the operations leading to the climactic land battle. As in perhaps every period, there were never enough landing craft available to make any plan a certainty – two centuries later Churchill would note plaintively about a more modern, specialized landing craft, the LST, that "the plans of two great empires … should be so much hamstrung and limited by a hundred or so of these particular vessels…."[2] Nevertheless a great many of the new oar-driven craft were present at Quebec, and their characteristics – as well as the skill of their crews - materially contributed to British success.

The success achieved by the British army once successfully landed adjacent to Quebec was dramatic, but not sufficient of itself. In the spring of 1760 resurgent French forces looked poised to overturn the verdict of the previous year's battle, and recapture the city of Quebec. It was at this point that the Royal Navy's influence proved truly decisive, when the first vessels to arrive that spring were British, not French. The many improvements in naval effectiveness introduced by the Royal Navy in this war allowed it to decisively defeat the French navy by late 1759, and it was the resulting high level of British sea control that sealed the fate of New France.

"A perfect good understanding between the Army and the Navy"

BRITISH SEAPOWER AND THE SIEGE OF QUEBEC

Donald E. Graves

Come cheer up, my lads, 'tis to glory we steer,
To add something more to this wonderful year,
To honour we call you, not press you like slaves,
For who are so free as the sons of the waves?

Heart of oak are our ships,
Jolly tars are our men,
We always are ready,
Steady, boys, steady,
We'll fight and we'll conquer, again and again.[1]

FOR NEARLY TWO AND A HALF CENTURIES the fall of Quebec in 1759 has fascinated historians. This is not surprising because of the high drama in the event and the glorious death in battle of the two opposing army commanders. While much ink has been spilled describing land operations during the famous siege – and particularly concerning British Major-General James Wolfe – much less attention has been devoted to the naval side of the operation.[2] Most historians of the siege pay lip service to the contribution made by Vice-Admiral Charles Saunders and his fleet to the successful outcome of the operation but few discuss the Royal Navy's role in any great detail. The following examination of the navy's contribution to the siege of Quebec is intended to correct this oversight.

Quebec was a masterpiece of amphibious operations, which have been defined by one knowledgeable observer as "the ability to project force from the sea onto land, into a hostile or potentially hostile environment, in a tactical posture, without any reliance on ports."[3] Amphibious landings, be they in the 18th or the 21st century, are among the most difficult of all naval operations to mount, but in June 1759 when the crews of Vice-Admiral Charles Saunders's ships saw the town of Quebec for the first time, the Royal Navy had a workable doctrine for amphibious warfare. It included special command procedures; combat loading of transports; fleet organization at sea; landing place reconnaissance; specialized landing craft; signals and communications procedures; the formation and control of landing craft; the use of separate assault echelons; naval gunfire support; and continuing long term logistical support of an army once it was ashore.

This doctrine did not derive from theory – the first British text on amphibious warfare was only published in 1759 – but from recent operational experience in the War of the Austrian Succession (1739-48).[4] Before that conflict, Britain did not emphasize large overseas expeditions in her strategic planning and when she attempted them, they often failed. The attacks made by the belligerents on each other's colonial possessions during the Nine Years' War of 1688-97 were little more than raids amounting to what one historian has termed a "medieval process of cross-ravaging."[5] During the War of the Spanish Succession in 1702-13 Britain largely favoured continental, as opposed to colonial commitment, and the one large-scale amphibious operation mounted – Rear-Admiral Hovenden Walker's attack on Quebec in 1711 – was a conspicuous failure.

The development of amphibious doctrine was accelerated by the operational demands placed on the Royal Navy during the War of the Austrian Succession. British strategy focused on the Caribbean because it was an area that offered "chances both of large-scale ship actions and assaults on territory."[6] Memories of the exploits of the 17th-century freebooters and buccaneers fuelled the government's expectation of maximum gain with minimum risk, and this was reinforced by Vice-Admiral Edward Vernon's successful raid on Porto Bello, undertaken at the outbreak of the conflict in 1739. In 1741, therefore, the government dispatched a large expeditionary force to the West Indies but conditions in the Caribbean had changed since the days of the freebooters as Spanish overseas policy now emphasized the fortified defence of major ports. Attacks on such places demanded elaborate preparations, large assault forces and lengthy operations, all of which were vulnerable to the vagaries of climate, disease and weather. The failure of the British attacks at Chagre and St. Augustine in 1740 presaged many of the problems experienced by the attack on Cartagena in 1741, the largest such operation mounted by Britain until that time.[7]

The siege of Cartagena, from March until May 1741, was a textbook example in the problems of joint command in amphibious operations. It was marred and ultimately doomed by the acrimonious relations between Vice-Admiral Edward Vernon, the naval commander, and General Thomas Wentworth, the army commander. In the words of a period commentator:

> The Admiral and General had contracted a hearty Contempt for each other, and took all Opportunities of expressing their mutual Dislike; far from acting vigorously in concert, for the Advantage of the Community, they maintained a mutual Reserve, and separate Cabals, each proved more eager for the disgrace of his Rival, than zealous for the Honour of the Nation.[8]

Cartagena was also notable because a large number of junior naval officers who would play major roles in similar expeditions in the future were present (among them Captain Edward Boscawen and Lieutenant Philip Durell) but it demonstrated that, in the area of amphibious operations, the Royal Navy was low on its learning curve. It was clear from the failure at Cartagena that *ad hoc* raiding expeditions in the spirit of Morgan and Drake were no longer enough; a successful attack on the colonial possessions of European powers needed lengthy preparation and considerable military and naval forces.[9]

For the Royal Navy, these operations provided valuable lessons. **First and foremost** was the question of joint command – never again would relations between the two services deteriorate to the level reached by Vernon and Wentworth. At La Guira and Porto Caballo in 1742, Commodore Charles Knowles shared command but was senior to his army counterpart. As a rear admiral, Knowles exercised supreme command of the attack on Port Louis in 1748 as did Boscawen in the expedition against Pondicherry in the same year. At Louisbourg in 1745, Commodore Peter Warren and General William Pepperell exercised joint command and, although there was some friction, they achieved a good level of interservice cooperation. At Port L'Orient in 1746, however, the operation came under the command of the senior military officer, Lieutenant-General James St. Clair, perhaps an unwise choice as it failed miserably. Of the ten major amphibious expeditions mounted during the war of the Austrian Succession, only one (Louisbourg, 1745) was an overwhelming success but, as the Duke of Wellington would later remark about his own experience in a bungled military campaign, he had "learned what one ought not do, and that is always something."[10] Another often overlooked result of the conflict was that the Royal Marines were established on a permanent basis and placed under the jurisdiction of the Admiralty, giving the navy, in embryonic form, a force of soldiers potentially useful for amphibious operations.[11]

In the mid-1750s British-French rivalry in North America was a major cause of the Seven Years' War, which had its formal beginning in January 1756 and ultimately involved every major European power. Prime Minister William Pitt, who came to office late that year, instituted a strategy of using Britain's naval strength to attack its European enemies overseas rather than on the continent. The success of this strategy depended, of course, on the navy's ability to carry out amphibious operations, and that service got off to a shaky start in such matters when its first major combined operation, an attack on the French naval base at Rochefort in September 1757, was a total failure.

Under the joint command of Vice-Admiral Edward Hawke and Major-General John Mordaunt, the Rochefort operation went wrong from the outset. Based on inaccurate intelligence that the defences of Rochefort were too weak to withstand a determined attack, it was also marred by indecision among the commanders and adverse winds and tides. After dithering about for some time, the assault force returned to England having accomplished almost nothing and both Hawke and Mordaunt were court-martialed for their actions but acquitted. It is interesting to note that Hawke, who has been criticized for his indecision and lack of aggression at Rochefort, was one of the few senior officers in the navy without any previous experience in amphibious operations.[12]

Rochefort, however, did have some positive results. It provided useful experience to a number of officers who were to play prominent roles in subsequent amphibious operations and who "at least, learned what one ought not to do." Among them was Lieutenant-Colonel James Wolfe, who had commanded an infantry battalion during the expedition, and who summarized the lessons he had learned:

> an Admiral should endeavour to run into an enemy's port immediately after he appears before it; that he should anchor the transport ships and frigates as close as he can to the land; that he should reconnoitre and observe it as quick as possible, and lose no time in getting the troops on shore; that previous directions should be given in respect to landing the troops, and a proper disposition made for boats of all sorts, appointing leaders and fit persons for conducting the different divisions.[13]

In a nutshell, these were the basics of amphibious assault technique as it was known in Britain by the mid-18th century.

Rochefort also gave rise to an important technical development — the construction of specialized landing craft — because the failure "was in some measure attributed to the want of proper boats to land a sufficient number of troops at once."[14] The "Sluggishness, Awkardness, and different sizes of the Transport-boats, were so apparent" that it was clear improvements would have to be made and they were, in a commendably short period of time. In early April 1758, less than six months after the abortive attack, the Admiralty approved a design for a new flat-bottomed boat to be used in landing operations and by the following month they were being constructed at Portsmouth.[15] This was the first specialized landing craft in British service and it was not without its faults:

> It differed only in these respects from the common Boats of the Fleet; it was constructed go in shallower Water and being all of size, they contained the like numbers. Each had two Sails, and was full of Benches; one (if not two) was made along the whole length of the Center of the Boat, with little ones branching to the right and left, like so many ribs, with little Benches also round the edge. There were ten Rowers on each side. Between every Rower and the edge of the Boat, sat a Musketeer to defend him; by which Method each was deprived of the Liberty necessary in his Occupation, that a few Soldiers on the sides might be in a position to fire very bad, the Rowers were obliged only to paddle. The Contrivance of this piece of Mechanism seemed, as if one main aim had been, to render it as difficult as possible for the Soldiers, when they reached the shore, to get out of it: During which Performance, the Oars being tied with Cordage sloped down the outside of the Boat like the fins of a Fish; which was the ingenious part of the Construction. Each Boat when freighted to the utmost, contained 70 Soldiers, besides the 20 Rowers.[16]

Failure at Rochefort in 1757 was offset by victory at Louisbourg in 1758. In no small measure this success was due to the qualities and experience of both the military and naval commanders. Vice-Admiral Edward Boscawen had been present at Porto Bello and Cartagena and had commanded the unsuccessful operation against Pondicherry in 1748. One of his principal subordinates, Commodore Philip Durell, was a veteran of not only the ill-fated Caribbean operations of the previous war but also the first siege of Louisbourg in 1745. The army component, under the overall command of Major-General Jeffrey Amherst, benefited from the organizational talents of the newly-promoted Brigadier General James Wolfe, the observant veteran of Rochefort.

Amherst insisted on properly preparing his troops for the operation. In the month prior to the landing, after the fleet and the landing force had rendez-voused at Halifax, his four subordinate brigadier generals did not fail to accustom the Troops to what they were soon to encounter. Some Military Operations were dayly carried on. They frequently landed in the boats of the Transports and practised in the woods, the different Manuvres …… In all these operations you may imagine that Gen. Wolfe was remarkably active. The Scene afforded Scope for his Military Genius. We found it possible to land 3500 men in the Boats belonging to the Transports, and when the Boats from Men of War assisted, 5000 men could be landed.[17]

The hard work and attention to detail paid off. Although the French defenders of Louisbourg were on the alert, and weather and surf conditions were against the attackers, they landed in the face of determined resistance. Boscawen, who had witnessed at first hand the antagonism between Vernon and Wentworth at Cartagena, was resolved that the fleet would properly support the army. He sent naval shore parties to assist with the siege and ordered each one of his captains to take a daily turn supervising the unloading of stores.[18]

Another factor that contributed to the success at Louisbourg as well as future large amphibious operations was the great improvement in the Royal Navy's ability to supply fleets at long distances. Advances in naval victualling in the first half of the 18th century permitted the Admiralty to keep ships at sea for longer periods of time. As Boscawen commented in 1756: "This ship has now been at sea twelve weeks, which is longer than I ever knew any first-rate at sea" whereas, just a few decades earlier, our "cruisers would not keep the sea above a fortnight, till one or two of them were broken for it, now three months is but a common cruise."[19]

By the fall of Louisbourg in 1758, the Royal Navy had, through a process of trial and error, developed an amphibious capability to meet the strategic demands that Pitt's ministry placed on it. As no other European state made similar demands on its navy, no European nation could match Britain's capacity to project seapower onto land at great distances. As a case in point, the major French amphibious operational success of the Seven Years' War, the spring 1756 attack on Minorca, was mounted only 250 miles from the large French naval base at Toulon and does not compare favourably with the Louisbourg operation which saw a fleet of 39 warships and 127 merchant vessels, assembled on both sides of the Atlantic, transport, land and supply an army of 13,000 troops. The French military and naval forces at Minorca were only marginally supplied and their success derived more from the gross misconduct of the British naval commander, Admiral John Byng – who was court-martialled and executed for his faults – than it does from any demonstrable level of amphibious expertise.[20]

Pitt turns to Quebec

Louisbourg secured, Pitt now turned to Quebec, the capital of New France. Three previous British attempts been made to take Quebec, the most successful being the first in 1629 when the Kirke brothers, privateers in English service, captured the infant settlement. It was returned to France in 1632 and in 1690 a second attempt was mounted when Sir William Phips commanded a large expedition organized by the New England colonies, consisting of 32 ships and 2,000 Massachusetts militia, which found its way, more by good luck than good navigation, up the St. Lawrence in September 1690. This was far too late in the season to mount a serious siege and, baffled by a determined defence and devastated by sickness, Phips was forced to sail away. The third and most serious attempt came in the summer of 1711 when a fleet under the command of Rear-Admiral Hovenden Walker, consisting of 11 warships and 60 transports, carrying 7,500 troops (including five regiments from Marlborough's army), ascended the St. Lawrence. This operation ended in disaster, however, on the foggy night of 14 August 1711 when eight of Walker's transports ran aground on the north shore of the river and 900 men drowned. Even if Walker had reached Quebec, it would have still been too late in the year to commence siege operations and in any case Walker was short of provisions.[21]

This sorry record made it clear to the Pitt ministry that success at Quebec was dependant on a number of factors. First, an operation against this objective would have to be mounted as early in the season as possible as the St. Lawrence would become hazardous in late October or early November, trapping a fleet in the river. Second, it would require a force powerful enough to prevail against Quebec's fortifications, its strong natural defences and its garrison of regular

Vice Admiral Charles Saunders (1713-1775). A veteran naval officer with 32 years of service, the competent but taciturn Saunders was very much the senior partner in the Quebec expedition. Without his efforts and those of his fleet, Wolfe would not have taken the city but, in contrast to Wolfe, who has been immortalized, Saunders has been nearly forgotten. (Library and Archives Canada, C-69298)

Rear Admiral Phillip Durell (1707-1766), shown here as a captain in 1746, was Saunders's second-in-command during the siege. Durell was a veteran of the Louisbourg campaign of 1758. (Library and Archives Canada, C-117939)

troops. Third, the attacking forces would have to be well supplied and this would require the presence of large seaborne logistical element, or "fleet train" in modern parlance. Fourth, and most important, to be successful, an operation against Quebec would require good leadership, particularly good naval leadership.

To command the naval forces in the Quebec expedition, the government chose 44-year-old Vice-Admiral Charles Saunders, a veteran of 32 years of service. Saunders had circumnavigated the globe with Anson in the *Centurion* in 1740-44, had a good record as a frigate captain and had commanded the Mediterranean fleet in 1756-57. A protégé of Admiral George Anson, First Lord of the Admiralty, Saunders was a highly competent officer renowned – in a service that did not encourage idle chatter – for his taciturn manner. Chosen as second-in-command to Saunders was 52-year-old Rear-Admiral Philip Durell, a veteran of many years experience in North American waters including both sieges of Louisbourg. After the fall of that place in 1758, Durell had remained in Halifax with that part of Boscawen's fleet chosen to winter in Nova Scotia. The third senior naval officer was Rear-Admiral Charles Holmes, who also had 32 years

Rear Admiral Charles Holmes (1711-1761) was a veteran of several amphibious operations, including the Louisbourg campaign of 1758. Holmes was Saunders's third-in-command during the Quebec campaign. To Holmes fell the task of organizing the actual landing at the Anse au Foulon. (Author's collection)

Major General James Wolfe (1727-1759).
This portrait by Joseph Highbourne, the most
pleasing likeness of the man, shows him as a
21-year-old major. The operation against Quebec was Wolfe's first major independent command. A professional soldier with a reputation
as a tactician, he would be hampered throughout the summer of 1759 by ill health and tensions with his subordinate generals. (Library
and Archives Canada, C-3916)

of service, including three years in
North American waters where he had
commanded a squadron during the
latest siege of Louisbourg.[22]

These were veteran officers with
proven records but Pitt's choice of
the man to command the army was
somewhat of a gamble. It fell on 32-year-old James Wolfe, a substantive colonel
who was given a temporary promotion to major-general for the operation. Wolfe
had a superb record as a regimental commander and had performed well as a
brigade commander at Louisbourg but had never before held an important independent command. As his most recent biographer remarks, for the first time in
his career, Wolfe would be obliged "to assume responsibility for the more cerebral
business of strategy" by planning and commanding in a separate campaign.[23]
The young general seemed somewhat daunted by his appointment, remarking
to Lieutenant-General Jeffrey Amherst, the senior military commander in North
America, that "they have put this heavy task upon my Shoulders, and I find nothing encouraging in the undertaking."[24] Wolfe could take satisfaction, however, in
the fact that he was given a small but superb army consisting of about 9,000 well
trained troops.

Pitt's plan envisaged a two-pronged attack. Saunders and Wolfe would sail
up the St. Lawrence and besiege Quebec while Amherst would move on Montreal by land from Lake Champlain. The 1759 Quebec expedition was the largest
combined operation yet mounted by Britain during the current conflict and the
largest since the disastrous attack on Cartagena in 1741. With that unfortunate
experience perhaps in mind, the cabinet emphasized the need for interservice
co-operation between the army and navy. Wolfe was instructed that, "Whereas
the Success of this Expedition will very much depend upon an entire Good Un-

derstanding between our Land and Sea Officers," he was strictly required "to maintain and cultivate such a good Understanding and Agreement" while, for his part, Saunders was directed to "cultivate the same good Understanding and Agreement."[25]

In December 1758, the government began active preparations for the expedition. In all, 49 warships, comprising 22 ships of the line, 13 frigates, 5 sloops and 9 smaller vessels manned by 13,500 officers and seamen were placed under Saunders's command, as well as 2,100 marines – a force amounting to about one-fifth the Royal Navy's strength in terms of warships and personnel. Twenty thousand tons of commercial shipping were chartered in Britain and 6,000 more obtained in the American colonies – some 140 vessels.[26] It took time to gather this force, but on 14 February 1759 Holmes sailed from Portsmouth with part of the warships and transports while Saunders and Wolfe left three days later with the remainder. Their destination was Louisbourg, the staging point for the expedition where they would rendezvous with Durell's squadron.[27]

Durell had an important task to perform. The assembly of an expedition of this size could not be kept secret and there was some concern that the French would reinforce Quebec before Saunders's fleet arrived before it. For this reason, Pitt directed Durell that as soon as "the Navigation of the Gulph and River St. Lawrence shall be practicable," he was to "repair with the Squadron under your Command to the River St. Lawrence," where he was to station his ships "in such a manner as may most effectually prevent any Succours whatever passing up that River to Quebec: and you are to remain in the Station abovementioned, till you shall receive further orders from Admiral Saunders."[28] This was a wise precaution as in March 1759 France did dispatch a convoy of 17 merchantmen escorted by a number of frigates to the capital of New France.[29]

It was therefore with some surprise that Saunders and Wolfe – having spent eight hard days trying to enter Louisbourg's icebound harbour – arrived at Halifax on 30 April to find Durell and his squadron at anchor. Wolfe, who, it seems, had a personal dislike of Durell, expressed himself "astonished" to find, in view of Pitt's orders, that officer at Halifax but Saunders was not critical of his subordinate, reporting that "He waits only for a wind."[30] Durell sailed for the St. Lawrence on 5 May "to take such a station as will effectually cut off all succours" but Wolfe feared that "there is reason to think that some [French] store-ships have already got up" to Quebec.[31] In fact the French convoy had managed to ascend the river and its first ships arrived at Quebec on 10 May.[32]

Historians have generally been critical of Durell's failure to intercept the French ships although some have defended him.[33] It would seem, however, from the correspondence of both Durell and Saunders that the winter of 1758-1759 was unduly severe on the Atlantic coast and this adversely affected operations. In mid-March, Durell reported to the Admiralty from Halifax that the recent winter had

> proved the severest that has been known since the settling of the place – For these two Months past I have not heard from Louisbourg – many Vessels have attempted to go there, but have met with Ice eighteen or twenty Leagues from the Land, so were obliged to return, after having had some of their People froze to death, and others frost bitten to that degree, as to lose Legs and Arms.[34]

That this was no exaggeration is clear from Saunders's correspondence with the Admiralty. On 17 April, at the end of his voyage from Portsmouth, he had tried to enter Louisbourg but had been "stopped by a Body of Ice" and finally forced to make for Halifax.[35] Even Wolfe, no friend of Durell, reported that "the Fogs upon this Coast are so frequent and lasting, and the Climate in every Respect so unfavourable to Military Operations," that, if the expedition had sailed from Britain earlier, it would still have experienced delays.[36] For his part, Durell did dispatch the small ship of the line HMS *Sutherland* (50 guns) under Massachusetts-born Captain John Rous, an officer of great experience in the local waters, to patrol off Canso in early May, and Rous managed to snap up one French ship, but the remainder of the convoy, escorted by *Capitaines* Jacques Kanon and Jean Vauquelin, officers very familiar with the St. Lawrence, managed to evade capture and reach Quebec.[37]

Winter continued to plague the expedition. Holmes, who had left Spithead before Saunders, did not even reach Halifax until 13 May. The two admirals then sailed for Louisbourg where they arrived on 15 May but the weather remained uncooperative – two days after Holmes and Saunders anchored at Louisbourg, a convoy of troop ships from Boston, Halifax and New York was prevented from entering Louisbourg "by the vast Quantity of ice surrounding the Harbor, with constant thick Fogs" and forced to wait in open sea off the harbor mouth for ten days. Even at this late date, Saunders recorded, the harbour was so full of ice that "it was not Practicable for Boats to Pass" through it. "This Severity of the Winter," he advised the Admiralty, "has greatly retarded our Sailing from Louisbourg, and has, by much, exceeded any that can be remembered by the Oldest Inhabitants of this Part of the World."[38] It was not just ice that made matters difficult; it was also cold temperatures as "running Ropes freese in the Blocks; the Sails are still like sheets of Tin; and the Men cannot expose their Hands long enough to the Cold, to do their Duty aloft; so that Topsails are not easily handled."[39]

At Louisbourg Saunders and Wolfe made their final preparations for the voyage up the St. Lawrence. Saunders issued sailing orders for all naval and merchant captains providing signals for orders and movement and places of rendezvous for ships that might get separated. The merchant vessels were organized in three divisions – Blue, Red and White – containing respectively the 1st, 2nd and 3rd Brigades of the army, and each division was under the command of a naval officer who was responsible for conducting it safely to the objective. With the wry humour which seems to have been one of his hallmarks, Saunders warned the merchant captains that

> the Master of every transport is hereby strictly enjoined to look out for, and punctually to obey, all such signals as shall be made by the Commanding Officers of the division he belongs to: and, in case of neglect in any one, the Captains of his Majesty's ships are directed to compel them to a stricter observance of their duty by firing shot at them, and to give me an account thereof, which I shall transmit to the Navy board, in order to their charging the [cost of the] same against the hire of these vessels, for whose neglect his Majesty's stores are so unnecessarily expended.[40]

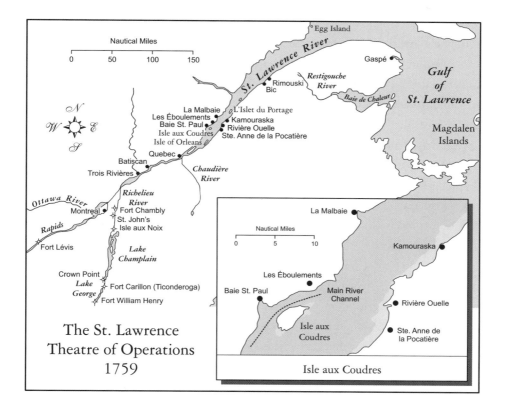

The St. Lawrence Theatre of Operations 1759

For his part, Wolfe directed the army officers under his command that, as "the Navigation in the River St. Lawrence may in some place be difficult, the troops are to be as useful as possible in working their ships in obedience to the Admiral's commands and attentive to all signals."[41]

Particular attention was paid to the landing craft for the operation. Saunders's fleet was provided with 83 craft for transporting soldiers and artillery from ship to shore. These were of three types: flat boats, whale boats and cutters. Flat boats were the true landing craft and there never seemed to be enough of them. Wolfe complained of shortages and he was strict about them being misused; naval and merchant marine officers were not to take them "for watering their ships, or other purposes; they are solely intended for the use of the troops."[42] Whale boats were larger, more robustly built and deeper craft with a pointed bow and stern so that they could be easily beached and were normally used to transport artillery and stores. Cutters were clinker-built craft, between 24 and 32 feet in length, equipped with a rudder and two masts, which could perform a variety of tasks. Augmenting these craft were about 70 boats that could be furnished by the warships and merchantmen. Brigadier-General Robert Monckton, Wolfe's second-in-command, calculated that the expedition possessed enough craft of various types to carry 3,319 troops in one lift.[43]

Quebec's "most effective rampart"

On 4 June 1759, Saunders weighed anchor for Quebec although bad weather imposed a delay of two days before all his ships cleared Louisbourg harbour. The naval commander now faced the problem of taking his fleet and merchant vessels, with little accurate navigational information and of course no navigational aids, nearly a thousand statute miles up the greatest tidal river in the world. The most dangerous stretch of the journey was the last 376 miles from the Gaspé Peninsula to Quebec as the fleet would have to ascend the St. Lawrence proper (see map, page 15). This stretch was regarded as extremely perilous and, as one French officer put it, "bristling with reefs and unparalleled for danger and difficulty of navigation" it was Quebec's "most effective rampart."[44] Although the government of New France had made attempts to chart the river, the French navy and merchant marine preferred to put their reliance on skilled pilots.[45]

On this head, Saunders had asked the Admiralty to send him "any Pilots for the River Saint Lawrence, they shall be well paid, and I shall be greatly obliged to you for them."[46] It is unlikely that the Admiralty could comply with this request but, in the event, 17 French pilots captured at Louisbourg the previous year were impressed into service in 1759 to guide the fleet. Despite later claims that they

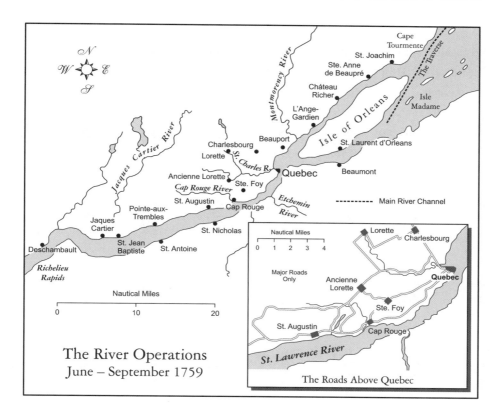

The River Operations
June – September 1759

The Roads Above Quebec

served reluctantly, most actually seemed to have been quite willing and the Royal Navy paid them well for their services – indeed, they proved "clamorous" for more money.[47]

Some of these men were on board Durell's squadron when it sailed from Halifax on 5 May and Durell made good progress. By 21 May he had reached the Île Bic, near modern Rimouski, the point in the mouth of the St. Lawrence where ocean navigation changed to river pilotage. Here he captured more French pilots who, unwillingly, accompanied him to Isle aux Coudres, some 45 miles below Quebec, which he reached on 28 May (see inset map, page 15). Nearby was an excellent anchorage that was to function as the rendezvous for the main fleet coming up from Louisbourg.[48]

While he waited for Saunders and Holmes, Durell did some useful work by sounding the Passage de la Traverse, 30 miles upriver from Isle aux Coudres and near the eastern end of the Île d'Orléans (see map above). This part of the St. Lawrence is divided by a number of low-lying islands into three channels of which the least dangerous was the northern channel running for about 15 miles from Cap Brule to the eastern tip of the Île d'Orléans. Known as the Traverse and still the modern route for ocean-going vessels, it is narrow and twisting, and subject to continuous silting which even today requires constant dredging. The ebb tide in the Traverse is very strong, running between 5 and 6 knots, and often creating a tide rip of short, violent waves, which is hazardous for small vessels; this is particularly true in the summer when the prevailing winds are westerly, and there is a danger of being forced aground.[49]

British naval officers had expected that the French would defend this difficult stretch of the river. In fact, when Marquis de Montcalm, the French commander at Quebec, had been advised on 2 June of the arrival of British warships at Bic, he had immediately invoked a council of war to discuss what measures could be taken to hamper the progress of the fleet. It was decided to sink eight ships in the Traverse but as the French had never made accurate soundings of it, these had to be done first so as to ascertain the best positions to place these obstacles. This intention was dropped, however, when local mariners assured Montcalm that the Traverse itself "would be a sufficient obstacle to the enemy."[50] Unfortunately for Montcalm, this was extremely bad advice and when Saunders's fleet made its way through this seemingly impossible stretch of water, Montcalm not unnaturally concluded that his own seamen and pilots were "either liars or ignoramuses."[51]

Captain James Cook (1728-79). One of the most skilled navigators in the Royal Navy, Cook served as the master (pilot) of HMS *Pembroke* during the Louisbourg and Quebec expeditions. His professional skills were in great demand during the ascent of the uncharted St. Lawrence but Cook would go on to greater glory as an explorer. (Library and Archives Canada, C-034668)

The Royal Navy's conquest of the Traverse began on 8 June when Durell ordered Captain William Gordon of HMS *Devonshire* (70 guns) to proceed to its eastern extremity with three other warships (HM Ships *Centurion* (50), *Pembroke* (60) and *Squirrel* (20)) and three transports to sound and buoy a channel. The sounding was done by ships' boats, which clearly marked the navigable channel with buoys. This work was done well and quickly and much of the credit goes to Lieutenant James Cook, master of the *Pembroke* and a superb navigator who would go on to greater fame. Gordon's crews spent five days on this important task and when it was completed on 13 June, the Traverse no longer presented an obstacle to the fleet.[52]

For their part, Saunders and Wolfe reached Bic on 19 June and Isle aux Coudres on 23 June, 19 days out of Louisbourg, without the loss of a single ship. The progress of that large fleet up the river was an impressive sight, and on 7 June the logkeeper of HMS *Neptune* (90), Saunders's flagship, felt constrained to record that there were 136 sail visible from that ship's quarterdeck. Leaving his larger warships – the 70- to 90-gun vessels drawing 20 or more feet of water – under Durell's command, Saunders now transferred his flag from the *Neptune* to the smaller *Stirling Castle* (64). Over a period of three days between 25 and 27 June, utilizing the French pilots and six sounding vessels that went ahead of the fleet, the smaller warships and merchantmen of the Blue, Red and White Divisions, in sequence, ran the Traverse guided by anchored ships' boats in the most difficult parts. Despite the strong current, the passage was made at ebb tide, between 4 and 6 A.M., because low water revealed more of the hazards in the river.[53]

Thanks to Gordon's preliminary work, the passage of the dreaded Traverse was made smoothly and without incident. Army officer John Knox made it in the transport *Goodwill* of London with 179 officers and men of the 43rd Foot. Her master was Thomas Killick, a veteran Thames River pilot and a Younger Brother of the Corporation of Trinity House, the English Navigating Guild. Some days earlier, a French pilot had been put on board the *Goodwill* but proved to be a disagreeable person who "assumed great latitude in his conversation," confidently asserting that, although "some of the fleet would return to England … they would have a dismal tale to carry with them; for Canada would be the grave of the whole army, and he expected, in a short time, to see the walls of Quebec ornamented with scalps."[54] If it had not been for orders to treat this dismal individual well, Knox recalled, "we would have thrown him overboard."

By the time the *Goodwill* approached the "Traverse," Thomas Killick had lost all faith in the Frenchman and resolved to do his own navigating. Knox was a witness to what followed:

As soon as the [French] Pilot came on board today, he gave his instructions for the working of the ship, but the Master [Killick] would not permit him to speak; he fixed his Mate at the helm, charged him not to take orders from any person except himself, and, going forward with his trumpet to the forecastle, gave the necessary instructions. All that could be said by the Commanding-Officer, [of the 43rd] and the other Gentlemen on board, was to no purpose: the Pilot declared we should be lost, for that no French ship ever presumed to pass there without a Pilot. "Aye, aye, my dear" (replied our son of Neptune) "but d[amn] me I'll convince you, that an Englishman shall go where a Frenchman dare not show his nose." The *Richmond* frigate being close astern of us, the Commanding Officer called out to the Captain, and told him our case; he enquired who the Master was? – and was answered from the forecastle by the man himself, who him "he was old Killick, and that was enough."

I went forward with this experienced mariner, who pointed out the channel to me as we passed, shewing me, by the ripple and colour of the water, where there was any danger; and distinguishing the places where there were ledges of rocks (to me invisible) from banks of sand, mud or gravel. He gave his orders with great unconcern, joked with the sounding boats who lay off on each side, with different coloured flags for our guidance; and when any of them called to him, and pointed to the deepest water, he answered, "aye, aye, my dear, chalk it down, a d[amne]d dangerous navigation – eh, if you don't make a sputter about it, you'll get no credit for it in England ..."

After we had cleared this remarkable place, where the channel forms a complete zig-zag, the Master called to his Mate to give the helm to somebody else, saying, "D[amn] me, if there are not a thousand places in the Thames fifty times more hazardous than this: I am ashamed that Englishmen should make such a rout about it."[55]

It was no surprise that the *Goodwill* made it through the Traverse intact as did all British ships that attempted the passage. By 27 June, much of Saunders's fleet was anchored in the south channel of the St. Lawrence at the eastern end of the Île d'Orléans. The experience and knowledge gained by the conquest of the Traverse was put to good use and in the first two weeks of July, the larger warships were brought up the river to join the remainder of the fleet. The Royal Navy was now before Quebec.[56]

The navy's task was to assist the army to take the capital of New France, but to do so the fleet had to deal with the navigational realities of the Quebec basin. As much as Wolfe (himself a poor sailor) wished, they could not command the wind and the tide, and the support they could render was affected by the simple fact that for sailing vessels, the navigation around Quebec was extremely difficult. To comprehend the problems faced by Saunders and his captains, a brief discussion of the nautical characteristics of the Quebec basin is necessary.[57]

Quebec is perched on a rocky promontory known as Cape Diamond, which, at its highest point, is some 350 feet above the high-water mark of the St. Lawrence. The width of the river from Cape Diamond across to Pointe Lévis on the south shore is about 1,000 yards. Downriver to the east, the St. Lawrence is divided into two channels by the Île d'Orléans, a low, wooded island 17.5 miles in length and about 3.5 miles wide. The north channel, now known as the Chenal de l'Île d'Orléans, is fairly narrow, being about 1,000 yards at its widest point at high tide but only 100 navigable yards wide at low tide. This channel is not only shallow, being only 15 feet deep – which excluded its use by the larger warships (those carrying 50 or more guns) – but it is full of sand shoals which caused ex-

"General View of Quebec from Point Lévis" by Richard Short, 1759, depicts the capital of New France crouched on its rocky promontory above the St. Lawrence. This view shows the three waterfront batteries in the Lower Town (the low fortifications pierced by embrasures for artillery). Despite their location, these batteries were unable to prevent the British fleet from ascending the river past the town. (Library and Archives Canada, C-000355)

perienced mariners to avoid it. The south channel, now known as the Chenal des Grands Voilliers, is not only wider, averaging between 1,000 and 1,500 yards, but, with a minimum depth of 35 feet at low water, much deeper. In 1759 this channel provided a good, although somewhat crowded, anchorage for the fleet.

At Quebec, the northern and southern banks of the St. Lawrence are markedly different (see map, page 23). On the north shore, the St. Charles River flows into the St. Lawrence just east of the town, and from the confluence of these two streams eastward the north shore is formed by tidal flats (the Beauport flats) of mud and slate. At high water, these flats extend, on average, about 1,400 yards into the river and are fringed with shoal water between 6 and 15 feet deep. In the 18th century, the high water line was marked by a string of small boulders and stones deposited by the tidal currents. Below Quebec, the south shore of the river rises more steeply than the north but it is more accessible as the mud flats on this shore are not as wide.

The basin below Quebec is a tidal estuary. The ebb, or outgoing tide, flows for about 7.5 hours and then reverses to become the flood, or incoming tide, which flows for about 5 hours. Tides vary between 10 to 15 feet. The tidal current also affects the river above the town (Quebec is an aboriginal word meaning "narrow place") and the constriction of the St. Lawrence in this area, which actually begins at Sillery 3 miles above Quebec, causes an increase in the speed of the current from about 3 knots at flood tide to nearly 5 knots at ebb tide, strong enough to cause many ships to drag their anchors in the basin. As it passes Cape Diamond and moves into the basin proper, the ebb current flows strongly toward the north causing problems for unsuspecting vessels trying to avoid the shoals and mud flats of the Beauport bank.

The presence of shoals, mud flats, and tidal currents makes navigation in the Quebec basin difficult enough for sailing vessels, but it is made more so by the prevailing winds. Given their construction, 18th century sailing ships could not sail more than 6 points off the wind in any direction, which limited their ability to manoeuvre. To get up to Quebec and past the town, Saunders's captains needed a wind direction from northwest to south-southeast and, preferably, from north-northeast to east. Unfortunately, the prevailing winds in the Quebec area during the summer months are westerly – it has been calculated by the Canadian Hydrographic Service that there is a 45.1% chance of the wind being from the west between June and September as opposed to a 24.4% chance of it being from the east. These modern calculations have been confirmed by an examination of the logs of 24 of the warships in the fleet. These logs reveal that, during the 87 days between 18 June when the first British ships neared Quebec and 13 September, the day of the battle of the Plains of Abraham, the winds were westerly on 49 days and easterly on only 20 days. On the remaining days, it was either calm or the winds were northerly, southerly or extremely variable.[58]

In 1759 the natural defensive strength of Quebec was enhanced by man-made defences. Situated on its rocky promontory high above the river and nearly surrounded by cliffs and water, the town's major weakness was its land front facing west onto open country, the famous "Plains of Abraham." This was protected by a fortified and bastioned wall which all French military engineers agreed was ineffective as it could be dominated by higher ground to the west and enfiladed by artillery fire from across the St. Charles River. The east side of that waterway, the Beauport Shore, stretching some 4 miles to the Montmorency River, seemed to provide a more accessible (and thus more attractive) approach for a besieger and this is where Phips landed his army in 1690. Immediately across the St. Lawrence from Quebec was Pointe Lévis, which was slightly higher than Cape Diamond and dominated the town.[59]

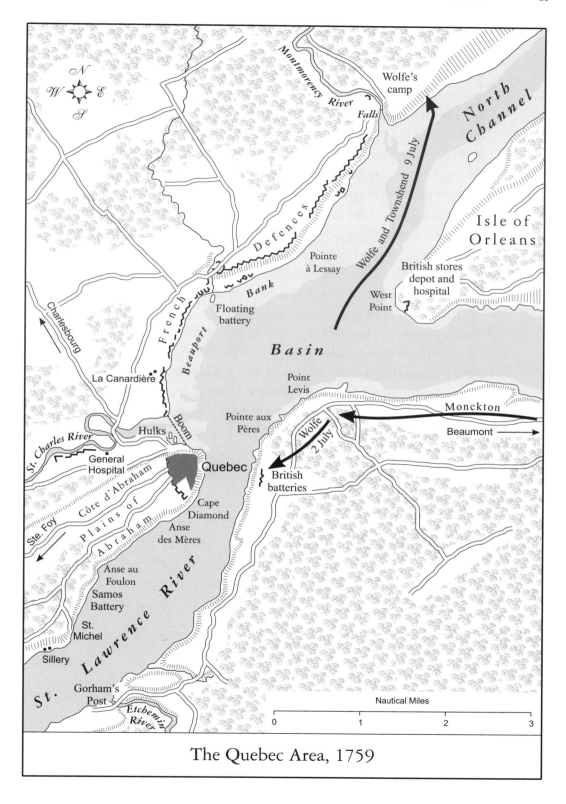

The Quebec Area, 1759

The opposing forces

To defend Quebec, Montcalm commanded 19,500 men comprising about 5,000 army or colonial regulars, about 2,000 sailors or merchant seamen, and about 12,500 militia with at least 290 pieces of artillery. Outside the town were large numbers of French-allied aboriginal warriors, whose strength fluctuated on a daily basis but whose presence was a matter of concern for Wolfe's troops. Although his numbers were impressive, the quality of Montcalm's army was poor and only his regular troops could be regarded as remotely comparable to the regiments under Wolfe's command. In addition, Montcalm's provisions were short and, although he possessed considerable artillery, his powder supplies were inadequate and of very poor quality, which restricted the effect of its fire. Well aware of the weaknesses of his command, Montcalm's plan was to deploy his forces in defensive positions, avoid pitched battle in the open against Wolfe's army, and simply wait out the besiegers, who would have to leave the area by early October or risk having their ships damaged by ice in the Gulf of St. Lawrence. The French commander therefore stationed the greater part of his troops in fortified earthworks along the vulnerable Beauport shore but neglected to defend Pointe Lévis across the river from the town, a serious error that would cost him dearly.[60]

Montcalm also gave some consideration to waterborne defences. Most of the ships that had arrived from France in May were taken up the river to Batiscan and left there with skeleton crews. The remainder of their sailors were sent to Quebec, where they were employed as gunners on the numerous batteries around the town. In addition,

> the Entrance of the River St. Charles was secured by a Boom, and this Boom defended by two Hulks with Cannon, wich were run ashore a little within the Chain; several Bateaux (or Boats) were put upon the Stocks, some of which were to carry a twelve, and others a fourteen Pounder[61] [gun]; a kind of floating Battery was likewise begun upon, of twelve Embrasures to carry Cannon of Twelve, eighteen, and twenty-four pounders, and ninety Men, and the Command given to Captain Duclos, of the Chezine, who was the Inventor of it. Batteries en Barbette[62] were erected on the Quay du Palais [in the Lower Town], and those on the Ramparts, and in the Lower Town, were repaired, completed, and considerably enlarged. Eight vessels were likewise fitted out as Fire-Ships .... Fire-Stages [rafts] were likewise built.[63]

Although some historians have stressed the fact that Wolfe's army was actually outnumbered by Montcalm's forces, this is not a realistic appraisal. Wolfe commanded just over 9,000 well-commanded, superbly-trained and disciplined

Louis-Joseph, Marquis de Montcalm (1712-1759). A professional soldier with considerable campaign experience in Europe, Montcalm assumed command of the forces in New France in 1756. He won some notable victories but by 1759 had become pessimistic about the outcome of the war in North America. (19th century copy of an original oil portrait, Library and Archives Canada, C-027665)

troops, with 163 pieces of artillery, plentifully supplied with ammunition and all other war materiel. He could also call on the services of an estimated 2,100 marines, not as well trained as his regular infantry but still useful to flesh out his strength. More importantly, there were at least 13,500 officers and seamen in Saunders's fleet who constituted a disciplined and very skilled force and not

only that, the British warships were armed with 1,800 guns and mortars, including many pieces of heavy 24 and 32-pdr. calibre. Finally, there was a useful auxiliary force of at least 4,500 merchant seamen in the civilian vessels. Overall, what arrived before Quebec in June 1759 was a massive military and naval assembly composed of about 29,000 soldiers, marines, sailors and merchant seamen equipped with just over 2,000 pieces of artillery. To put these figures in perspective, it has been estimated that the total population of all the French colonies in North America in 1759 was about 70,000 souls.[64] Wolfe's main problem was not strength but where to bring it to bear. He wanted to lure Montcalm out from behind his defences for, as Wolfe succinctly put it, Montcalm was "at the head of a great number of bad soldiers" while he "was at the head of a small number of good ones, that wish for nothing so much as to fight him."[65] The difficulty was that "the wary old fellow avoids an action doubtful of the behaviour of his army." Wolfe experienced many problems bringing his protagonist out in the open. This study is not a history of the siege of Quebec but a discussion of the naval aspects of the operation but a brief survey of events during the summer of 1759 is necessary to comprehend British command decisions and to provide proper context for the work of the fleet.[66]

Before his arrival on 27 June, Wolfe had hoped to seize the Beauport shore, the most useful access to the town, and was disappointed to find it heavily defended (see "Wolfe's different plans of attack," page 35). That same day his troops began landing on the Île d'Orléans, which became his main rear base area. On 29 June, Saunders suggested that Wolfe take the undefended Pointe Lévis, so as to forestall the French establishing batteries on this high ground that would prevent the admiral's ships operating in the basin. This was duly done on 1 July and on the following day, the erection of batteries began on Pointe aux Pères, the high ground immediately across from the town. Montcalm, realizing his error in not defending this vital area, made plans to attack it but, due to incompetence on the part of senior French officers, these plans came to naught. The possession of Pointe Lévis not only allowed Wolfe to bring the town under direct artillery bombardment, which both he and Saunders regarded as a necessary preliminary; it also permitted the fleet to get above Quebec and threaten Montcalm's vulnerable lines of communications.

After Saunders advised him that the fleet would not be able to provide any effective assistance in an attack on the Beauport shore, Wolfe began to actively plan for a landing above Quebec, with the village of St. Michael, on the north shore of the river about 3 miles west of the town, being the likely objective. To act as a diversion, however, on 9 July Wolfe landed one of this three brigades on the north shore of the St. Lawrence, just across the Montmorency River from the left or eastern flank of the French positions on the Beauport shore. The following day, however, Wolfe abandoned his plan to attack above the town, apparently because he feared that the landing force would be isolated and cut off by the enemy. While he contemplated his options, on 12 July the British batteries on Pointe aux Pères commenced firing on the town and were assisted by the three bomb vessels, HM Ships *Baltimore*, *Pelican* and *Racehorse*, armed with 10 and 13-inch mortars. This bombardment continued without cease for almost five weeks and left Quebec in ruins, although it had little real effect on military operations.

Brigadier-General Robert Monckton (1726-82), Wolfe's second-in-command, was a professional soldier who had served in North America since 1752, primarily in Nova Scotia, where he was lieutenant-governor from 1755 to 1758. Monckton gained an early British success when he captured Fort Beauséjour in 1755. (Library and Archives Canada, C-19118)

Brigadier-General George Townshend (1724-1807). A wealthy, aristocratic officer with less active service than Wolfe, Townshend was the government's, not Wolfe's, choice to command a brigade in the Quebec expedition and relations between the two were never cordial. Townshend assumed command after Wolfe's death. (Library and Archives Canada, C-8674)

On the night of 18 July, a small British naval force ran above Quebec, permitting Wolfe to reconnoitre west of the town. This done, he again began to plan to attack upriver but again abandoned the idea, possibly because Montcalm, nervous about his lines of communication, shifted troops from the Beauport shore to defend the area. For a brief time, Wolfe considered a waterborne assault directly against the Lower Town but he abandoned this project because it was too dangerous. Despite the reservations of Saunders, Holmes and Wolfe's three senior subordinates, Brigadier-Generals Robert Monckton, James Murray and George Townshend, Wolfe began to plan an attack on the Beauport shore. It was duly made on 31 July and was decisively rebuffed.

While he considered his next move, Wolfe initiated a punitive campaign against the French settlements around Quebec which lasted well into September. The smaller warships transported troops who despoiled the villages above and below the town and ultimately destroyed 1,400 farmhouses. On 5 August Saunders got 20 of the valuable flatboats past Quebec with comparative ease, which permit-

Brigadier-General James Murray (1721-94). This long-service professional served at Louisbourg in 1758. After the surrender of Quebec, he commanded the British garrison during the winter of 1759-60 and was defeated by Lévis the next spring at Ste. Foy. Murray served as governor of Quebec 1761-66 but his sympathies for the French-speaking population led to his recall. (Library and Archives Canada, C-2834)

ted amphibious operations above the town. This done, Wolfe dispatched Murray with his brigade to harass Montcalm's lines of communication, and although this caused much anxiety to the French commander, there seems to have been little positive result from these operations. Becoming nervous about the absence of nearly a third of his army, Wolfe recalled Murray, who returned on 25 August.

By the last days of August 1759 relations between Wolfe and his three subordinate brigadier-generals had become strained. This is not surprising as, after more than seven weeks of profitless activity, Monckton, Murray and Townshend were becoming exasperated by Wolfe's seeming inability to select a plan and maintain the aim necessary to accomplish it. To many, the army commander appeared unsure of himself – an intelligent observer complained about Wolfe's propensity to issue contradictory orders in succession, and also noted that "Every step he takes is wholly his own" and "he asks no one's opinion, and wants no advice" from his subordinates.[67] Matters were not helped when Wolfe (not a physically robust officer) fell sick on 19 August and was unable to command for nearly a week, and even when he did resume his duties, Saunders thought him weak and "still greatly out of Order."[68]

"Captious young general"

Unfortunately, the tension between senior army officers spilled over onto the navy. Although, with one minor exception (which will be discussed below), Saunders never recorded a single negative remark about his army counterpart, in the general air of peevishness present in the army, Wolfe and his staff criticized the navy's performance. The commanding general complained about the fleet's seeming inability to deal with the small French gunboats, which, although "paltry" craft, managed to "insult us."[69] He and his staff were also critical of the effectiveness of shore bombardment carried out by the warships and the three bomb vessels. Ignoring the fact that the army would not even be before Quebec without the fleet, one of Wolfe's staff officers went so far as to record that a general is "to be pity'd whose operations depend on Naval succour."[70]

These complaints are inaccurate and groundless as the navy had done its best to support the army. Concerning the French gunboats, these small but boldly-handled craft, sheltering within range of the French batteries on Cape Diamond and the Beauport shore, were quick to dart out and attack any sailing vessel that got into difficulty near the town. Since they were not dependent on the wind, the gunboats could lay off a British warship's quarter where her weapons could not reach them and pound its target to pieces at leisure. If another warship came to the rescue, they would withdraw within cover until the next opportunity. Trying to contain these irritants was an exercise in frustration and for this reason Saunders kept his larger ships back out of their range and protected the fleet with guardboats.[71]

These guardboats and the alert crews of the warships foiled two French attempts to destroy the fleet by using fireships or rafts. During the night of 28 June, just a day after the main body of the fleet arrived before Quebec, the French attacked and the log of HMS *Centurion* (50) records that "Att ½ past one A.M. saw 6 sail of French fire ships driving down the river, which we fired several guns in order to allarm the fleet that lay 3 or four mile below us: we were obliged to cut our best bower cable [to escape the flames]."[72] The experience of HMS *Pembroke* (60) was as follows:

> At midnight the enemy sent six fire ships down before the tide, all in flames; the *Sutherland, Centurion,* and *Porcupine* sloop got under sail and came down before them; sent all the boats ahead to take them in tow. At 2 A.M. two of the fire ships drove on shoar on the Island Orleans, and others was towed off clear of the ship.[73]

Twice during the siege (on 30 June and 27 July), the defenders deployed their secret weapon – fireships crammed with combustible material and set alight to destroy the British fleet. On both occasions, the boats of the fleet managed to tow them out of harm's way and little damage was done. The defenders expected great things of these weapons but they were disappointed. (Painting by Dominic Serres, Library and Archives Canada, C-004291)

On 28 July, the French mounted a second attack on the fleet when they sent a line of more than a hundred fire rafts, chained together, down the river. As the log of HMS *Dublin* (74) notes, Saunders had advance warning of this movement:

> At 10 Adm[ira]l. [Saunders] sent to acquaint all ye ships he had intiligence of some fire mesheanes would be sent amongst us; man[ne]d and arm[e]d 4 flatt[botto]m boats, on[e] of wh[ic]ch rowed guard wth an off[icer]; got all ye boats in rediness. At 12 P.M sign[a]l. was made for see[in]g ye fire mecheans com[in]g wth the tide, but by timely assistance of ye boats was tow[e]d cleare of ye [fire]ships.[74]

The frigate *Lowestoft* (28) was closer to the action:

> At 12 A.M the French sent down with the tide of ebb about 150 or 200 stages [rafts], each properly fited as fire ships, and all chained one to ye other in a line across ye river; they towed them as close to our ships as they durst venture, then set fire to the stages, they being so contrived that they were all in a flame in one minute. We sent all the guard boats with grapnels and chains fited a purpose to grapnel them and towe them clear of the fleet; ... towed the clear of all ye fleet and landed them on the Isle of Orlens.[75]

Wolfe and his staff also complained that the three bomb vessels, *Baltimore*, *Pelican* and *Racehorse*, did not move to within close range of the town and hinted that their mortar crews were incompetent, or at least very inaccurate.[76] Unfortunately, these vessels were armed with weapons that fired broadside and they thus had to be anchored with their sides to the target. They also required a stable anchorage so that they could use the cables by which their position was altered to shift the direction of their fire. Their captains found that, before Quebec, it was impossible to find a stable anchorage to fire effectively because the best positions were broadside to the river and tidal currents. As the log of HMS *Stirling Castle* (64) recorded on 13 July, the bomb vessels had to cease firing as the tide was "running too hot for their laying with a spring" or cable.[77]

Another problem was that, as soon as the bomb vessels found a good station from which to fire, the French would in turn bring them under fire – as the *Racehorse* noted on 10 July, just after she had taken station off Pointe Lévis and "fired 2 ten-inch shells and on[e] 6-pounder to see if our shot would rach a shore, which shott went on the shore," the French "began to fire a mortar that thea ad fixd at us and the rest of the ships" anchored nearby.[78] The result was that the bomb vessels had to constantly alter position and this had a negative effect on

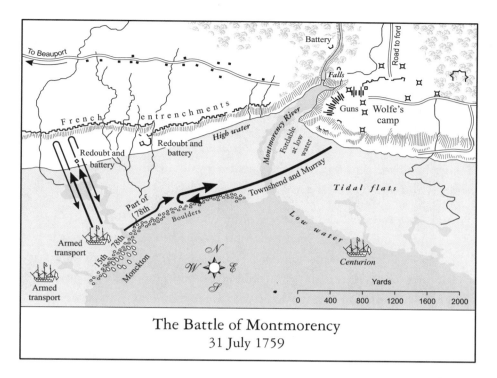

The Battle of Montmorency
31 July 1759

both the rate and the accuracy of their fire. A few decades later, the Royal Navy constructed bomb vessels that could fire forward and thus be anchored bow on to the direction of currents, improving the stability of the firing platform. In mid-July, realizing the three bomb vessels were ineffective, Saunders ordered them to put their heavy mortars and attendant crews ashore to join the heavy batteries at Pointe aux Pères.[79]

Throughout July and into August, the other vessels in the fleet carried out numerous shore bombardments at the request of the army. On 9 July, HM Ships *Captain* (64), *Racehorse* (bomb), *Richmond* (32), *Squirrel* (20) and *Sutherland* (50) bombarded the Beauport shore to cover the landing of troops near the Montmorency Falls but shallow water prevented them getting close enough to do much damage.[80] Six days later, HMS *Centurion* (50) engaged a partially-completed French shore battery on the Beauport shore and kept up her fire for two days:

> [16 July 1759] At 8 A.M began to cannonade a battery which was erecting; recd. sevl shot from dtto which cutt away a bobbstay and the clue of our maintopsl. At noon left off. ……

> [17 July 1759] At 5 P.M began to cannonade the above. At 7 left off. At 11 they began to fire from the battery. At 3 A.M unmoord and shier'd our ship nearer the battery, and began to cannonade. At 7 left off.[81]

Centurion was also involved in the attack on the Beauport shore that Wolfe made on 31 July (see map, page 31). Saunders had told his army counterpart that the navy could not render good support in this area as the mud flats prevented his large warships, with an armament powerful enough to damage the French defences (i.e, 24 and 32-pdr. guns), from getting within decent range. Wolfe persisted in the attack, however, and the navy did its best to assist him. *Centurion*, along with HM Ships *Pembroke* (60), *Richmond* (32) and *Trent* (28) carried out a shore bombardment but only the *Centurion* was able to get close enough to render useful support and she had a hot time of it:

> At ½ past 9 weigh'd and made sail to the estwrd. between the Isld of Orleans and the N[or]th shore, as did two arm'd catts. At 10 came too with the best bower in 4 f[atho]m water abreast 2 French batteries … began to cannonade batteries, as did the catts. …… Continued still to cannonad the batteries till 6 P.M when our army landed under the cover of our guns. At 7 P.M the whole body of the army was oblig'd to retreat, some in boats, and some repassing the fall of Mo[n]t[morency]; rec'd several wounded officers and soldiers; the

A view of the Montmorency Falls and the battle of 31 July 1759 by Hervey Smyth, who was present during the action. This interesting view, sketched from the British artillery position on the east side of the Montmorency River, shows HMS *Centurion* supporting the attack with her guns and the two "catts" deliberately grounded to provide inshore fire support. On the shore, directly opposite the *Centurion*, the grenadiers can be seen advancing to attack the redoubt, which is shrouded by smoke. (Library and Archives Canada, C-000782)

two catts was set on fire by our people. ...... During our firing the ship was aground and s[h]ewed about 3 inches [in the hold].[82]

The navy also provided the crews and armament for the two catts, the *Russell* and *Three Sisters*, which were run aground to provide close fire support.

This repulse was the cause of one of the few disagreements between Saunders and Wolfe. Both men had participated in the operation that had called for the two catts being run onshore, followed by one portion of the troops landing in small boats while another forded the Montmorency River to join them. Wolfe and Saunders had accompanied the landing craft although neither landed. After Wolfe had drafted his report to Pitt on the action, he showed a copy to Saunders, who objected to certain passages describing the position of the *Centurion* and the two catts which contained implied criticism of the effectiveness of their fire – perhaps feeling that Wolfe was trying to shift some of the responsibility for the defeat onto the navy. Wolfe was quick to assure the admiral that he meant to cast no aspersions on the handling of the ships during the landing and promised to leave out the offending passages. "I am sensible of my own errors in the course of the campaign," he assured Saunders and could "see clearly where I have been deficient."[83]

The complaints made by Wolfe and his staff about the navy at Quebec were frivolous and unfounded. They derive more from the irritability in the higher ranks of the army than they do from any shortcomings on the navy's part. In this respect, the opinion of C.P. Stacey, author of the most detailed study of the siege, is worth quoting:

> It is evident that Saunders, as the situation became increasingly plain, was prepared to support Wolfe to the limit of his strength, and did so. It may be that at certain points he could have acted somewhat more promptly, but there seems no sound basis for any important criticism of his action. Judging Wolfe by his own journal, it is clear that the impatient, sickly and captious young general was a difficult colleague, but Saunders shows no sign of having been influenced by such considerations. He had been sent to Canada to help the army take Quebec, and he carried out his mission.[84]

To which might be added that Saunders, unlike his army counterpart, carried out that mission without complaint.

It is notable that, although Wolfe and his staff complained about the navy's seeming inability to prevent "insult" from the French gunboats and to carry out effective shore bombardment, they did not comment on the multitude of other tasks performed by the fleet. A perusal of the logs of Saunders's warships reveals the extent of his sailors' contribution: providing and manning the boats that moved troops, guns, ammunition, supplies, the sick and the wounded; landing shore parties to haul heavy ordnance up the heights opposite the town or at Montmorency and gun crews to assist in firing them; providing guardboats to act as sentries for both the fleet and the seaward flanks of the land camps; landing marines to guard the army and provide labour details, thus freeing soldiers for operations; escorting transports up and down the river and often rescuing them when they experienced difficulties. Even the civilian seamen from the merchant vessels, Saunders reported, "exerted themselves in the Execution of their Duty" and had "willingly assisted me with Boats & people on landing the Troops and many other Services."[85] The navy and their merchant marine counterparts provided every biscuit eaten by the soldiers, every tot of rum they drank, every round they fired and the powder that propelled that round.[86]

During that long summer of 1759, the fleet was ever busy and, although the logs reveal the extent of that activity, space limitations preclude a more detailed mention. The log of HMS *Stirling Castle* (64) for 31 August records what was probably a typical day in the fleet:

> P.M Employed in making nippers [anchor bindings]; the *Trent* made the signl. for master of merchnt. ships. Att 9 received on bd 9 butts of spruce beer; longboat employed in bringing the artillery off from Montmorencey.
>
> A.M All the line of battle ships signl for a Lieutent. twice. At 11 do. made ye *Shrewsburgy*'s sigl. for a Lieutent; cleared hawse; reced on board 5 boatloads of wood; anchored here his Maj's ship *Trident*; carpenters employed on shore in repairing the longboat and some people in cutting wood.[87]

On the other hand some days were unusual – on 12 September, for example, the bomb vessel *Racehorse* recorded that her sailmaker was employed in mending sails "that was eatn by the squrels that swim on board."[88]

The crowded anchorage in the south channel of the Île d'Orléans did cause problems, particularly on 28 June. On that day, Saunders reported, "a very hard gale of wind came on, by which many Anchors and small Boats were lost and much damage received among the Transports by their running on board each other."[89] The frigate HMS *Lowestoft* (28) had a difficult time:

> At 2 P.M lost our gibb boom sprit sail yard, larboard bomkin, and reased ye larboard catthead by a transport that come foull of us; … let go the best bower anchor under foot. At 4 came foull of us another transport which carried

Wolfe's different plans of attack, May to September 1759

After studying the records of the 1759 siege, C.P. Stacey concluded that Wolfe considered no less than eight different plans to take Quebec.

The **First Plan** (numbered "1" on the map) was his original intention to land on the Beauport shore and attack across the St. Charles River. It was forestalled by Montcalm's fortification of the Beauport area.

Wolfe's **Second Plan**, contemplated between 5 and 10 July 1759, was to land at St. Michel above Quebec and force Montcalm to attack to preserve his line of communication. This plan was abandoned as too risky.

The **Third Plan** was a variant of the first. Forestalled from landing on the Beauport shore, Wolfe tried to position his army as close as possible and moved the greater part of it to the east side of the Montmorency River with a view to attacking directly across the river.

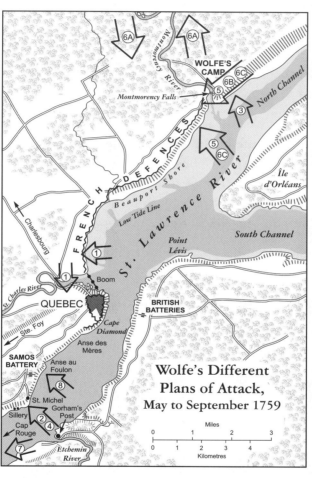

The third plan was forsaken momentarily in favour of the **Fourth Plan**, contemplated about 20 July. It was a variant of the second and again involved a landing above Quebec to seize an entrenched position but it was abandoned for reasons that are not certain.

Wolfe's **Fifth Plan** was an elaboration of the **Third Plan**, an attack directly across the Montmorency. It was carried out with disastrous results on 31 July.

The **Sixth Plan** included three different suggested attacks (numbered 6A to 6C in the map) in the Montmorency area contained in a letter probably dated 27 August from Wolfe to his three brigadiers. The first of these (6a) involved an encircling night march to ford the Montmorency 8-9 miles above its mouth and attack the rear of the Beauport defences. The second (6b) was a night attack directly across the Montmorency below the falls against the French entrenchments. The third (6c) was similar, a movement across the Montmorency followed by a major landing from boats on the Beauport shore.

In their reply dated 29 or 30 August, the brigadiers countered with the **Seventh Plan**, a landing above the city at the mouth of the Cap Rouge River to cut the French supply line and thus force Montcalm into battle.

Wolfe accepted his subordinates' suggestion to attack above the city but altered the objective in his **Eighth and Final Plan**, which was evolved on 9-10 September. The selected landing place was the Anse au Foulon, closer to the city. This plan was carried out on the night of 12/13 September 1759 and achieved victory.

MAP: CHRIS JOHNSON.

away our spare anchor, larboard main chaine, and our barge cutter, and one flatt bottom boat, all lost; we was obliged to cutt our small bower cable close by the inner end to clear us of ye transport, ... At 10 came ath[w]or[t] our hawse a scho[o]nar which carryd away our cutt water and rails of our head, veard away to a wholl cable on the sheet and the schoner cut away her main mast. ½ past 11 she got clear of us.[90]

The fleet also transported the troops that carried out Wolfe's punitive campaign against the surrounding area. Occasionally warships mounted their own raids using naval personnel as was the case with HMS *Scarborough* (20) on 1 September which carried out a raid on Mingan, downriver from Quebec on the north shore of the St. Lawrence. *Scarborough* landed

60 men and a Lieut all armed, where we found a little town in the skirt of a wood; it had 8 or 10 houses with barns and outhouses and a ridout [redoubt] where was ambrasures [embrasures] but no guns, and all the inhabitance was run in the woods where we searched strictly and found an Indian cannue and several other things, but most of their affects they carried with them, so we left 30 men in the ridout all night and had shooting parties out all day ...[91]

After destroying the village of Mingan and the redoubt, the *Scarborough* party learned from scouts that there were some vessels on the other side of the island.

At 2 [P.M] sent our boats mann'd and armed to take possession of them; sent a party of marines alongshore and a party of seamen cross the neck of land to support them, so they returned; sent the marines and seamen to ford the river, which they did, tho' up to the neck in water, in face of several of the enemy, who on our people approach run away; the boats being following them, we took possession of the vessels, found the enemy had killed their dogs to prevent being persued, the vessels being a large sloop, a schooner without masts, and 2 shallops [chaloups], being so far on the beach and it was neap tides we could not get them off, so set them all on fire, saw them burnt, and returnd without any damage, but looseing some small arms in fording the river.[92]

Operations above the town

The support of the fleet was essential for any operations above Quebec. The first British vessels to go above the town were HM Ships *Diana* (32), *Squirrel* (20) and *Sutherland* (50) which, with two merchant sloops, ran the obstacle during the flood tide on the night of 18 July. Although the channel between Quebec and Pointe Lévis opposite is fairly deep, it was not an easy passage. One of the merchant vessels ran afoul of the *Diana,* which got caught in a tidal eddy, and ran aground within gunshot of the town. As the frigate's log records, it was a perilous situation:

> we got down topg[allant] y[ar]ds and topm[as]ts and got the booms over board, did what we could to lighten her by send[in]g. the powder, shot, iron, ballast, &c. [overboard] The town and 5 floating batteries fir'd on [us], we recd some damage by them; hove 13 guns overboard. ...... the men of war's boats and people assisting us; the Adml. sent his Majt's ship *Pembroke,* who came near us and sent on board his stream cable and brought it too to his own capstan in order to heave us off, as did also the *Richmond;* we had anchors out both ahead and astern, the hawser brought too our own capstan; hove overboard 12 more of the guns. At ½ past 1 [P.M.] hove off in deep water.[93]

Disarmed, the *Diana* was useless for further operations and was sent back to Halifax.[94]

The British movement above Quebec worried Montcalm, who judged it to be of "the most alarming kind" as his enemy would "be able to steal up the south shore in their flat-bottomed boats, and cross over to the north even as far as Three Rivers."[95] He began to shift troops west of the town to guard the north shore of the St. Lawrence. On the night of 5 August, Saunders sent 20 flat-bottomed boats – about half of those with the fleet and capable of carrying 1,260 troops – above Quebec and they made the passage without difficulty.[96] When Wolfe ordered Brigadier-General Murray to carry out raids in the upper river, Saunders dispatched Rear-Admiral Holmes to command the ships above Quebec and "act in concert" with Murray and "give him all the Assistance the Ships and Boats could afford."[97] Murray's operations were not as productive as they might have been and Wolfe eventually recalled him.

On the night of 11 August, HM Ships *Hunter* (14) and *Lowestoft* (28), two transports, two armed sloops and a schooner attempted to run above Quebec but, with the exception of the schooner, were forced to turn back because of weak wind. It was nearly two weeks before another attempt was made and it came on the night of 27 August when *Hunter, Lowestoft* and *Seahorse* (20), two catts and an armed sloop got past the town without difficulty although, as the *Lowestoft* recorded, she

> received a smart fire of shot and shells, at which we fired 7 9-pounders. At 9 [P.M.] came abreast of another battery on the north shore which fired at us, at which we returned 4 9-pounders; in casting the ship lost overboard

20 fathoms of a hawser which we had for a slip rope; in passing the town received many shot holes through our sails and a deal of our running rigging shot away.[98]

On the night of 4 September, the remainder of the flat-bottomed boats, crowded with troops, went above.[99] Two days later, a small schooner somewhat whimsically named *The Terror of France* passed the town, and although the enemy "foolishly expended a number of shot at her," she made a safe passage "with her colours flying; and, coming to anchor in the upper river, she triumphantly saluted Admiral Holmes with a discharge" from her tiny swivel guns.[100]

During his illness in late August, Wolfe had asked Brigadier-Generals Monckton, Murray and Townshend "to consult together for the public utility" to consider "the best method of attacking the enemy."[101] After discussing the various options with Saunders, the three army officers were unanimous that

> the most probable method of striking an effectual blow, is to bring the Troops to the South Shore, and to direct the Operations above the Town: When we establish ourselves on the North Shore [west of Quebec], the French General must fight us on our own Terms; We shall be betwixt him and his provisions, and betwixt him and their Army opposing General Amherst.
>
> If he gives battle and we defeat him, Quebec and probably all Canada will be ours ......[102]

In effect, this was very nearly the plan Wolfe had contemplated nearly two months earlier before he became fixed on the Beauport shore. Wolfe agreed and, after further discussions with Saunders, began to concentrate the major part of his army on the south shore near the mouth of the Etchemin River, about 3.5 miles upriver from Pointe Lévis. On 1 September, Wolfe ordered the evacuation of the Montmorency position and the following day, the fleet transported and protected the troops as they were moved to the south shore of the St. Lawrence – not a man was lost in what was a very risky operation.[103]

Writing to Pitt on 2 September, Wolfe exhibited signs of both stress and indecision, cautioning the prime minister that:

> By the nature of the river, the most formidable part of this armament is deprived of the power of acting, yet we have almost the whole force of Canada to oppose. In this situation, there is such a choice of difficulties, that I own myself at a loss how to determine. The affairs of Great Britain, I know, require the most vigorous measures, but then the courage of a handful of brave men should be exerted only, where there is some hope of a favourable event.

Cap Rouge by Hervey Smyth, 1759. By early August, British warships had ascended above the town and were threatening its supply lines. This view shows a large warship (probably HMS *Sutherland*, 50 guns) attended by a smaller vessel. Note the two small French gunboats, one of which is firing at the *Sutherland*. Also note the buoy in the river in the right foreground – wherever the Royal Navy went in the vicinity of Quebec, it sounded, charted and marked the waterways. (Library and Archives Canada, C-000783)

However, you may be assured, Sir, that the small part of the campaign which remains shall be employed (as far as I am able) for the honour of his majesty, and the interest of the nation, in which I am sure of being well seconded by the admiral, and the generals.[104]

For his part, Saunders informed Pitt that Wolfe had resolved to "quit the Camp at Montmorenci, and go above the Town, in hopes getting between the Enemy and their Provision, supposed to be in the Ships there, and by that means force them to an Action."[105]

The stage was now set for the final act.

"The most hazardous & difficult Task"

In the first week of September 1759 the navy made ready for major operations above Quebec. Overall command of this task was vested in Holmes, who was pleased with the change of direction as the besiegers "had met with nothing but Disappointments at Montmorency; and all Our Attempts below the Town."[106] "A Plan was immediately set on foot," Holmes continued, to attempt a Landing about four Leagues [roughly 12 miles] above the Town, and it was ready to be put in execution when General Wolfe reconnoitred down the River and fixed upon Foulon, a Spot adjacent to the Citadel, which, tho' a very strong Ground, being a Steep Hill with Abbatis laid across the accessible parts of it and a Guard on the Summit, He never-the-less thought that a Sudden brisk Attack, a little before day break, would bring his Army on the plain, within two miles of the Town.[107]

The anonymous officer on Wolfe's personal staff records how this change of plan came about:

The 6th [September 1759] Mr. Wolfe went up [the river]. The ships moved up to Carrouge [Cap Rouge, about 10 miles above Quebec], the following day the General went in his Barge, and reconnoitred the Coast at less than 200 yards distance all the way up [river] to Pointe de Trempe [Pointe aux Trembles, nearly 30 miles above the town] and there fixt on a place for the Descent, and gave his orders in consequence; Heavy rain delay'd this operation, and the General fearing for the Health of the soldiers, who were much crowd'd aboard the ships, order'd half of them to be landed on the southern shore and Cantoon'd [cantoned] in the Village of St. Nicolas: During this Interval [9 September] the General went in Captn. Leaske's[108] schooner and reconnoitered close by the shore from Carrouge down to the Town of Quebec.[109]

According to this officer, during his reconnaissance Wolfe had spotted a landing place closer to the town at a small cove known as the Anse au Foulon and "having observ'd the Foulon, thought it practicable and fixed on it for the Descent."

Wolfe's latest change of plan was not popular. As Holmes commented:

This alteration of the Plan of Operations was not, I believe approved of by many, beside [Wolfe] himself. It had been proposed to him a Month before, when the first Ships passed the Town, and when it [the Anse au Foulon] was entirely defenceless and unguarded: but Montmorency was then his favourite Scheme, and he rejected it. He now laid hold of it when it was highly improbable he should succeed, from every Circumstance that had happend since: But the Season was far spent, and it was ncessary to strike some Stroke, to ballance the Campaign upon one side or another.[110]

There were several important objections to the choice of the Anse au Foulon as a landing place. First, it was much closer to the town than the previous objective at Cap Rouge and thus, even if the landing was successful, it would not cut

French communications to the west, potentially allowing the garrison of Quebec to escape and fight another day. Second, Montcalm had recently reinforced the defences at the Foulon by posting troops on top of the cliffs that overlooked it. Third, given the tidal current and the difficulty of river navigation, a landing force risked either overshooting the Foulon or missing it entirely.

But Wolfe's mind was made up. On 10 September, he made a second reconnaisance of the Foulon, taking with him a group of officers including Rear-Admiral Holmes; Lieutenant James Chads (sometimes Shads), the naval officer charged with conducting the water movement of the landing; Brigadier Generals Murray and Townshend; and Colonel William Howe, commander of the light infantry that would make the initial assault landing. Wolfe did not carry out this reconnaissance by water but simply conducted this party to the post of Goreham's Rangers, located at the junction of the Etchemin River and the St. Lawrence, nearly, but not quite, across from the Foulon. According to the anonymous staff officer, Wolfe did not indicate to these officers the exact location for the planned landing but simply pointed out "the Places He thought most accessable."[111] That the landing place and the time of the landing, however, were clear in Wolfe's mind on 10 September is evident from a letter he wrote that same day to a friend that stated that he intended to make "a powerful effort" at the Anse au Foulon about four in the morning of 13 September 1759.[112]

Unfortunately, Wolfe was not so forthcoming with his three subordinate generals, who remained unclear about the exact objective even after orders were issued on 11 September for the landing at the Anse au Foulon. These orders specified that the assault troops were to be embarked in their boats at St. Nicholas on the south shore by 9 P.M. on the night of 12 September.[113] During the morning of that day, the three generals sent a letter to Wolfe with a request for information as to the exact landing place. Shortly afterward (perhaps he brought the letter with him) Brigadier-General Robert Monckton, who had not been present at the officers' reconnaissance of 10 September, visited Wolfe's headquarters, possibly to seek clarification about the objective. After Monckton departed, Wolfe commented to his personal staff that his "Brigadiers had brought him up the River and now flinch'd" and "He did not hesitate to say that two of them were Cowards and one a Villain" – an indication of just how bad relations were between Wolfe and his generals.[114] As it was, Wolfe did not see fit to respond to his subordinates' written request for clarification until 8.30 P.M. that evening, just 30 minutes before the troops were to embark. He addressed his reply to Monckton, informing him that the landing site was at a place

called the Foulon distant upon two miles, or two miles & a half from Quebec, where you remarked an encampement of 12 or 13 Tents, & an Abbatis,
below it – you mentioned to day [i.e. 12 September] that You had perceived a
breast-work there, which made me imagine you as well acquainted w[it]h. the
Place, as the nature of things will admit off.[115]

As they were responsible for getting the army on shore, Wolfe had, of course,
informed Holmes and Chads of the exact landing place. Holmes believed that
the proposed operation was

> The most hazardous & difficult Task I was ever engaged in – For the distance
> of the landing place; the impetuosity of the Tide; the darkness of the Night:
> & the great Chance of exactly hitting the very spot intended, without discov-
> ery or alarm, made the whole extremely difficult: And the failing in any part
> of my Disposition, as it might have overset the Generals Plan, would have
> brought upon me an imputation of being the Cause of the miscarriage of the
> attack, & all the Misfortune that might happen to the Troops in attempting
> it.[116]

Lieutenant James Chads had the primary responsibility for the landing. Very
little is known about this officer beyond the fact that he was the commander of the
fireship HMS *Vesuvius* and seems to have been a specialist in small boat work as
he had commanded the landing craft during the attack of 31 July on the Beauport
shore.[117] Chads had some serious concerns about the proposed landing site and, in
the words of the anonymous but highly-partisan officer on Wolfe's staff,

> on the Eve of the attack made many frivolous objections such as that the Heat
> [force] of the Tide wou'd hurry the boats beyond the object &c. &c. which
> gave reason to suspect some one had tamper'd with him: The General told
> him he shou'd have made his objections earlyer, that shou'd the disembarka-
> tion miscarry, that He wou'd shelter him from any Blame, that all that cou'd
> be done was to do his utmost. That if Captn. Shads wou'd write any thing
> to testify that the miscarriage was G. Wolfe's and not Captn. Shads that he
> wou'd sign it. Shads still persisting in his absurdity, the General told him He
> cou'd do no more than lay his head to the block to save Shads, then left the
> Cabin.[118]

Again, this is unfair comment about a naval officer who was only advising Wolfe
of the dangers of moving on a dark night down an uncharted river flooded by
recent rains on a strong ebb tide. The statement that, by doing his duty, Chads

"gave reason to suspect some one had tamper'd with him" or, in other words, had been influenced by a senior officer who disliked Wolfe, is no reflection on Chads but yet another indication of the bad relations between Wolfe and his subordinates. As it is, these words exhibit a tinge of paranoia on the part of their author.

The objections raised by Chads were entirely reasonable. Most simply put, Wolfe's plan was to move 1,700 troops in 36 small craft, of which 31 were flat-bottomed boats, 8.7 statute miles down the current of an ebb tide which would give the landing force a maximum speed of 5 knots and about 4 knots by the time they reached the Foulon.[119] There was a very great likelihood that this speed would cause the assault force to overshoot the Foulon and the path which led from it to the top of the cliffs. The fact that there was a French picket at the Foulon, a garrison on the cliffs above it and a battery nearby that commanded the river were not the concern of Chads; that lay within the sphere of Wolfe and Colonel William Howe, the commander of the assault force. Both Chads and his superior, Holmes, felt that a landing at the Foulon was a risky business as its success depended entirely on two things: luck and the skill of the naval boat crews. To a greater or lesser extent, there is an element of chance in all operations of war but prudent commanders do their best by careful planning to reduce that element. What Wolfe was proposing was to carry out an operation that depended very much on luck to be successful.[120]

Orders being orders, however, Chads carried out his duty and closely regulated the organization of the troops and craft for the Foulon landing. The number of landing craft available, and their loads, for the initial flight were:[121]

Wave	Unit	Strength	Craft
1st	Light infantry	c. 400	8 flat boats
2nd	28th Foot	c. 300	6 flat boats
3rd	43rd Foot	c. 200	4 flat boats
4th	47th Foot	c. 250	5 flat boats
	Louisbourg Grenadiers	c. 50	1 flat boat
5th	58th Foot	c. 300	6 flat boats
6th	78th	c. 200	1 flat boat
			3 ships' longboats
			1 ship's barge
			1 ship's cutter

The second, or follow-up flight, was to be transported in warships or transports and then landed as boats became available after the first flight had got ashore. The details of the second flight were as follows:[122]

Vessel	Type	Unit	Strength
HMS *Lowestoft*	frigate (28)	15th Foot	300
HMS *Squirrel*	frigate (20)	Louisbourg Grenadiers	240
HMS *Seahorse*	frigate (20)	78th Foot	250
HMS *Hunter*	sloop (14)	78th Foot	120
	three armed sloops[123]	light infantry	200
Laurel	transport ship	60th Foot	400
Adventure	transport ship	35th Foot	400

Despite Chads's concerns, Wolfe gave him full responsibility and issued orders that

> Captain [Lieutenant] Shads has received the Genl.'s direction in respect to the order in which the troops move and are to land, and no officer must attempt to make the least alteration or interfere with Capt. Shad's particular province, least as the boats move in the night there be confusion and disorder among them.[124]

The landing at the Anse au Foulon in the early hours of 13 September 1759 is such a well known event that it is only necessary to briefly summarize it. The assembly and loading of the troops – after much practice on the navy's part over the last 15 weeks – went without a hitch. The first flight pushed off on time at about 2 A.M. and reached the Foulon at about 4 A.M. but, just as Chads had predicted, the current caused the lead boats to overshoot the landing place and they grounded some distance below the cove. There was a French garrison at the Foulon but

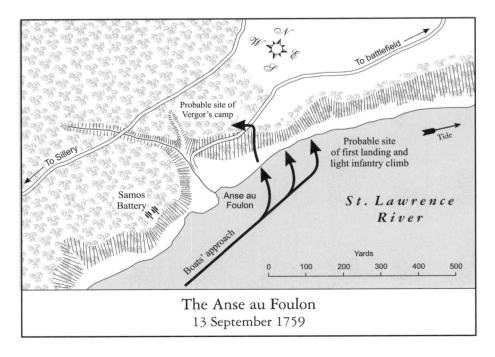

The Anse au Foulon
13 September 1759

An aerial photograph taken in the 1950s shows the Anse au Foulon, the intended landing place on the night of 12/13 September 1759. The breakwater is a modern construction. Wolfe was warned by the naval officers that the assault force risked the danger of overshooting the landing place and they did. (C.P. Stacey, *Quebec, 1759: The Siege and the Battle*)

they were unable to prevent the landing. Their commander, *Capitaine* Vergor of the *compagnies franches de la marine*, or colonial regulars, recorded that:

> On the 12th [of September 1759], he received an order to allow the passage of a convoy of boats that were to go down [the river] during the night of the 12th to 13th, to take provisions to Quebec, and he gave orders accordingly and had the sentries warned to allow the vessels in question to pass through, after, however, having recognized [identified] them. 3 hours before daybreak, the sentries gave notice that the vessels were in sight and they [the sentries] immediately demanded "where the boats had come from, from which regiment, and where they were going," to which they [the boats] replied "France. Marine"[125] and that "they were on their way to Quebec with provisions." Continuing to observe these boats, which had already passed by their post, the sentries noticed that they were coming round and trying to enter the [Foulon] Cove and assumed that they were the enemy and did not hesitate a moment to make arrangements to fire upon them.[126]

Having overshot the cove and its upward path, Howe's light infantry were forced to climb the cliffs to get at Vergor's troops above. Lieutenant Gordon Skelly of HMS *Devonshire*, who was with the lead boat, remembered it this way:

The navy delivers the army to the field of battle. Shortly after 2 a.m. on 13 September, flatboats land Colonel William Howe's light infantry, the advance guard of Wolfe's army, just below the Anse au Foulon, in this scene by noted marine artist Peter Rindlisbacher. (Courtesy of the artist)

the signal was made to proceed to land, the orders being for the four vanguard boats to land, and ours to lead the four, the rest of the boats keeping astern us at as small a distance as possible. ......

By the time we had run our boats ashore at the foot of the eminence, which seemed inaccessible, and the troops began to draw up, the enemy were no longer in doubt, and now began to fire irregularly from above, into us in the boats, which were scarce perceptible, it being extremely dark, but they killed one or two officers of the army, one midshipman, and several men in a little time. ......

Whilst our people were attempting to mount the eminence [cliffs], we kept our boats close to the foot of it in case of a retreat, and we for some time heard nothing but a war whoop or cry made use of [by the French] to make their numbers seem greater, as they fired upon us from the bushes. But presently we were apprised of our troops having found a way up by hearing their voices as they gave a loud Huzza! and fell amongst the enemy, by which means the musketry ceased, and at daybreak we found that they had got possession of the grounds above.[127]

The Anse au Foulon and the heights above being secured, the second and subsequent flights of troops began to arrive. As Holmes noted, the army landed

> undiscovered by the Enemy a little before Day; but not without hazard of being drove by the Currant, below the Town – The Sloops drew close in; and the Men of War & Transports got to their Station at Day break. …..
>
> The Boats were now employed in landing Troops from the Ships with the utmost diligence, and in carrying over 1200 Men, which Mr. Wolfe had ordered to march from Point Levi over Night and cover themselves in the Wood on the South Shore opposite Foulon, and in dragging the Artillery up the Hill.[128]

"In this manner," Holmes concluded, Wolfe got "his Army on the Enemies Shore, within two Miles of the Town, before his arrival was well known at their Head Quarters." By 7 A.M. on 13 September, Wolfe had 4,400 troops and two pieces of field artillery deployed on the Plains of Abraham west of Quebec.

The gamble had paid off and an operation that depended almost entirely on luck had been successful. With his army safely on shore and successfully deployed, the British commander "could face the next stage with comparative confidence."[129] Not a little share of the credit must go to the naval crews manning the boats, who were able to battle the current and get the first assault flight ashore, albeit downstream of the intended landing place. Credit also belongs to Colonel William Howe's light infantry who, being unable to use the path leading from the Foulon up the cliffs, had simply climbed straight up the heights in front of them. No officer was more aware of how risky the entire venture was than Saunders, who commented that

> considering the darkness of the Night and the Rapidity of the Current, this was a very critical Operation, and very properly and successfully conducted. When General Wolfe and the troops with him had landed, the Difficulty of gaining the Top of the Hill, is scarcely credible; it was very steep in its ascent and hight, and had no path where two could go abreast, but they were obliged to pull themselves up, by the Stumps and boughs of Trees that covered the Declivity.[130]

Getting the troops ashore near the Foulon was not all the Royal Navy did on that memorable night. The fleet contributed toward the success of the landing by mounting a feint with ships' boats to distract Montcalm's attention. Saunders ordered all the ships' boats that could be spared from the vessels in the basin to rendezvous off Pointe Lévis as darkness fell and then make for the Beauport shore in a simulated assault landing. The log of HMS *Stirling Castle* (64) records that on 12 September her boats were employed laying

> several buoys off Bowport to draw the enemy's attention that way. ….. At 11 observed a French cannon cutting away the aforesaid buoys, att whom the *Richmond* fired several shott …… sent several longboats above the town with cannon, mortar, shot and shells, &c.; the rest of the boats of the fleet assembled off Point Levee mann'd and arm'd. Att 11 [P.M.] the put off from thence and keept rowing between Bowport and mouth of Charles River. Att 1 A.M. heard some vollies of sm[al]l. arms off Bowport. Att ½ past 4 heard the report of sev[era]l. vollies of sml. arms and cannon above the town, which we afterwards found to be occasioned by Genl. Wolf's landing our army about a mile above Cape Diamond where he succeeded and got a footing on the riseing ground, during which time the town and our battery [on Pointe aux Pères] cannonaded each other very briskly.[131]

Holmes believed that this diversion caused Montcalm to think "our grand Aim was still below Quebec & pointed towards Beauport."[132]

Now came the finale. At about 10 A.M. on 13 September, Montcalm, with a force of about 4,500 regulars and militia, attacked Wolfe's army on the Plains of Abraham – exactly what the British commander had desired for more than two months. The French advanced with great spirit and opened a ragged musketry while the British line held their fire. On command, the redcoats then loosed off several disciplined volleys and charged with the bayonet, putting their enemy to flight. The entire action took less than 30 minutes and at the end of it Montcalm was mortally wounded and his army in full retreat into Quebec. He was not the only senior officer to be hit during the battle. The log of HMS *Lowestoft* records that at "½ past 10 was brought on board General Moncton wounded" and "At 11 was brought on board ye corps of General Wolf."[133]

Victory

Montcalm died that evening, even as the greater part of his forces withdrew from Quebec. Using the northern roads through the villages of Charlesbourg, Lorette and Ancienne Lorette, they were able to evade the British army, now commanded by Townshend, and escape to the west, bringing to pass what had been feared when the Foulon was chosen as a landing site – that it was too close to Quebec to cut off the enemy's retreat. In Quebec, the Chevalier de Ramezay commanded a dispirited and half-starved garrison of 2,200, comprising mainly sailors and militia.

The Royal Navy now assisted the army in preparing for a formal siege of the town. As Holmes recorded,

> From the 13th to the 17th [of September] the Army and Fleet were incessantly employed in getting every thing ready for opening Batteries against the Town – It was now getting very late in the Season and the utmost diligence was used and the greatest fatigues undergone, with Spirit and Cheerfulness by every Body, to bring the Campaign soon to an End: We had got up 17 pieces of battering Cannon from below, besides others of a Smaller Calibre, Mortars, Shells, Shots, Powder, Plank etc. all in readiness when the Batteries should be formed, to continue our progress against the Town.[134]

By 17 September, the Chevalier de Ramezay was beginning to contemplate surrender and his decision to do so may have been prompted by Saunders, who that day

> made a Motion with the Squadron below, which added to the operations of the General [Townshend] at land, put the Enemy in the utmost Consternation.
>
> He [Saunders] moved seven of the best Line of Battle Ships within random Gun Shot of the Town which struck them with the Apprehension of their coming up along side of the lower Town with the Night Tide, and that they would be stormed by Sea and Land. – Under these Terrors, they made offers of Capitulation about 3 in the Afternoon and it was perfected and signed this morning [of 18 September].[135]

Among the first British parties to enter Quebec that evening were a party of seamen from HMS *Shrewsbury* under the command of Captain Hugh Palliser. A month later on 18 October – actually very late in the season – Saunders sailed from Quebec. He left behind HMS *Porcupine* (14) and the bomb vessel HMS *Racehorse* to winter in the town along with the British garrison under Brigadier-General James Murray.[136]

There was, of course an epilogue. Once the Royal Navy was out of the St. Lawrence, the Chevalier de Lévis, who had succeeded Montcalm as French commander, attempted to run ships past Quebec. Most were destroyed but some made it to France carrying a request for a fleet and reinforcements to help recapture the capital of New France. The French navy, which had suffered a disastrous defeat at Quiberon Bay in November 1759, was in no position to help and only a small squadron was sent, but it was caught by the Royal Navy and destroyed downriver from Quebec in July 1760. Meanwhile, Lévis assembled almost every

A jolly tar and two soldiers visit the ruined interior of the Jesuits Church after the battle while grenadiers talk to the local ladies. When the British fleet sailed from Quebec in October, a garrison was left in the city under the command of Brigadier General James Murray. (Library and Archives Canada, C-000351)

regular soldier under his command and in April marched on Quebec. Brigadier-General James Murray, although badly outnumbered, somewhat foolishly emerged from the town to engage Lévis on almost the same ground as the battle of 13 September 1759. The result was the single bloodiest day's fighting ever done on what is now Canadian soil and a decisive defeat for the British army.[137]

Lévis, still convinced that help was on the way from France, lay siege to Quebec and both sides waited for the first ship to come up the St. Lawrence as the nationality would determine the outcome. On 9 May that vessel arrived – it was the frigate HMS *Lowestoft*, so prominent in the operations of the previous year. Lévis abandon the siege and withdrew to Montreal, where, after the arrival of a large army under Amherst in September, New France surrendered to overwhelming British force.[138]

"Justice to the Admirals"
In conclusion, it must be stated that, far from the rather waspish comment made by Wolfe's staff officer that "much is the General to be pity'd whose operations

depend on Naval succour" being valid, the Royal Navy rendered outstanding support to the army during the summer of 1759 at Quebec. The navy was really the senior partner in the operation and the importance of its contribution to the successful outcome was better expressed by Brigadier-General George Townshend in his dispatch to Prime Minister Pitt, dated 18 September, reporting the fall of the town. "I should not do justice to the Admirals, and the naval service," Townshend commented,

> if I neglected this occasion of acknowledging how much we are indebted, for our success, to the constant assistance and support received from them, and the perfect harmony and correspondence which has prevailed throughout all our operations, in uncommon difficulties which the nature of this country, in particular, presents to military operations of a great extent, and which no army itself can solely supply; the immense labour in artillery, stores, and provisions; the long watchings and attendance in boats; the drawing up [of] our artillery by the seamen, even in the heat of action; it is my duty, short as my command has been, to acknowledge, for that time, how great a share the navy has had in this successful campaign.[139]

Success was only made possible by good relations between the two services, despite the claims of later commentators. Vice-Admiral Charles Saunders emphasized this point when he assured Pitt after the surrender of Quebec that "during this tedious Campaign, there has continued a perfect good understanding between the Army and the Navy."[140]

Tribute to a hero. Major-General James Wolfe has several monuments and memorials in both Britain and Canada including this statue at Greenwich overlooking the former Royal Naval College. Vice-Admiral Saunders and his sailors, whose efforts actually secured victory at Quebec in 1759, have never received the same attention. (Photo by Dianne Graves)

The Naval War of 1812

THE WAR OF 1812 has been described variously as many things, including being a strange or incredible conflict. At sea two very different navies were pitted against each other. The Royal Navy had been fighting France for close to two decades before the United States declared war. The long period of fighting had seen the British fleet grow in size and competency so that in 1812 it was by far the largest and most powerful in the world. The United States Navy was much smaller and had much less of a tradition of victory at this point. What might have seemed a David and Goliath contest soon proved galling in the extreme to the Royal Navy, as the formidably large and gallant frigates of the United States Navy won a series of single-ship victories over their British counterparts. These successes could not reverse the marked inferiority at sea of the United States, but they served to highlight the limitations of the much more stretched Royal Navy when faced with a determined and highly capable opponent.

The Royal Navy responded by seeking to confine these U.S. frigates to port by imposing a blockade and by adapting some of its frigates to better challenge the upstart U.S. designs. The blockade's success was uneven, as the United States Navy still managed to get some frigates to sea where they could threaten British shipping, while on other occasions the blockaders succeeded in bringing the escaping U.S. frigates to battle.

The small size of the United States Navy might suggest that these battles were of little importance. This is wrong, however, as the continued existence and occasional successes of U.S. warships as the conflict dragged on for over two years proved significant for a number of reasons. The U.S. Navy realized quite well that achieving command of the sea could never be a realistic objective, given the great overall strength of the Royal Navy. Nonetheless, so long as U.S. ships continued to successfully get to sea and pose a threat to British merchant ships, the costs inflicted by losses and by heightened insurance charges would ensure that British businessmen would lobby strenuously for an end to the war. National pride was also at stake, as the outcome of relatively small naval battles could be used for important propaganda purposes. To a modern reader this might seem of little importance, but it is interesting to note – as one author does – how the results of one

A short and deadly fight

The hands drafted off USS *Constitution* to USS *Chesapeake* in 1813 took part in the most intense and evenly matched frigate action of the war. Patrolling off Boston on 1 June of that year, Captain Philip B.V. Broke, HMS *Shannon*, anticipated that USS *Chesapeake* would soon sail. He even sent in a letter of challenge, which went unread as *Chesapeake*'s commander, Captain James Lawrence, had already sailed. As *Shannon* lay to off Boston under easy sail, the aggressive Lawrence did not hesitate to close and engage an almost identical opponent. (*Shannon* carried 52 guns, while *Chesapeake* mounted 50 – their respective weights of broadside were 550 pounds for the British warship, 542 for the American.)

The action opened at 5:50 P.M. on 1 June. Lawrence rashly steered directly for a broadside engagement, forsaking an opportunity to rake the British vessel, and Broke's highly trained gunners made every one of their first rounds tell. Within ten minutes the two ships had collided and Captain Broke led his crew on board the American vessel. Mortally wounded, Captain Lawrence exclaimed, "Don't give up the ship!" as he was carried below. These inspiring words proved insufficient, as Broke – also wounded during the fighting – and his crew forced the American vessel, with its members from several different ships sailing together for the first time, to surrender at 6:05 P.M.

Although it lasted only 15 short minutes the butcher's bill for the battle was long. At least 60 were killed and more than 90 wounded on the American ship, while about half as many died and about 60 were wounded aboard the British vessel. The decisive factor in this action was Broke's painstaking dedication to gunnery and his acute tactical judgment. He had exercised his gun crews regularly, insisting not only on speed but – unusually for that time – accuracy. His efforts proved devastatingly effective, with *Shannon* scoring more than twice the number of hits as *Chesapeake*.

The aquatint of USS *Chesapeake* and HMS *Shannon* by Robert Dodd was published in 1813. (Library of Congress) battle proved useful diplomatically many years later, and how names from that war are still very much alive today in both navies, sometimes in surprising ways!

The comparatively few naval battles of this war, which altogether involved fewer men than fought at the single battle of Trafalgar, were therefore highly important to both sides. The failure of American armies to conquer Canada further magnified the importance and significance of naval successes to the United States. The continued threat posed by U.S. warships as negotiations dragged on in Ghent was one of the few positive forces U.S. diplomats could apply.

The next two chapters look at two key naval battles in the latter part of the war between U.S. frigates and their Royal Navy opponents. The two studies are different yet complementary in almost every sense of the word. The first covers perhaps the most famous naval battle of the most famous sailing warship in the United States Navy, the USS *Constitution*. This ship succeeded in slipping past blockading forces at the end of 1814 and proceeded on a successful sea cruise during which it encountered a British convoy and defeated two British escorts in a fascinating sea battle. This success seemed doomed to be reversed when a strong British squadron that had long chased the *Constitution* overtook her and her prizes, but the luck and skill of the American ship allowed it – and one of its prizes – to escape yet again. Both the initial battle and the subsequent escape are studied in detail in William Dudley's compelling assessment.

The second battle study reviews a more successful effort by the British to stop U.S. frigates. This battle witnessed a day-long chase by a British squadron of USS *President*. The key vessel in the chase was HMS *Endymion*, which was specifically adapted to counter the U.S. frigates. The result of the battle was the capture of the U.S. frigate, but the manner of that capture has caused enduring controversy. Andrew Lambert's vivid account makes a persuasive argument that *President*'s defeat was not due to overwhelming force but rather to the calculated efforts of *Endymion*. The many adjustments made by the British ship, and the successful tactics used to foil the desperate efforts of the U.S. vessel to avoid its final defeat, are carefully examined. Finally, the continued importance of this event in Royal Navy history is demonstrated in a way that shows that its effects resonate even today.

The fact that these two battles have much in common, aside from the dramatically different results, should be obvious. However, the different perspective of the two authors, one American and the other British, highlights intriguing contrasts. Both studies are scholarly yet approach the conflict with views that clearly originate on opposite sides of the Atlantic Ocean. The result is two complementary studies that work together to help contemporary readers understand naval events in this now-distant war.

"Old Ironsides'" Last Battle

USS *CONSTITUTION* VERSUS
HM SHIPS *CYANE* AND *LEVANT*

William S. Dudley

IN THE EARLY AFTERNOON of a chilly winter's day near the island of Madeira in the eastern Atlantic, the lookout at USS *Constitution*'s masthead excitedly hailed the main deck. He had sighted a distant ship on the starboard bow that within minutes changed course and headed for *Constitution*. Soon, the lookout hailed the deck a second time, another ship sighted – this one to leeward, on the port bow. The first ship was closing *Constitution*, at the same time hoisting signal flags, evidently trying to identify the stranger as friend or foe. Not receiving a satisfactory answer by 1430 hours she suddenly reversed course, hoisted studding sails and headed toward her consort, signals flying and guns firing, trying to warn her of a very present danger. *Constitution*, an American 44-gun frigate, commanded by Captain Charles Stewart, had been at sea since 17 December 1814, and found her long sought quarry, a British Mediterranean convoy heading for the West Indies from Gibraltar. The ships just sighted were HMS *Levant* (20 guns), commanded by Captain George Douglas, and HMS *Cyane* (34), commanded by Captain Gordon Falcon. They were rear guard escorts of the convoy that they were now forming up to defend.

The ensuing battle would become one of the best known of the frigate victories in the American annals of the War of 1812. It would make the reputation of Captain Stewart as one of the U.S. Navy's premier ship-handlers. From the British point of view, Stewart's victory was denigrated as an unquestionably better manned, more powerful ship overwhelming two stoutly fought, lightly built ships, in an uneven match. There was also a fascinating sequel to the battle as a squadron of powerful British frigates caught up with *Constitution*. They had been tracking the ship since her escape from Boston through the British blockade two months earlier and hoped to settle scores, especially after *Constitution*'s defeats of HMS *Guerriere* and HMS *Java*. Before proceeding, it is important to put these actions in proper perspective by reviewing the earlier relevant events of the war and their international context, with respect to the United States Navy.

The War of 1812 fought between the early American republic and the time-tested, though battle-weary, British Empire has been variously described as unnecessary, avoidable, an affair of honour, a stalemate, a second war of in-dependence, a victory for one side or the other, and probably all of the forego-ing, depending on one's nationality or political viewpoint. Another interpreta-tion would be to see it as representing the growing pains of nationhood for the United States, Canada and Great Britain, all of whom were at different stages of evolution. Imperial Britain and her allies had been at war with Revolution-ary France since 1792. The Royal Navy, then still the world's most powerful maritime power, had recovered from its setback during the War of the American Revolution. But the French navy was but a shadow of the naval force that had fought the British to a stalemate in the battle off the Virginia Capes, enabling the French and American armies to force the surrender of General Cornwallis's army at Yorktown in 1781.

Meanwhile, in the United States, the new government and its relatively inex-perienced political and military leaders struggled to make the engine of govern-ment function in a world at war. To a great extent the Americans, blessed with a flexible constitution and tremendous untapped resources, were inventing govern-ment as they went along. The military force at its disposal was small, weak and under the command of an aging leadership. The Continental Navy created dur-ing the Revolutionary War had been allowed to disband and for more than ten years after the Peace of Paris, the American navy did not exist. Only after news reached home that American merchant ships and sailors had been captured in the Mediterranean did the federal government act to "create a naval armament" that would in 1798 become the United States Navy under a Navy Department with Secretary Benjamin Stoddert as its chief executive. A few of the former naval officers from the Continental Navy and some with experience in Revolutionary War privateers survived to lead the navy in its first contests with the French in the Quasi-War and immediately thereafter in the Barbary Wars on the North African coast. By the time of the War of 1812, nearly all of the navy's officers and most of its enlisted sailors were those of a younger generation, born in the 1770s and 1780s, who joined the navy in 1798.[1]

Trade – and rising tensions

American merchant trade burgeoned during the years after the Peace of Paris in 1783, despite the fact that American trading ships could no longer claim the protection of the Royal Navy. Merchant ships, such as the *Empress of China* out of Philadelphia, reached the Far East as early as 1785.[2] Others penetrated the Mediterranean, and if they could avoid the Barbary corsairs, returned with valuable cargoes, particularly after 1805 when an American naval squadron had subdued, at least temporarily, the pirates of Tripoli. Exports from the United States surged from the early 1790s when the annual figure was approximately $20 million to $94 million by 1801 and to $108 million by 1807. Imports followed a similar course, from $23 million in 1790, to $110 million in 1801, and rising to $138 million in 1807.[3]

With Admiral Nelson's victory over the French and Spanish fleets in 1805 and Napoleon's victories in 1806 and 1807, the Royal Navy was supreme at sea and the French armies dominant on the Continent. Anticipating further conquests and fearing no opposing armies, Napoleon pressed on into the Iberian Peninsula and Russia, where fierce opposition strained his coalition army to the breaking point.

During these years, each power also resorted to economic warfare, attempting to weaken its enemy through denial of commodities and manufactures. To counter the Emperor Napoleon Bonaparte's "Continental System," the Royal Navy had devoted itself to a rigorous blockade of French ports in the Mediterranean, the Bay of Biscay, and the French channel ports. By the Orders in Council of 1807, the British prohibited neutrals from trading with France and her puppet state allies. In the Berlin Decree of December 1807, the French government declared any ship that had traded at British ports would be "denationalized" and subject to seizure as British property. Thus, neutral American merchants were caught in a bind and suffered losses of ships and cargoes to both Britain and France.

President Jefferson believed that if the United States withheld trade from Britain and France, their need for American grain and other commodities would prompt them, eventually, to relax their trade regulations. He proposed, and Congress passed, the Embargo Act of 1807, which prohibited American exports to Europe and limited imports. This experiment in economic coercion failed. Jefferson and his supporters underestimated the European belligerents' determination to seal off each other's trade and did not account for their alternate sources of supply. In the United States, the Embargo (ridiculed as the "Ograbme Act") infuriated New England's shipping interests and the merchants who depended on overseas trade for their livelihoods, sent the region's economy into steep decline and stimulated smuggling with Canada and other British colonies. Two years later, after Jefferson left the presidency, the Republican-controlled Congress repealed the Embargo, substituted less restrictive trade limits, and trade with Great Britain resumed, but without any concessions on the part of the British. However, there had been enough concern among English manufacturers that a revocation of the Orders in Council was under consideration in Parliament in the early months of 1812.

European warfare at sea had taken a further turn that vexed the American political leadership. The desperate need for sailors to man hundreds of ships led to impressment of sailors from the merchant ships the Royal Navy halted on the high seas. Their policy dictated that any English-speaking sailors in American ships who could not prove they were native born Americans would be presumed to be British on the basis "once an Englishman always an Englishman." Some 6,000 American sailors were pressed into British service during the years 1803-1812.[4] Royal Navy ships were also subject to desertions when in the same port with American merchant ships because of higher wages and lighter discipline. American diplomats had raised this issue with their British counterparts, to no avail. Anglo-American tension grew, particularly after the *Chesapeake–Leopard* Affair in 1807 that had erupted when a British warship fired on the frigate USS *Chesapeake* and forcibly removed four Royal Navy deserters whom the Americans had recruited for service in *Chesapeake*. From that point on, U.S. Navy captains were alert to the possibility of combat with the Royal Navy whether their nations were at war or not. In his message to Congress of 1 June 1812, President Madison proposed a debate on war or peace with Great Britain. He pronounced that "thousands of American citizens, under the safeguard of public law, and of their national flag, have been torn from their country, and from everything dear to them; have been dragged on board ships of war of a foreign nation; and exposed, under the severities of their discipline, to be exiled to the most distant and deadly climes, to risk their lives in the battles of their oppressors, and to be the melancholy instruments of taking away those of their own brethren."[5]

There were, of course, other causes of war, although there is no space here to discuss them at length. Americans on the western and northern frontiers complained of British agents provoking Indians to attack settlements and discourage migration farther westward; many felt a desire to vindicate national honour sullied by British insults to the flag and interference with commerce, a general feeling that United States had fought for independence from Britain, yet were still subject to its domineering presence; and others believed that the great experiment in democratic government was in danger if the United States did not assert itself when its national interests were being threatened. The "War Hawks" were the most forward of the groups in Congress advocating war, but they were not, as one might have expected, from the seacoast. They were a small group of vociferous men of the South and West, who persuaded a number of less enthusiastic representatives to vote with them.[6] Federalists from New England and the Middle Atlantic states opposed war with Great Britain despite their region's shipping losses and the impressment of sailors. In their view, Napoleonic France was the preferred enemy, representing a threat to liberty, property and good order. This was not, however, the view of the Republican majority in Congress. They controlled 75% of the seats in the House of Representatives and 82% of the Senate seats. The 17 June Congressional vote on war was 79 to 49 in the House and 19 to 13 in the Senate.[7]

America goes to war

The U.S. Navy, while not in any way a match for the Royal Navy in the number and variety of its warships, had a small number of stout frigates and smaller vessels under the command of experienced officers. Navy Department Secretary Paul Hamilton had polled his leading captains as to the best strategy to use in opposing the British. Commodore John Rodgers had argued in favour of concentrating most of the larger ships in squadrons and to let the lighter vessels

annoy the enemy's West Indies and coasting trade. The larger frigates and sloops of war would cruise in a small squadron of two or three on the coasts of England, Ireland and Scotland, to "menace them in the very teeth." He thought that

Secretary of the Navy Paul Hamilton initially accepted Commodore John Rodgers's proposal that U.S. warships cruise in squadrons, but altered this policy to allow single ships more latitude in the fall of 1812. (Naval Historical Center, NH 54757-KN)

since the British North American Station controlled roughly the same number of ships, the Americans would stand a fair chance of defeating them, if they met. Rodgers won Hamilton's approval, but his plan did not work out in the way expected. The British ships based on Halifax formed a squadron under the command of Commodore Philip B.V. Broke, who concentrated his ships off the New Jersey coast after the Americans departed.

Commodore Rodgers set sail in his flagship USS *President* immediately after learning of Congress's declaration of war and headed eastward to intercept a merchant convoy rumoured to be on its return to England from the West Indies. Commodore Rodgers returned with his squadron at the end of August 1812 with disappointing results. Although he had captured a few prizes, he failed to intercept the West Indies convoy. Also, owing to unwise tactics, a failure of nerve or both, he lost the opportunity to bring to action HMS *Belvidera* (36 guns), which his squadron encountered while in pursuit of the convoy.[8] Captain Stephen Decatur, Jr. had advised that "the plan best calculated for our little Navy to annoy the Trade of Great Britain, in the greatest extent, & at the same time to expose it least, to the immense force of that Government, would be to send them out with as large a supply of provisions as they can carry, distant from our coast, & singly, or not more than two frigates in company, without giving them specific instructions as to place of cruising but to rely on the enterprise of the officers."[9] This was the strategy that Secretary Hamilton adopted for the navy's campaign in the fall of 1812.

Two American frigates of interest, USS *Constellation* and USS *Constitution*, did not sail with Rodgers but their adventures have a direct bearing on this study because of Charles Stewart, who commanded both in succession. USS *Constellation* (38 guns), one of the first six frigates built and the second to be launched in 1797, had a notable war record. Under Captain Thomas Truxtun's command during the Quasi-War with France, she had captured the French frigate *L'Insurgente* (36) and had engaged *La Vengeance* (50), but was unable to capture her due to battle damage. In 1801, after a period of repairs, Navy Secretary Robert Smith ordered *Constellation* to the Mediterranean under the command of Captain Alexander Murray. The Bashaw of Tripoli had declared war against the United States for failure to pay tribute. Tripolitan gunboats attacked and captured American merchantmen as they entered the Mediterranean on trading voyages. *Constellation*'s primary tasks there were to escort merchant ships and to blockade off Tripoli. She returned to the United States in 1803 under orders to be laid up in ordinary, but the situation in the Mediterranean demanded even more ships than before, so back she went, now under the command of Captain Hugh Campbell.

During 1805, *Constellation* served again on blockade but prolonged service and lack of maintenance resulted in severe leakage.

A board of officers including Captain Stephen Decatur, Jr., Captain John Shaw and Master Commandant Charles Stewart surveyed the frigate, found the ship's hull to be in a weakened condition and recommended her immediate return to the United States for a complete overhaul. Commodore John Rodgers gave command to Master Commandant Charles Stewart and ordered him to return to the Washington Navy Yard.[10] Yard commandant Captain Thomas Tingey gave the ship a light repair, then placed her in ordinary. This lack of timely overhaul would have a lasting effect when she was finally brought back to service. This was the last Stewart saw of his ship until ordered back to command in the summer of 1812.

USS *Constitution*, then under the command of Captain Isaac Hull, had been ordered to join Rodgers's squadron but departed Chesapeake Bay three weeks later than expected and did not make the rendezvous off Sandy Hook. Hull had been busy working his green crew up to the level of performance required of a ship going into action and was awaiting the arrival of his Marine detachment and extra spars before departing on a cruise. Anxious to get under way lest he be caught by blockaders off the Virginia Capes, he sailed from Annapolis on Sunday, 5 July.[11] Hull dropped down the bay and on 12 July left Cape Charles astern, with no blockaders in sight. He headed north, on a light and variable southwest breeze. He made slow progress; by late afternoon of 16 July he was still only off the coast of New Jersey at about the latitude of Egg Harbor when a lookout reported a single ship advancing from the northeast. At about the same time, Hull sighted several ships toward shore that he thought might be Rodgers's squadron. But when the approaching ship did not answer recognition signals, it became apparent that the inshore squadron might be the enemy. It later proved to have been Commodore Broke's squadron, comprising *Shannon* (38 guns), *Africa* (64), *Belvidera* (36) and *Aeolus* (32). By the next morning, the ship closing from the northeast, *Guerriere* (38) had joined with the squadron in pursuit of *Constitution*.

Over the next several days, beleaguered by light winds, Hull bent every effort to keeping ahead of the British, wetting the sails, manning the ships' boats to tow ahead, kedging – dropping the ships' anchors ahead then winding on the anchor windlass to pull the ship forward – and finally, taking advantage of an approaching squall to bend on as much canvas as possible. Hull's ship sped on, hidden by the squall and then nightfall, and was so far ahead the next day that the enemy gave up the chase. This was the first of *Constitution*'s fortunate adventures during the War of 1812, an escape by means of skill, luck and hard work. One month

On 17 July 1812 USS *Constitution* encountered a squadron of five British warships off New York. The superior force of British ships, under the command of Commodore Philip Broke in HMS *Shannon*, chased their quarry for nearly three days. Light winds resulted in desperate measures on both sides, including kedging and the use of ships' boats to tow the ships, as seen here with *Constitution*'s crews pulling urgently at their oars. The U.S. warship would meet one of her pursuers, HMS *Guerriere*, in very different circumstances almost exactly a month later. (Painting by Anton Otto Fischer, Naval Historical Center, NH 85542-KN)

later Hull met *Guerriere* again under different circumstances, both sailing singly off the Grand Banks, where *Constitution* gained her first frigate victory of the war, by reducing her opponent to a mastless, sinking hulk.

Returning to Boston, throngs greeted Hull as the first naval hero of the war at the same time that news arrived of an American military setback in the Northwest. Having declared war in June, President James Madison ordered plans for the conquest of Canada to be set in motion on three fronts, at Detroit, Niagara and Lower Canada north of Lake Champlain. Unfortunately for him, American armies were poorly prepared and badly led by generals whose best days were long past. Led by the capable General Isaac Brock, the outnumbered British troops and Canadian militia put up a flexible and energetic defence at Detroit and Niagara. General William Hull (Captain Isaac Hull's uncle) surrendered his army and the fort at Detroit in mid-August. British and Canadian troops repulsed General Stephen Van Rensselaer's invasion across the Niagara River, although some units temporarily reached Queenston Heights on 13 October, where General Brock met his death at the hands of an American rifleman.

After several months of delay, General Henry Dearborn's Albany-based troops faltered at Champlain as they headed into Canada along the Richelieu River. The Canadian Voltigeurs under Major Charles-Michel d'Irumberry de Salaberry and Embodied Militia under Captain Joseph-Françoise Perrault had deployed along the Lacolle River. At the first sign of opposition, on 19 November, the Americans retired in confusion and retreated to Plattsburgh. That marked the end of the inept campaign to invade Canada in 1812.[12]

In early September, Secretary Hamilton, having considered the generally successful results of the past months, ordered his ships to prepare for another cruise on the basis earlier suggested by Captain Decatur. Three small squadrons would sail, each having two frigates and one smaller vessel; then at the discretion of the commodore, the ships would separate and cruise on an agreed-upon course to rendezvous later if possible. These commands were arranged by seniority,

On 19 August 1812, the 44-gun frigate USS *Constitution*, commanded by Captain Isaac Hull, decisively defeated HMS *Guerriere* in a short but intense battle off Newfoundland. This would be the first of three successful battles by that famous warship. It was during this battle that *Constitution's* crew began to call their ship "Old Ironsides." (Library of Congress)

Commodore William Bainbridge commanded *Constitution* on her second war cruise, winning plaudits for the United States Navy when he defeated HMS *Java*. Later as commandant of the Boston Navy Yard, his prickly personality led him to call for a court of inquiry on Captain Charles Stewart when that officer returned from an unsuccessful war cruise in early 1814. The court cleared Stewart of any wrongdoing. (Naval Historical Center, NH 56072)

thus Commodore Rodgers in *President* would have as consorts USS *Congress* (36 guns), commanded by Captain John Smith, and sloop of war *Wasp* (18), commanded by Master Commandant Jacob Jones. Commodore William Bainbridge was to have command of *Constitution*, sailing in company with Captain David Porter in the frigate *Essex* (32) and Master Commandant James Lawrence in the sloop of war *Hornet* (18). Commodore Decatur was to have sailed in the frigate *United States* (44), accompanied by Captain Samuel Evans in the frigate *Chesapeake* (38) and Master Commandant Arthur Sinclair's brig *Argus* (18). *Chesapeake*, however, lacked sufficient crew for the cruise and would not be ready for sea for two months. Rodgers's and Decatur's squadrons sortied from Boston together on 8 October, sailed in company for four days and then separated, with Decatur's two vessels heading southeast for the Azores islands.

Bainbridge's squadron never did sail together as *Essex* suffered delays in preparing for sea. *Constitution* got underway from Boston with *Hornet* on 27 October. Bainbridge followed his plan to sail southeast for the Cape Verde Islands, crossing the North Atlantic trade routes, and then to bear away for Brazil with the blessing of the northeasterly trade winds and to work back up the South American coast in search of targets of opportunity. He paused at the Brazilian island of Fernao de Noronha in hopes of finding news of David Porter at that agreed-upon rendezvous, but found nothing. In implementing this strategy, good fortune attended both Bainbridge and Lawrence. On 29 December, HMS *Java*, Captain Henry Lambert commanding, bound from England to India, approached *Constitution* and offered battle off the coast of Brazil near the port of Salvador da Bahia. It was a hard-fought three-hour battle in which *Java* shot away *Constitution*'s helm and wounded Bainbridge twice while *Constitution*'s guns tore away at *Java*'s rigging, took down two of her masts and fatally wounded Captain Lambert. Unable to manoeuvre his damaged ship and threatened by raking, *Java*'s First Lieutenant Henry Ducie Chads hauled down his flag in surrender.[13]

Meanwhile, Captain Porter had arrived at the Fernao de Noronha rendez-vous, picked up a message Bainbridge had left for him and departed for the next rendezvous at Cabo Frio farther south along the Brazilian coast. Bainbridge had planned to meet Porter there, but now *Constitution* had battle damage and with his own wounds to tend he decided to return to Boston for repairs and recu-peration. Lawrence, in *Hornet*, had been patrolling off Bahia hoping to capture a specie-laden British sloop of war, *La Bonne Citoyenne* (20), that had been about to depart but refused when *Hornet* made her appearance. He knew that HMS *Montague* (74), based at Rio de Janeiro, might make her appearance any day hav-ing heard of the loss of *Java*. Not anxious to tempt fate in this way, Lawrence gave up this opportunity and sailed north. He was off the Demerara River in British Guiana when, on 24 February, he encountered HMS *Peacock*, a smart brig under

Constitution is shown here delivering yet another devastating broadside into the dismasted and battered HMS *Java*, soon before the latter surrendered. This defeat, on 29 December 1812, resulting in the death of *Java*'s commanding officer, Captain Henry Lambert, and 47 others proved particularly galling to the Royal Navy. (Library of Congress)

Commander William Peake's command. *Hornet* emerged victorious after a swift engagement in which Peake died and *Peacock* sank. Lawrence followed Bainbridge back to the United States, while Captain Porter sailed *Essex* around Cape Horn to meet his destiny in the Pacific.

Captain Charles Stewart: "Proceed to sea…"

Captain Charles Stewart, born in 1778, had spent his youth in merchant sail, joined the U.S. Navy with the rank of lieutenant in 1798, served with distinction in the Quasi-War with France and Barbary Wars, and received his promotion to captain in April 1806 at the age of 27. Returning from the Mediterranean, he took up his next assignment, the supervision of gunboat construction. He worked with Adam and Noah Brown, Christian Bergh and Henry Eckford, all well-known New York shipbuilders, to bring President Jefferson's gunboat flotillas into being. By November 1806, he considered the job completed and applied to the Navy Department for a furlough to renew his interests in overseas trade.[14]

With the declaration of war with Great Britain, Captain Stewart reported ready for duty at Washington, and Secretary Paul Hamilton did not hesitate. He ordered him to the brig *Argus* on 22 June with orders to "proceed to sea and scour the West Indies and Gulf Stream – consider yourself as possessing every belligerent right of attack, capture and defense of and against any of the public or private ships of the Kingdom of Great Britain, Ireland and other dependencies."[15] On his return, Hamilton ordered Stewart to command USS *Constellation*, then undergoing an extensive rebuild at the Washington Navy Yard. As will be recalled, the ship had been scheduled for a thorough overhaul in 1806 when she returned from the Mediterranean, but this must have been postponed because it was not until March 1812 that Commodore Thomas Tingey, commandant of the Washington Navy Yard, commenced the overdue repairs. Even then, the lack of timber, equipment and ordnance delayed completion of the yard's work until at least 11 October when a gunboat captain noticed her ensign hoisted, indicating that *Constellation* had been recommissioned. During the weeks that followed gunboats brought anchors, guns and provisions for *Constellation*, apparently still in need of filling its complement as evidenced by the opening of a recruiting rendezvous in Alexandria. While these preparations were going forward, Secretary Hamilton asked Stewart for help persuading Congress to authorize a new round of ship-building to which he complied with a letter of 12 November, arguing for the construction of heavy frigates and ships of the line.[16] This communication evidently had a positive effect. Congress authorized the building of four 74-gun ships of the line and six 44-gun frigates on 23 December 1812.

Captain Charles Stewart gained experience as merchant sailor and then joined the newly formed U.S. Navy in 1798. Commissioned as a 20-year-old lieutenant, Stewart served under Commodores Thomas Truxtun and Edward Preble in wars with France and the Barbary States. He would eventually command USS *Constitution* in her last battle of the War of 1812, gaining fame for himself and the young U.S. Navy. (Library of Congress)."

Stewart, concerned that ice forming on the Potomac might soon prevent his departure, dropped down the river with his frigate still not completely ready for sea in late December. He put in at Annapolis to continue preparations in accord with instructions from William Jones, the newly-appointed Secretary of the Navy. Still not satisfied with his tests of gunpowder and lacking spare sails and other equipment, but anxious to escape the Chesapeake before British blockaders took position off the Virginia Capes, Stewart got underway for Hampton Roads. Unbeknownst to Stewart, the British Admiralty on 26 December 1812 gave orders for its ships to commence a blockade of Chesapeake and Delaware Bays.[17]

To his great chagrin, Stewart discovered on 2 February, the day he brought his ship into Hampton Roads, that seven Royal Navy war vessels (two ships of the line, three frigates, a brig and a schooner) had arrived off Cape Henry, making their way into Chesapeake Bay. He had little choice other than to lighten ship and haul her up the Elizabeth River above (south of) Craney Island, where she could be protected by the guns of Fort Norfolk and the navy gunboat flotilla. In the three months that followed, Stewart exerted efforts to strengthen the gunboats' crews and armaments, a step that would pay dividends. The enemy ships remained in control of the waters of Hampton Roads for the next two years, although at certain times Baltimore privateers and the corvette USS *Adams* found it possible to slip out. *Constellation* remained where she was for the rest of the war. Her finest hour occurred on 22 June 1813, when Admiral Warren's forces made a deliberate attempt to thrust past the gunboats and the Craney Island defences. Having anticipated the attack, Stewart's successor Captain Joseph Tarbell ordered *Constellation*'s guns, sailors and marines landed to join the artillery batteries of Virginia militia on the island. With a coordinated defence, they repulsed the British attack in what was the one of the few defeats inflicted on the British during their Chesapeake campaigns of 1813 and 1814.[18]

Secretary Jones finally decided there was a more valuable duty for the capable Stewart than commanding an inert frigate trapped in Chesapeake Bay. He ordered him to Boston to take command of *Constitution,* then refitting at Boston, having returned from South American waters and Bainbridge's successful battle with HMS *Java.* Disconcerting things can happen to ships that are in navy yards for repairs. In this case, the Navy Department had need of experienced hands to serve in the squadron then about to begin operations on Lake Ontario under Commodore Isaac Chauncey. The ship was in an interim between commanding officers, with Bainbridge nominally commanding and Stewart not yet arrived. Yielding to pleas from his friend Chauncey, Bainbridge sent 150 *Constitution* sailors to the Sackets Harbor naval base in New York State. The deficit would presumably be made up by opening a recruiting rendezvous in Boston; it was also the case that USS *Chesapeake*, Captain James Lawrence's next command, was in need of additional hands, soon to be met by a draft of more men from *Constitution.* Both calls on the crew of his new ship would not likely have met with Stewart's approval.[19] Only by the end of September 1813 was the ship was in a condition to depart. Yet, bad weather and the Royal Navy's continuing close blockade of Massachusetts Bay made escape difficult for Stewart, who finally succeeded in getting out on the last day of the year.[20]

When *Constitution* made her sortie from Boston, there were no blockaders in sight and Stewart made haste for southern waters, entering the Caribbean from the east, and then proceeded west to the Guyanas. There he chased several vessels but none was a proper prize and two evaded capture by entering shallows along the coast. Cursed with lack of wind, Stewart tried to change his luck by altering his ship's paint scheme. He painted over the stripe along the gun deck ports with ochre to make his ship resemble a British frigate. He soon took two prizes, the small merchantman *Lovely Ann,* and HM Schooner *Pictou* (16), which surrendered without a fight off Barbados.[21] *Constitution* now headed north toward the Leeward Islands, where she captured two more prizes, the schooner *Phoenix* and the brig *Catherine,* both merchantmen without valuable cargoes. Stewart then had a chance to take a British frigate later identified as HMS *Pique* but failed to come up with her in calm. The next day *Pique* had disappeared, and so he sailed for Bermuda, rounding the island south and east, but still found nothing. In fact, if Stewart had not run into bad luck he would have had no luck at all.

On 26 March, he learned that his lower mainmast had developed a dangerous crack. At this point, he realized that the ship needed a new mainmast, which
could not be supplied at sea, and so he made for Boston in the hope the Navy
Yard could give him a quick turnaround. But he had one more serious obstacle –
just off Cape Ann his lookouts sighted two ships in the southeast heading at him.
Correctly surmising they were enemy warships, Stewart set course for Salem,
just inside Cape Ann, whose entrance featured a narrow rockbound channel. In
a light breeze, with the 38-gun frigates HMS *Junon* and *Tenedos* gaining, he set
skysails and royal stunsails despite the sprung mainmast and was lightening ship
at about the time a steady breeze came to favour him. With the guidance of a local Marblehead seaman, Stewart headed north at Halfway Rock and slipped into
narrow Marblehead Harbor, under the guns of the fort. His pursuers, only six
miles behind, hauled off at that point, unsure of the inshore navigation and the
danger of gun batteries. Days later he shifted into nearby Salem, and when the
enemy frigates moved offshore, he brought *Constitution* safely into Boston.[22]

For the next eight months *Constitution* obtained the needed repairs while
Captain Stewart endured a court of inquiry induced by Commodore Bainbridge,
then in charge of the Boston Navy Yard to answer for his "early" return to port
and for certain irregularities that the prickly Bainbridge raised about an officer
whose reputation he sought to bring down to the level of his own. Though he had
enjoyed fame as the captor of *Java*, Bainbridge's loss of the frigate *Philadelphia*
and the subsequent period of captivity in Tripoli ten years earlier still rankled his
soul. Stewart's straightforward explanations won the day as well as establishing
the fact that, far from curtailing his cruise, he had not returned to port for 92
days, almost the longest absence of any American frigate until that time.[23]

The principal obstacle Stewart faced in getting his ship to sea again was the
strengthening blockade off Boston. The military and naval situation had changed
significantly since his last sortie in December 1813. By the mid-summer of 1814,
the British and allied armies had succeeded in weakening Napoleon's armies on
the Continent, driving him back from Russia and invading France. He abdicated
on 11 April, agreeing to retire to the island of Elba. In the southwest of France, the
Duke of Wellington's army, fighting its way north from Spain, captured Bordeaux
and then Toulouse before receiving news of the Emperor's abdication. These developments released military and naval units for service in North America. British troops that had seen service against Napoleon's armies were loaded into transports for service in Canada, in the Chesapeake and ultimately at New Orleans.

The Admiralty sent more ships of the line to reinforce the blockade off the
coast and raid United States' ports and harbour fortifications. In one of its most

interesting moves, the Admiralty ordered certain of their three-decked ships of the line to be "razeed" (to be cut down by one gun deck) and fully armed, thus creating stronger, but lighter and faster ships that could better contend with the U.S. Navy's "super frigates." But they also ordered the building of more powerful frigates to cope with the American 44s. Anticipating that Stewart would try to break through the blockade as blustery fall and winter weather came on, the Admiralty bolstered the blockade off Boston with three ships especially sent to defeat *Constitution*. According to British historian Rif Winfield, "this need led the Admiralty to call on [21 April 1813] for two draughts for 50-gun spar deck frigates, designs being produced three days later." Two of the ships so designed, HMS *Leander* (50), Captain Sir George Collier, and HMS *Newcastle* (50), Captain Lord George Stuart, were rapidly built and commissioned within ten months. Winfield describes them as "large 24 pdr. frigates of 50 guns (Fourth Rates)." In addition to these, the Admiralty provided *Acasta* (40), commissioned in 1797, commanded by Captain Alexander Kerr.[24] In point of fact, both *Newcastle* and *Leander* were very heavily-armed, with thirty 24-pdr. long guns on their gun decks and twenty-eight 42-pdr. carronades on their upper decks, carrying at least ten more guns than rated. For the sake of comparison, at this time *Constitution* carried thirty 24-pdrs., twenty 32-pdr. carronades, and two 24-pdr. gunades, totalling 52 guns in a frigate rated at 44 guns.[25]

Halifax – rumours of the enemy at sea

Rear Admiral Edward Griffith, based in Halifax, was in charge of the Royal Navy's ships on the North American station. The senior officer afloat in charge of the blockade was the highly energetic Rear Admiral Henry Hotham, who in December 1814 positioned his flagship, HMS *Superb* (74), off New London. Biding his time, Stewart held his ship in readiness in port until the blockaders were no longer in sight from Boston Harbor. Then, on Sunday, 17 December, *Constitution* departed to the cheers of the crowd on Long Wharf and the crew of the privateer *Prince of Neufchatel*. The British did not learn of her absence until the next day when HM Brig *Arab* (18) came in as close as she could and found the quarry had bolted. With this bad news she hastened to Provincetown, where *Newcastle* and *Acasta* had anchored for protection from the weather. On Christmas Eve, *Leander* arrived from Halifax with the squadron commodore, Captain Sir George Collier, in command. He had received reports that Stewart intended to rendezvous with USS *President* and USS *Congress* to wreak havoc in the English Channel. The rumours were false. Captain Stewart headed south to the Atlantic shipping lanes west of Bermuda. Commodore Collier took the time

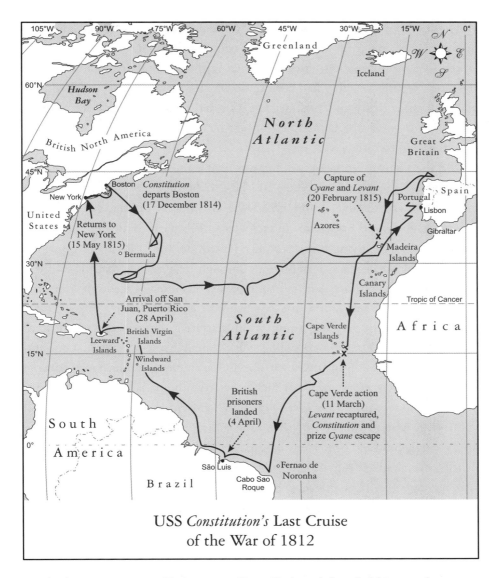

USS *Constitution's* Last Cruise
of the War of 1812

to plot his strategy over Christmas at Cape Cod, and then led his squadron east-ward toward the Azores and beyond.

On 29 December, Hotham reported to the Admiralty that "American Frig-ates *Constitution* and *Congress*, sailed from Boston and Portsmouth about the 17th Instant. *Leander, Newcastle* and *Acasta*, gone in search towards Western Islands, Madeira, Canaries, Barbados, Carolinas, and to Halifax. *Newcastle* has been on shore near Cape Cod; not much damaged."[26] Stewart discovered the *Lord Nelson*, a British merchant schooner in distress, and took her as a prize, removing her welcome cargo of food and supplies intended for Bermuda and the Windward Islands.[27] Vainly searching for a convoy, *Constitution* skirted the Tropic of Cancer and proceeded east for several days, then northeast to find the shipping lanes that linked Europe to the South Atlantic. Nearing Portugal, Stewart halted several vessels to gain information. From a Russian brig on 8 February, he learned that the peace negotiations that had been underway for many months in Ghent had yielded a peace treaty on Christmas Eve, but that ratifications had not yet been exchanged between the United States and Great Britain. With the war not yet over, the hunt for prizes went on.

Off the Tagus River, he sighted but was unable to exchange information with a Portuguese ship that later entered Lisbon harbour. There, word spread that *Constitution* was at large. HMS *Elizabeth* (74), then at Lisbon, got underway immediately but did not make contact. On 19 February, following a southerly track, *Constitution*, flying British colours, halted the *Josef*, a Russian ship whose captain knew Stewart and instantly saw through his pretence that his ship was HMS *Endymion* (38). *Josef*'s master soon passed this intelligence to the pursuing *Leander*, *Newcastle* and *Acasta*, just days behind *Constitution*.[28]

To return to the very beginning of this narrative, shortly after 1:00 P.M. of the following day, *Constitution*, sailing southwesterly, came up with first one and then a second warship of unknown size and armament, sailing loose cruising formation some 5 miles abeam of each other. Two hours later after having approached and recognized *Constitution* as a probable enemy, the first ship made all sail to close with its consort and then took station about 100 yards astern of her on a westerly heading. British naval historian William James states that both the British commanders "fully believed that she was the American frigate *Constitution*; having received intelligence, before leaving port, of her being in their intended track."[29] That being the case, one suspects they might well have devised a plan in case their courses intersected. They did hope to postpone the action until nightfall, but *Constitution* closed too rapidly.[30] In the meantime, Stewart, sensing a battle in the offing, made sail to overtake but suffered a crack in his main royal mast and had to reduce sail to effect repairs.

Within an hour, his ship was gaining. By 5:00, he had taken the position he wanted, to windward and off to one side yet equidistant between the two ships, at 300 yards distance (see tactical diagram, page 76). He broke out the American ensign; the two British ships, *Cyane* and *Levant*, hoisted their red ensigns and waited for *Constitution* to fire first.[31] In these two warships, Stewart, as experienced a sea officer that he was, faced men of considerably more fighting experience than his own. Captain Gordon Thomas Falcon, *Cyane*'s commanding officer, received his first commission in May 1800, served in *Leopard* (50) and *Barfleur* (98), was promoted to commander in *Melpomene*, advanced to post captain in

1813 and was given command of *Cyane* in March 1814. As a lieutenant in *Leopard*, Falcon was the officer who boarded the frigate *Chesapeake* in 1807, after she was fired on, and demanded to examine the crew for deserters.[32] Captain George Douglas, *Levant*'s commanding officer, was senior to Captain Falcon and as such he bore responsibility for conducting their operations. Douglas had entered the Royal Navy as a midshipman in *Excellent* (74), transferred to the frigate *Castor* (32), then to *Spencer* (74), and in that ship took part in the Battle of Trafalgar. Made post captain on 28 February 1812, he assumed command of *Levant* in April 1814 at 27 years of age.

Constitution vs. *Cyane* and *Levant*

In recounting details of the *Constitution* vs. *Cyane* and *Levant* engagement, there are factual disagreements on both sides that are difficult to reconcile. According to Stewart, he fired first and immediately received fire from both ships. According to the contemporary British naval historian William James, the distance was "a full three quarters of a mile" between the opposing ships. James's statements, however, are open to question since his writings, subject to exaggeration, frequently show an anti-American bias. Three quarters of a nautical mile is about 1,500 yards, five times as far as the 300 yards that Stewart stipulated in his report. This is not an idle question, for it brings into play the effective range of the British guns that in the case of *Cyane* and *Levant* were almost entirely carronades. *Cyane* carried twenty-two 32-pdr. carronades, eight 18-pdr. carronades and two long 9-pdrs. Thus, she had 30 carronades of different strengths and throw weights plus the two long nines. *Levant* mounted 20 guns, including eighteen 32-pdr. carronades, and two long nines. At 300 yards, the carronades would have had telling effect, but at the range of three quarters of a nautical mile, carronades, being short-barrelled, would have lost much of their power. The British estimates of distance suggest that *Constitution*'s preferred tactic was to stand off at greater-than-carronade range and fire away with her long 24-pdrs.

Stewart's report indicates he put *Constitution* well within carronade range from the outset. If so, why would he have done that? To give the smaller British ships a sporting chance? This is unlikely, so the question is: why did he state that he brought *Constitution* that close, within carronade range? We do not know the answer, but the point of battle is to win, not to flirt recklessly with danger when so many other things can go wrong. The only logical conclusions are that it was either a mistaken judgment of distance or an attempt to impress others with the possible risks he was taking in choosing to fight two enemy ships at once. If he was so close, one wonders why the butcher's bill was so low. An answer to this may be

This dramatic Thomas Birch painting captures the close engagement of *Constitution*, *Cyane* and *Levant* as the three ships manoeuvred for advantage in the moonlight in the eastern Atlantic near the island of Madeira. (New York Historical Society)

seen in James's comment that "both ships returned her fire, but having only car-ronades, their shot all fell short, while *Constitution*'s 24 pound shot were cutting to pieces their sails and rigging." The opposing ships' strength in terms of manpower was also of great disparity. *Constitution*'s normal complement was 450 officers and enlisted, more than those on board *Cyane* (171) and *Levant* (131) combined, 302 men and boys. James claims that *Constitution*'s actual complement on the day of battle was 369 men and three boys. He also states that *Constitution*'s "original complement was 480 at least" although he gives no source for that statement.[33]

As the battle proceeded, *Constitution* fired at both ships, but at first concen-trated on *Cyane* on his port quarter. With considerable smoke and a setting sun, the light was diminishing. After ceasing fire to clear the air, Stewart saw Captain Falcon moving his ship to starboard to gain a raking position. Stewart gave orders to back the main and mizzen topsails and loosed his jibs to back the ship, making sternway in the direction of *Cyane* to give her a broadside. *Constitution*'s carron-ades cut up *Cyane*'s rigging, sails and spars. Captain Falcon's *Levant* attempted to close *Constitution* and cross her bow in a raking position. Stewart ordered his sails braced around, and when they filled away, the towering frigate surged forward and raked *Levant* as she wore. As *Levant* withdrew to replace rigging and spars,

Stewart saw *Cyane* about to renew battle. He manoeuvred his ship to a position to rake the ship's larboard quarter and at 6:50 P.M. *Cyane* struck her colours.[34]

Captain Falcon's after-action report described his difficult position. "It was my constant endeavor to close with the enemy, finding we were too far distant for the carronades, at the same time exposed to the full effect of his long guns, & obtain a position on his quarter, in this however, I was only partially successful, as the situation and superior sailing of the enemy's ship enabled him to keep the *Cyane* generally on his broadside, consequently exposed to a heavy fire from which in the early part of the action the Ship suffered very much in the rigging and latterly in the hull." When later Falcon tried to wear ship to conform to *Levant*'s movements, he found *Cyane*'s rigging to be so cut up as to be unmanageable. "I had the mortification to find that not a brace or a bowline, except the larboard fore brace were left, but observing the *Levant* was exposed to a heavy raking fire, the *Cyane* was brought to the wind on the larboard tack, unfortunately with all the sails aback in which situation the action was maintained so long as gun would bear, the smoke under the lee preventing my discovering for some time that *Levant* was continuing before the wind."

When the firing halted for a short time, Falcon ordered the reeving of fresh braces but could not finish before *Constitution* took position close to his larboard quarter, firing into *Cyane*. This put Falcon in an impossible situation, "nearly the whole of the standing and all of the running rigging cut, the sails very much shot and torn, – all the lower masts severely wounded, particularly the main and mizzen masts both of which were tottering. Foreyard, Fore and Main topmasts, Gaff and Driver boom, Main topgallant yard and foretopgallant mast shot away or severely wounded, – a number of shot in the hull, eight or nine between wind and water, – six guns dismounted or otherwise disabled by shot, drawing of bolts, etc. – with a considerable reduction of our strength in killed and wounded." He consulted his officers and found their views, like his own, to be that the crippled state of the ship and the situation prevented any prospect of success against a force considerably more than double their own and that as further resistance would only be attended with loss of lives equally unavailing and unnecessary, he felt it his duty to strike the colours.[35]

Stewart sent an officer and some marines to take possession of *Cyane* and remove her officers. During this interlude, Captain Douglas drew *Levant* off to reeve new braces and at about the time Stewart directed his ship toward him, Douglas bravely headed back into the fray. The two ships passed starboard to starboard and exchanged broadsides. Douglas wore ship while Stewart manoeuvred *Constitution* to rake *Levant*. In Douglas's own words, "At 9:10 P.M. finding

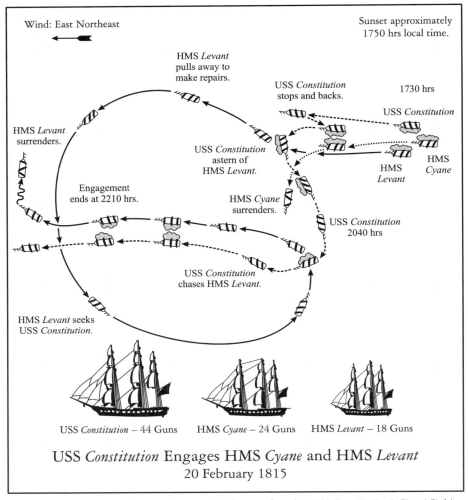

Wind: East Northeast

Sunset approximately 1750 hrs local time.

HMS *Levant* pulls away to make repairs.

USS *Constitution* stops and backs.

1730 hrs

USS *Constitution*

HMS *Levant* surrenders.

USS *Constitution* astern of HMS *Levant*.

HMS *Levant*

HMS *Cyane*

Engagement ends at 2210 hrs.

HMS *Cyane* surrenders.

USS *Constitution* 2040 hrs

USS *Constitution* chases HMS *Levant*.

HMS *Levant* seeks USS *Constitution*.

USS *Constitution* – 44 Guns   HMS *Cyane* – 24 Guns   HMS *Levant* – 18 Guns

USS *Constitution* Engages HMS *Cyane* and HMS *Levant*
20 February 1815

Drawing by Christopher Johnson, after a similar diagram found in *USS Constitution's Finest Fight: The Journal of Acting Chaplain Assheton Humphreys, U.S. Navy*, Tyrone G. Martin, editor.

it was out of my power to weather him, pass'd close under his lee and gave him our starboard broadsides, as long as the guns would bear, receiving at the same time a most heavy and destructive fire from the Enemy both in the rigging and in the Hull, at 9:30 P.M. finding that the *Cyane* had undoubtedly been obliged to strike her colours, the *Levant* was again put before the wind with hopes of saving the Ship, receiving several heavy raking broadsides in wearing from the enemy who were in chase of us. Every effort was now to make all sail, but owing to the crippled state the ship was again in, the whole of the lower and running rigging, the wheel, Main topgallant yard, Mizzen topmast and Starboard foretopmast Studding sail boom being shot away, the lower Masts much wounded and the sail shot and torn to pieces, caused an unavoidable delay." Stewart followed, gaining slowly while firing two long 24s from the bow, through his bridle ports. These wreaked havoc on *Levant's* quarterdeck. Douglas at last capitulated, firing a gun to leeward at 10:10 P.M., by the American version or at 10:40 P.M. by the British version.[36] The losses on each side were as follows: *Levant*, 6 seamen and marines killed, and 1 officer and 14 seamen and marines wounded; *Cyane*, 6 killed and 13 wounded. Thus, the British suffered 12 killed and 28 wounded. The Americans lost 4 dead and 14 wounded.

Compared to single-ship naval actions earlier in the war and considering the armament available, the casualties were surprisingly light on both sides. In tabular form, this is how these results appear (men who died of wounds are combined with those killed):

Battle Casualties – Navy Vessels, Atlantic, 1812-1813

Engagement	Americans	British
Constitution vs. *Guerriere*	K-7 W-8	K-6 W-56
Wasp vs. *Frolic*	K-5 W-5	K-30 W-60
United States vs. *Macedonian*	K-6 W-5	K-43 W-61
Constitution vs. *Java*	K-11 W-22	K-48 W-102
Hornet vs. *Peacock*	K-1 W-2	K-8 W-30
Chesapeake vs. *Shannon*	K-61 W-85	K-33 W-50
Enterprise vs. *Boxer*	K-2 W-10	K-7 W-8
Constitution vs. *Cyane* and *Levant*	K-4 W-14	K-12 W-28

While it may be seem insensitive to reduce battle casualties to statistics, in just these eight selected actions from 1812 and 1813, the American losses in dead (including mortally wounded) averaged 12.1 and wounded 18.6, while British losses averaged 23.3 dead and 49.4 wounded. Americans in battle suffered only half as many killed as the British and only one third as many wounded as the British. This ratio almost exactly matches the *Constitution* vs. *Cyane* and *Levant* casualties. By these measures, all three ships suffered casualties far lower than might have been expected, considering the disparity in ships' sizes, personnel and weaponry. Generally it appears American gunnery was more deadly (in fatalities) than the British in these contests by at least 50%. Possible conclusions are that *Constitution* kept her distance from the smaller ships' powerful carronades and that her long 24-pdrs. were either concentrating on spars and rigging rather than British personnel and hull or were less accurate than could have been expected. The onset of darkness and difficulty in aiming might have influenced the result, with only slight assistance from moonlight. As the American historian Theodore Roosevelt wrote: "It was now moonlight and an immense column of smoke formed under the lee of the *Constitution*, shrouding from sight her foes; and as the fire of the latter had almost ceased, Captain Stewart also ordered his men to stop so as to find out the positions of the ships."[37]

While both *Cyane* and *Levant* were somewhat smaller, more lightly built, and possessed less numerous crews than *Constitution*, they did together possess a greater weight of shot and more powerful carronade batteries. Although *Constitution* had the advantage in larger crew and number of long 24-pdrs., the two British ships could, if manoeuvred differently, have posed difficult problems for their opponent in fighting two enemies at once. Sailing, as they were at first, in line-ahead formation, they presented a most convenient target as *Constitution* overtook them to windward and commenced the gunnery duel. Whether this was British battle doctrine at work is unknown, but one could imagine a sudden breakaway at some earlier point to lead *Constitution* out of her dominant (windward) position. One British ship might have fired on the beam while the other aggressively sought a close raking position astern or ahead. But this was not the case, and the two ships remained in relatively the same positions.

Let us also consider differing views on Stewart's famous battle manoeuvre, in Roosevelt's words, "giving a broadside to the sloop (*Levant*), Stewart braced aback his main and mizzen top-sails, with top gallant sails set, shook all forward, and backed rapidly astern under cover of the smoke, abreast the corvette (*Cyane*), forcing the latter fill again to avoid being raked." This, combined with skilful ship handling led Roosevelt to proclaim that "as regards *Constitution*, her maneuvering was as brilliant as any recorded in naval annals, and it would have been simply impossible to surpass the consummate skill with which she was handled in the smoke, always keeping both of her antagonists to leeward, and while raking both of them not once being raked herself."[38] James's account does not mention this manoeuvre. Historian Tyrone Martin states that Stewart "threw his main and mizzen sails flat aback, with top gallants still set, shook all forward, let fly his jib, braked his ship, and unleashed a heavy fire." He does not assert that Stewart actually "backed" his ship. In a footnote, he states his opinion that "this seems unlikely given the swift sequence of events. What is more likely is that *Constitution* rapidly decelerated, causing her surprised foe to move forward on her larboard side, well within lethal range of the American batteries."[39] On the other hand, Captain Stewart positively asserts in his report to the Secretary of the Navy that he did in fact achieve sternway, stating, " we found ourselves abreast of the headmost ship, the stern most ship luffing up for our larboard quarter, – we poured a broad side into the headmost ship, and then braced aback our main and mizzen topsails, and backed astern under cover of the smoke, abreast of the stern most ship, when the action was continued with spirit and considerable effect...."[40] I accept the latter version since it is well documented. After all, it is perfectly well known that square-rigged ships were capable of and often had to perform the backing manoeuvre while navigating in narrow straits, tortuous channels and other situations. There is no reason why it could not have been used in battle, especially as a surprise by a well-trained crew under cover of darkness and obscured by smoke. That Stewart used the tactic of backing in battle is testimony to his coolness under stress and confidence in his ship and crew. It is a virtually unique instance of this manoeuvre in the annals of fighting sail.

With the battle over, Stewart ordered the British officers and seamen taken prisoner and transferred to *Constitution.* His next task was to make repairs on the two prizes and get underway as quickly as possible. He realized that other British warships were on the prowl; since he had reduced his crew by the necessity of manning the prizes, his ship was now more vulnerable. He headed south under easy sail while the crew travelled back and forth to *Cyane* and *Levant* with materials to repair sails, blocks, and the running and standing rigging for the long voyage ahead.

The newly captured British officers, it was reported, treated each other with insults and contempt as they argued about what had gone wrong in the battle and who was most at fault. Even Captains Douglas and Falcon were at loggerheads, until finally Stewart told them that they would have lost no matter what they had done differently, and that if they doubted it, he would put them back on board their ships to try it all over again. Apparently this silenced the argument.[41] They might, however, have comforted themselves with the thought that their ships, acting in concert, had saved the convoy it was their mission to protect. It is a legitimate function of escorting vessels to sacrifice themselves, if necessary, to save a convoy. Numerous examples of this exist, of which perhaps the most dramatic was the action of Captain Richard Pearson in HMS *Serapis* in 1779. He saved a Baltic convoy in the North Sea though he surrendered his ship, when the American Captain John Paul Jones and his squadron attacked *Serapis* off Flamborough Head in a fierce evening engagement. The king knighted Pearson for his conduct and the merchants whose ships he saved celebrated and rewarded his bravery.

Escape from Cape Verde

Stewart led his small squadron of prizes toward the Portuguese-controlled Cape Verde Islands, close by the west coast of Africa, often the last port of call for ships heading for the Caribbean or South America (see map, page 71). There he could land his prisoners and take on water and other provisions, while plotting a course for the return to the United States. While ostensibly neutral toward the United States, the Portuguese were friendly toward the British with whom they had been

allied in the Peninsular War. Stewart would have to take care not to remain any
longer than necessary. As he neared the islands, fog banks reduced visibility and
slowed his progress, but he dropped anchor in the harbour of Porto Praya on 10
March. On the very next day, while *Constitution*'s crew was engaged in transfer-
ring prisoners and taking on stores, Lieutenant William Shubrick, the officer of
the deck, noticed a British midshipman being reprimanded by a British lieutenant for exclaiming excitedly that ships, shrouded in fog except for upper masts,
were approaching the anchorage. Shubrick called Stewart, who instantly came
on deck and saw he was in imminent danger. Commodore Sir George Collier's
three-ship squadron had at last found its prey and was about to pounce. Stewart
quickly ordered his ships to cut their anchor cables and leave the harbour without delay.[42] That the two prize masters and crew could comply indicates their
readiness to react to this no-doubt-anticipated possibility. *Constitution, Cyane*
and *Levant* were under way in four minutes, clearing East Point, as the pursuing
ships, *Leander, Newcastle* and *Acasta*, set out in chase and the Portuguese fort's
batteries came alive to fire at the fleeing Americans.[43]

Although Stewart's course initially was eastward, it took only an hour and a
half of fast sailing to discover that while *Constitution* was outdistancing her consorts and the British squadron, Captain Kerr's weatherly *Acasta* was beginning to
cut the distance to the American prizes, who were still in a line-ahead formation.
Fearing that *Cyane*, the slower of the two prizes, would soon come with range of
Acasta's guns, Stewart hoisted signals ordering the prizemaster, Lieutenant Beek-
man Hoffman, to tack to the northwest. He did so with alacrity and much to
Stewart's surprise none of the British ships tacked to pursue *Cyane*. They remained stubbornly in the wake of *Constitution* and *Levant*. By 2:30, it appeared
that *Levant* would soon fall victim to the overtaking enemy. Stewart made signals
to prizemaster Lieutenant Henry E. Ballard, instructing him, likewise, to tack.
Again, Commodore Collier surprised Stewart. Instead of dividing the squadron,
perhaps sending two ships to continue the chase after *Constitution*, all three ships
tacked after *Levant*. At that point, Ballard understood that his ship, being under-
manned, outgunned and outnumbered, would soon be taken. He headed back
toward Porto Praya and was tempted run his ship aground, but changed his mind
because of the number of prisoners who might have drowned. He anchored inshore, and when subjected to gunfire, hauled down his colours. The Cape Verdes'
"neutrality" had mattered not a whit; Ballard and his shipmates now experienced
turnabout as they became prisoners of the British.[44]

With the entire British squadron committed to the capture of *Levant*, Stewart
was left free to continue on his way, trusting that Lieutenant Hoffman could sail

Cyane to the United States without escort. After running south to the latitude of Guinea, Stewart shaped a course for Brazil, hoping to intercept the British frigate *Inconstant* (36), which he had learned was en route carrying specie, one million pounds sterling in bullion, from the Rio de la Plata to England. Despite keeping a sharp lookout, this was not to be, and two weeks later *Constitution* made landfall near Cabo São Roque and headed toward Maranhao, an island seaport in the mouth of the Itapecuru River. Here he allowed the British officers, on their parole, to make arrangements with local officials to find transportation for their crew to return to England (see map, page 71), but Stewart still did not know whether the Treaty of Ghent had been ratified. Stewart's homeward-bound voyage finally commenced on 15 April. Thirteen uneventful days later, off San Juan, Puerto Rico, Lieutenant William Hunter went ashore and obtained the good news that peace was at hand, having been in effect since ratifications became official in late February. Captain Stewart arrived in New York harbour on 15 May, firing a 15-gun salute. New Yorkers received him with much celebration. They were

After the battle, Stewart and crew took the better part of a week to replace shattered spars and severed running rigging of the two prizes. They then proceeded, sometimes in open order, sometimes in close formation, toward the Cape Verde Islands, where Stewart hoped to land prisoners and re-provision. Stewart had to keep a close watch on *Cyane*, the worse damaged and slower of the two. Though this artist, as did others, indicated that the American ship towed her prize, as yet no written record of when, or if, this actually happened has been found. (Library of Congress)

prepared because news of his victory had preceded him more than a month earlier with Lieutenant Hoffman's arrival in *Cyane.* The U.S. Congress rewarded Stewart handsomely with a gold medal struck in his honour and prize money provided for officers and enlisted men for the value of *Cyane* and *Levant. Constitution*'s odyssey came to a close with her return to her home port of Boston, but the fame of the ship's three victories under the command of three successive captains became a legend of great pride in the United States.

Why *Levant?*

This was not the end of the story. Questions remained about Commodore Collier's curious decision to pursue *Levant* instead of *Constitution,* the ship he had blockaded at Boston and pursued across the entire Atlantic Ocean. William James, so frequently critical of American naval practices, showed himself at a loss to explain this event. As he stated, "when the forces of those ships, (each two of which threw a heavier broadside than the *Constitution,*) and the distinguished character of the officers commanding them, come under consideration, it absolves the British from anything like an unwillingness to fight: at the same time, we must all regret, that it should have been deemed expedient to withhold from the public eye, those "untoward circumstances which led to *Constitution*'s, – as it now appears, – most unaccountable escape.""[45] Roosevelt thought that "had the *Newcastle* and *Acasta* kept on after the *Constitution* there was a fair chance of overtaking her, for *Acasta* had weathered on her and the chase could not bear up for fear of being cut off by the *Newcastle.* At any rate, the pursuit should not have been given up so early." He also quotes James's opinion, "it is the most blundering piece of business recorded in [his history of the war]."[46] Yet, James neither states precisely what the "blunder" was nor who committed it, although it may be assumed to have been that of the squadron commodore for he bore the full responsibility.

John Marshall's account provides more detail as to what happened. He quotes a letter of Commander I. McDougall, former first lieutenant in *Leander,* who explained what took place that day on the flagship's quarterdeck: "Shortly after this conversation, the other ship [*Levant*] tacked and Sir George Collier ordered the *Acasta*'s signal be made to tack after her. In making the signal, the *Acasta*'s distinguishing pendants got foul, and before they could be cleared the *Newcastle* mistook it for a general signal. Fearing the consequences of such a mistake, Sir George desired the optional signal to be hoisted with the *Newcastle*'s pendant, and I am positive he never intended for her to tack. When the *Acasta* had filled on the starboard tack, I observed to Sir George, that if the ships standing in shore were really frigates, which it was impossible to ascertain, owing to the haziness of the weather, they would be more than a match for the *Acasta*. He replied, 'It is true Kerr can do wonders, but not impossibilities; and I believe I must go round as when the ship that tacked first hears *Acasta* engaged she will naturally come to her consort's assistance.' Sir George then asked if I saw the headmost ship [*Constitution*] and the *Newcastle*. I went with my glass to look and observed the latter but could not see the former. He then, after looking through his glass, ordered the helm to be put down; and shortly after we had filled the *Newcastle* was observed to tack, which circumstance displeased him very much; but he remarked that he was satisfied if she had been gaining upon the enemy's ship and keeping her in sight, Lord George would never have discontinued the chace." After *Leander* had closed on *Levant* and fired on her, where she was anchored, close to shore, "and to our great mortification [we] observed she was a corvette or 20 gun ship … on leaving the anchorage [the next day] Sir George Collier displayed the greatest zeal and anxiety to meet the *Constitution*; and if we had not fallen in with an American vessel that gave us authentic information of the peace, there is little doubt that *Leander* would have met her singly, having taken up the exact position that would have caused a junction."[47]

In support of Collier's account, the lack of visibility at sea that day must have been highly frustrating. It was not only the fog that may have been usual at that time of year, but in addition a very fine sand was blowing west from the coast of Africa. As Collier wrote, "on the 8th the weather was, in addition being very hazy [undecipherable] from the fine sand with which the air was impregnated, so that a quantity was collected from the Sails & Rigging of this circumstance assisted so materially that the land was not possible been seen more than a mile off." He remarked that his navigation was also affected and he made sure to stand off the Cape Verdes until he could better see the coast and found himself to the leeward of Porto Praya on the Island of St. Iago. At that point he approached the harbour and at a distance of about five miles, "my mortification was increased to the highest degree by observing three ships apparently Frigates in the act of making sail from the anchorage of Porto Praya & hauling by the wind on the Larboard Tack without waiting to cross Top Gallant Yards!"[48] From this account, one can readily see that Collier was not yet aware that *Constitution* had defeated and captured *Cyane* and *Levant*. He was then under the impression that the ships in company with *Constitution* were American frigates, not the less powerful ships they turned out to be.

In the 17 June 1815 issue of *Niles Weekly Register*, a Baltimore newspaper of national scope, another explanation appeared. After the *Levant* tacked, *"Leander,*

Sir George Collier, who was most astern then made signal [to] the *Acasta* to tack, and the *Newcastle*, Lord George Stuart, to continue the chase. The *Acasta* sailed faster than the *Constitution* and was gaining on her, the *Newcastle* about the same rate of sailing, and the latter fired several broadsides, but the shot fell short one to two hundred yards. After the other ships tacked [to follow *Levant*], the *Newcastle* made a signal that her foretopsail yard was sprung, and tacked also. The British officers on board, who had expressed the most perfect confidence that the *Constitution* would be taken in an hour, felt the greatest vexation and disappointment, which they expressed in very emphatic terms… (The springing of the foretopsail yard of the *Newcastle* was the subject of much joking at Barbadoes where the squadron went after the chase. It was Lord George Stuart's *heart* that was unfit for service)."[49] This implication undermines the "blunder" hypothesis that James advanced and puts into question the "character" argument he uttered in defence of the mere thought that a British commander might exhibit faintness of heart. If Captain Stuart's reason for not following orders was merely a false excuse, then it fell upon Commodore Collier to shoulder the burden of not capturing or destroying *Constitution,* unless he were to publicly accuse a brother officer of cowardice, something he apparently did not wish to do. In a recent British utterance on the matter, Winfield states, " *Leander* was not as weatherly as most British frigates, but was decidedly fast and might have caught the *Constitution* in March 1815 had not a signaling blunder diverted her squadron."[50]

In a pathetic epilogue to the Cape Verde action, the Admiralty did not hold an inquiry or court martial into the matter, nor did it make known what really happened in any other way. Naval historian Andrew Lambert has explained the aftermath. "Mortified by William James's account of his action with *Constitution* in his *Naval Occurrences of the War of 1812* which implied incompetence at the very least, Collier requested an Admiralty rebuttal. When this was denied, 'in a fit of insanity,' he cut his throat on 24 March 1824. American accounts agree with Marshall that the weather was too thick for identification of the ships…."[51] It can be agreed that the fog and haze was a major factor in explaining Collier's decision to tack and assist *Acasta*, but it does not explain the behaviour of Lord George Stuart's *Newcastle*. The question of the signal sent regarding a sprung foretopsail yard, raises the issue – was this what caused *Newcastle* to tack or was it because Stuart chose to interpret Collier's signal to *Acasta* as a general signal to tack? It seems Stuart may have suddenly realized that if he continued pursuit of *Constitution*, it could end being a single-ship action, rather than two or three frigates against one. For this historian, it remains an open question.

"Never has she failed us!"

Thus ended the saga of "Old Ironsides" in her last battle. As intriguing as it was, the engagement had no impact on the War of 1812. It was, for many Americans, a symbolic coda that reminded them of the importance of national pride, pointing the way to true independence and freedom on the seas. This was not *Constitution*'s last cruise, for she continued to serve the United States Navy in peacetime, protecting commerce, showing the flag, as a training ship, and finally as a national naval icon, similar to Admiral Lord Nelson's flagship HMS *Victory*. "Old Ironsides" remains in commission in the United States Navy as the oldest serving warship afloat and is open for visitors at the Charlestown Navy Yard in Boston, Massachusetts. Her achievement, attaining this level of prominence and protection, has not been a smooth path. There have been moments in her career when the expense of repairing and rebuilding endangered her existence. It was at such a time in 1830 that Oliver Wendell Holmes wrote his memorable poem "Old Ironsides," and Commodore Bainbridge one year later offered his toast "The Ship! Never has she failed us!"[52] Commodore Charles Stewart went on to become the grand old man of the United States Navy. In 1861, during the first year of President Abraham Lincoln's administration, Congress honoured Stewart, placing him at the head of those honoured with the rank of rear admiral (the first) on the retired list, at the age of 82 with more than 63 years of active service in the United States Navy.[53]

The U.S. Congress rewarded Commodore Stewart handsomely with a gold medal struck in his honour and prize money provided for officers and enlisted men for the value of *Cyane* and *Levant*. (Naval Historical Center)

Taking the *President*

HMS *ENDYMION* AND THE USS *PRESIDENT*[1]

Andrew Lambert

THE WAR OF 1812 WAS A CURIOUS CONFLICT, and for the United States, as Henry Adams observed:

> The worst disaster of the naval war occurred January 15, when the frigate *President* – one of three American forty-fours, under Stephen Decatur, the favourite ocean hero of the American service – suffered defeat and capture within fifty miles of Sandy Hook. No naval battle of the war was more disputed in its merits, although its occurrence in the darkest moments of national depression was almost immediately forgotten in the elation of the peace a few days later.[2]

He may not have realized quite how long a shadow this event would cast.

For all the skill and courage involved, single-ship actions rarely exercise a profound influence on international relations. But the War of 1812 was an unusual war. It has often been defined as little more than a succession of such Homeric contests, contests in which the new boys took on the old masters and whipped them. Of course if the Americans had been as successful as their propaganda claimed then Canada would long ago have become a part of the United States.

USS *President*

In 1794 the United States decided to build a new navy, specifically to deal with the Algerine corsairs. The corsairs regularly preyed on American merchant shipping in the Mediterranean and held her sailors for ransom. In the long term it would be more economical, dignified and appropriate to combat the pirates than pay them off with warships and stores. The biggest ships of the new fleet would carry 44 guns, the same force as the largest Algerine vessels. Shipbuilder Joshua

Humphreys exploited his brief to design a super-frigate significantly larger than existing European frigates. With the hull strength of a battleship and a large battery of heavy guns, the American ships would be ensured of success in battle. Humphreys designed a long, broad-beamed ship mounting thirty 24-pdr. cannon on the gun deck, with space for up to twenty more cannon, or by 1812 32-pdr. carronades, on the extended quarter deck and forecastle. Equipped with an enormous spread of canvas on unusually strong masts, these ships had the potential for high speed and superior performance in the hands of experienced seamen, despite their size and weight. Although the 44s would go on to achieve iconic status, they attracted little attention before 1812. No-one considered them the equal of a battleship, and none of them had seen action with another frigate.

President was the last of three 44-gun frigates to be completed, launched from Forman Cheeseman's yard in New York on 10 April 1800. She entered service carrying an elaborate bust of George Washington as a figurehead – replaced by a simple scroll head before 1812. Once in service it was clear that *President* was the best sailing ship among the 44s, very fast going large and before the wind, steady in a seaway and rarely straining her standing rigging. By contrast the sluggish

The best sailing frigate and pride of the United States Navy, the USS *President* is seen here in profile with a fair wind behind her. (Naval Historical Center, NH 592)

United States was known as the "Old Wagon." *Constitution* was considered her superior, but one captain reputedly offered $5,000 to exchange *Constitution* for *President*.[4] *President* was the pride of the United States Navy, the chosen command of her senior seagoing officer, and the British knew her name.

In 1811, after extensive service in the West Indies and the Mediterranean, *President*, Captain John Rodgers, was off New York looking for a warship reported to be interfering with American merchant shipping. Anxious to avenge the humiliation of the *Leopard–Chesapeake* incident of 1807 Rodgers was looking for a fight; he painted the name of the ship in the largest letters the topsails could carry, to ensure no-one mistook her for a British warship. On the night of 14-15 May Rodgers encountered the small British sloop *Lille Belt* and opened fire. The British lost 11 dead and 21 wounded, the Americans one wounded. Despite Rodgers profuse apologies, his actions might have caused a war under any other circumstances; but Britain and the United States were already on the verge of war and the bloodshed had little effect.

Not that Rodgers was troubled by such issues. He was more concerned to prove the fight had been worthy of his mighty vessel, but attempts to magnify a diminutive ex-Danish sloop into a frigate only served to highlight his poor eyesight, a failing that would dog his wartime career and that of the *President*. Unlike her sisters *Constitution* and *United States*, *President* did not fall in with a Brit-

Dimensions and Armament		
	USS *President*	HMS *Endymion*
Length overall	204'	not available
Length lower deck	173' 3"	159' 3⅜"
Length of keel	145'	132' 3"
Extreme beam	44' 4"	42' 7"
Depth in hold	13' 11"	12' 4"
Draft	22' 6"	18' 11"
Tonnage	1,533.5 tons	1,277 tons
Armament (1815)		
Main deck	30 24-pdrs.	26 24-pdrs.
Upper deck	20 42-pdr. carronades	20 32-pdr. carronades
	2 long 24-pdrs.	1 long 18-pdr.
	1 24-pdr. howitzer	
Fore top	2 brass 6-pdrs.	
Main top	2 brass 6-pdrs.	
Mizzen top	2 smaller guns.[3]	
Broadside weight of fire	916 pounds	676 pounds
Crew	480	346

On the night of 14/15 May 1811, USS *President*, Captain John Rodgers, overtook and exchanged gunfire with the small British sloop *Lille Belt*. In this image the British ship is represented as being the same size as her opponent, a far bigger and heavier ship. (Naval Historical Center, NH 58942)

ish frigate willing to fight, the futile pursuit of HMS *Belvidera* on 23 June 1812 being the nearest she came to an enemy warship. Rodgers was both unlucky and over-cautious. This mattered because in 1812 *Constitution* captured the 18-pdr. frigates HMS *Guerriere* and *Java*, while *United States* took the similar British-built HMS *Macedonian*. In all three cases the Americans were quick to call these frigate actions "equal," which they were not, rather than celebrating the skill that Isaac Hull and Stephen Decatur had shown in their engagements. It was powerful propaganda to represent these battles as fair tests of the two navies, and by extension the two nations. Such claims mattered because the original American war plan, the military conquest of Canada, had gone dreadfully wrong.

"Fair and equal"
These single-ship actions were affairs of "honour," the theatre of battle being governed by an unwritten Anglo-American chivalric code that determined the manner in which actions opened and when it was "proper" to surrender.[5] Ulti-

mately victory and defeat were less significant than honour and dignity. Yet such niceties were more often observed in the breach. The American victories of 1812 were powerful propaganda because they were claimed to be "fair and equal" contests between ships of the same rate, actions in which superior skill and the merit of the national cause were the key to success. If the Americans defeated British ships of the same or superior force, they could claim to be the masters of the ocean. If their triumphs were secured by superior power, they were of no political or propaganda value whatsoever.

While excitable journalists and anxious Republican politicians accepted the "fair and equal" argument, there were problems. Under American Prize Law the crew of a ship that captured one of equal or greater force was entitled to receive the entire sum the state paid for the prize vessel. If the prize was of inferior rate, then half the prize went to the Navy Pension Fund. Consequently "fair and equal" became a question of money. After taking HMS *Macedonian* in October 1812, Commodore Stephen Decatur boldly claimed the prize was of equal rate to his ship, the USS *United States*. Although well aware that the claim was false, he still claimed the full prize payout and the additional laurels that such an achievement would warrant. Since anyone with an eye could see the truth when the prize entered port, one of his officers, who also stood to benefit from the deception, told a national newspaper that the claim of equality reflected the fact that the British chose to fight in such ships. American Treasury Secretary Albert Gallatin was unimpressed. He declared *Macedonian* "was a vessel inferior both in Caliber and number of guns, as well as ... men," and refused to pay the full amount. Decatur gave way; it would not be his last defeat.[6]

Both "unlucky and overcautious," Captain John Rodgers, USN, failed in his goal of capturing a British frigate in the War of 1812. (Lossing, *Pictorial Field-Book of the War of 1812*)

Although based on fraudulent prize claims, the American argument of "equal" combat was quickly turned to account by a remarkable American propaganda offensive, a sudden outburst of government-inspired broadsheets and hagiographies. This English-language material was doubly dangerous, soon finding its way into British newspapers. The claim would be subjected to forensic examination by British lawyer William James, who demonstrated that the actions between *Constitution* and *Guerriere*, *United States* and *Macedonian*, and *Constitution* and *Java* were only "fair and equal" in the overheated atmosphere of

The one-sided action between the larger USS *United States* and HMS *Macedonian* on 25 October 1812 made Commodore Stephen Decatur a national hero. His claim that the British ship was of equal force reflected the curious character of American Prize regulations, not the reality. (Library of Congress)

Republican Washington in late 1812. On hearing that the *Macedonian* had been captured by the *United States,* British frigate captain Philip Broke, who had seen the 44s up close, observed:

> We must catch one of these great American ships with our squadron, to send her home for a show, that people may see *what a great creature it is,* and that our frigates have fought very well, though so unlucky.[7]

Once an American 44 was placed alongside the sister ships of *Guerriere, Macedonian* and *Java* the obvious advantage of one-third in size, firepower and crew became obvious.[8] While most American historians accept James's verdict, Henry Adams aside, few are prepared to acknowledge his role in the process.[9]

An inconclusive war

On 14 February 1814 *President* put into New York after another uneventful cruise, and there she remained for precisely 11 months.[10] The British blockade and the threat to the city kept her tied up in harbour. John Rodgers relinquished command, perhaps hoping his luck would change. By late 1814 the war, begun so lightly with an invasion of Canada, had turned sour. The United States' capital had been burned, her coastal and overseas trade annihilated, bankruptcy loomed and British armies were poised to invade from north and south. New York was threatened, and only one service could soothe public alarm.

The navy was the one American institution to emerge from the war with credit. Morale-boosting victories at sea and on the Great Lakes made it the mainstay of government propaganda. Naval heroics served to obscure the repeated failures of the army. The naval icons were the super frigates *Constitution, United States* and *President*, and a handful of successful captains. Among them, Stephen Decatur was the celebrity icon of a vainglorious age. Cheap newspapers and mass literacy spread his name and his story across an entire nation.

Stephen Decatur

Thirty-six-year-old Stephen Decatur (1779-1820) burst onto the national stage in 1804 with a daring raid on Tripoli harbour. Immediately promoted captain, he sustained his reputation whenever the opportunity admitted. A considerate and caring officer, Decatur was popular with officers and men alike, ensuring his ships were well manned.[11] When the war broke out, Decatur commanded the *United States*. On 25 October he captured HMS *Macedonian* off the Azores in a one-sided battle dominated by his superior tactical skill and the superior firepower of his powerful 44. He doubled the effect by bringing the captured ship home as a prize. When he attempted to return to sea, he found a superior British force, commanded by Captain Sir Thomas Hardy, blockading the eastern end of Long Island Sound. His three ships were driven into New London, Connecticut, and remained there. Decatur was not going to fight such a superior force – he was not that kind of hero. His circumspection did not please everyone. James Biddle, commander of the USS *Hornet*, became thoroughly disenchanted: "Decatur has lost very much of his reputation by his continuance in port. Indeed he has certainly lost all his energy and enterprise."[12] Biddle might have been a hot-headed 21-year-old, but there was more than a grain of truth in his indictment. Decatur, acutely conscious of his public image, would have sensed as much.

For all the bravado of his public persona, Decatur was an innovative and reflective officer. His tactics were carefully thought out, and he was prepared to try novel weapons. In 1811 he successfully tested exploding shells, but decided not to use them in a war with the British "as he means to have fair play with them."[13] Such noble sentiments did not last long. Frustrated by his detention at New London, Decatur asked torpedo pioneer Robert Fulton to help remove the blockading force. Eventually a small American coaster fitted with an improvised

Captain Stephen Decatur, USN, displayed bravado in public but his tactical actions were thoughtful and innovative. While the press proclaimed Decatur an American naval hero, fellow officers were not convinced, and by early 1815 he had a lot to prove. His personal vanity would contribute to his early death in a duel 22 March 1820. (Library of Congress)

explosive device was run into the blockade to be captured, before exploding alongside Hardy's flagship. Hardy was incandescent, threatening to demolish the nearby town if any more underhand "tricks" were attempted.[14] However, the British had not seen the last of Decatur's devices. In early 1814 he finally admitted defeat, paying off *United States*. By contrast Biddle's *Hornet* slipped out and got clean away. Biddle rejoined Decatur, who had shifted key personnel to the *President* at New York. Anxious to get to sea again, Decatur had turned down a new frigate, preferring the fast-sailing *President*.

New York

New York was already a major commercial centre, possessing an enormous harbour, sheltered from the ocean and linked to major inland waterways, including the Hudson River, and coastwise to New England through the sheltered waterway of Long Island Sound. However, the seaward approaches were less than ideal as numerous sand banks lay beneath the 6 miles of water stretching between Sandy Hook and Coney Island. Pilots were essential for ocean-going vessels attempting the narrow navigable channel. These were dangerous waters, littered with shipwrecks. The tidal range of over 4 feet made the choice of time significant for ships wishing to pass the shallows. Twice-monthly spring tides added a foot more water. The ebb tide swept past Sandy Hook at 3 knots, speeding outgoing ships on their way, but making their return all but impossible. Such information was equally well known to blockaded and blockader alike, and more useful for those lying in wait outside.

The bar rendered it unsafe to enter New York in poor visibility. At least there

Stephen Decatur, frustrated by the British blockade of New London, sought the assistance of inventor Robert Fulton. In due course a vessel fitted with an improvised explosive device sailed out, exploding once the blockading ships came alongside to inspect it. (Library of Congress)

were soundings to be had 20 miles out to sea, to warn experienced mariners that the coast was near. Once at sea, ships heading for Europe hugged the Atlantic coast of Long Island, which stretched 104 miles from Coney Island to Montauk Point without a single harbour large enough to take a frigate. Once past Sandy Hook a sailing ship had only two options, press on or put back, a choice determined by wind, tide and weather.[15]

The first year of the war barely affected New York; both merchant ships and privateers were able to get to sea for their lucrative endeavours. But the British soon clamped down on these trades. From the summer of 1813 the British blockaded off Sandy Hook and at the entrance to Long Island Sound. Commerce quickly dried up as ocean-going and coastal vessels were unable to enter the harbour. Import prices doubled, while export goods found few buyers. New York's commercial community became heartily sick of the unnecessary war that Mr. Madison had dragged them into.[16] For much of 1814 New Yorkers feared a British attack, batteries were built, along with Robert Fulton's steamship *Demologos*. *President* was tied to the city as Decatur commanded the harbour defences.[17] Only in late 1814, with the British committed to amphibious operations elsewhere, did Decatur's thoughts turn to the sea.

1814 had not been a good year for American seafarers. Much of the American commercial marine was either laid up or in enemy hands, while the remaining naval vessels were rarely at sea. At the close of the year New York was full of prime seamen, as were many other ports along the Atlantic seaboard. The Navy Department decided to employ these men and such ships as could be got to sea in one last attempt to influence the result of war. Unfortunately for the navy's plans, the British had excellent intelligence sources in New York, and American officers proved uncommonly garrulous.

In November 1814 Secretary of the Navy William Jones and Decatur discussed how to hurt the British. Neither man considered using naval force on the Ameri-

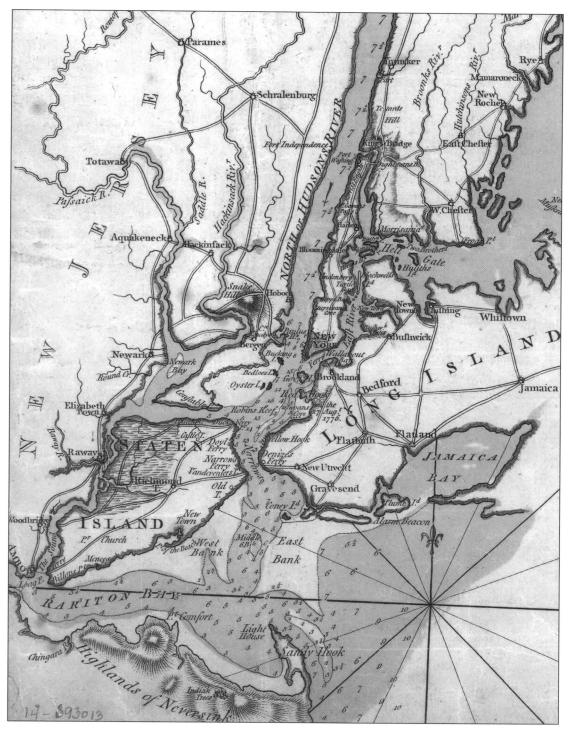

New York harbour and its approaches are shown in this detail from "A Topographical Map of Hudsons River…" by Claude Joseph Saucier, engraved 1777 by William Faden. (Library of Congress)

can coast to defend other coastal cities or protect coasting. Decatur favoured a distant cruise, and Jones, an experienced seafarer and ship-owner,[18] came up with the most ambitious strategic concept yet advanced by an American statesman – the capture of the British East India Company's annual China trade fleet. The target was very attractive. Not only did the fleet carry wealth beyond the dreams of avarice, but it was critical to the entire British commercial system. The stock market panics caused by the American privateers in 1812 would be as nothing to bankrupting the East India Company. After such a disaster, holding Canada would become remarkably unimportant.

Relying on his "pretty intimate knowledge of the navigation of those seas ... and of the trade resources and force of the enemy in that quarter," Jones promised Decatur "more honor and advantage, and your country more reputation" from this plan than any other option. With *President, Constitution* and a small sloop under his command, Decatur would rendezvous at Tristan da Cunha to replenish water and obtain refreshments. Then the squadron would set course for the Sunda Strait, picking up the North West Monsoon between Madagascar and the Island of Bourbon (Mauritius). The next refreshment stop would be Pulo Aor or Pulo Condore at the Straits of Malacca. This strategic choke point was the ideal place to "intercept all the trade from and to China." Capturing Indian Country ships would provide Decatur with the latest intelligence on the East India Fleet. Jones included a recent letter from Canton, explaining how easy it would be to capture "the whole China Fleet," and the effect this would have on Britain. He added a French hydrographic treatise on the area. Did the plan appeal to Decatur? The only fly in the ointment was Jones's impending resignation.[19]

After two years locked up in port, his name slowly slipping down the list of American heroes, Decatur was hooked. He advised hiring a store ship, to be fitted as a privateer once emptied.[20] Jones agreed and detached *Constitution*, replacing her with the newly arrived sloops *Hornet* and *Peacock*.[21] While James Biddle considered the change "a most infamous arrangement," he was in the minority.[22] Decatur purchased the 260-ton schooner *Tom Bowline*, and although rather small for his needs, "she promises to be a fast sailer," especially after her hull had been freshly coppered. North Pacific fur trade magnate John Jacob Astor offered to take another 200 or 300 barrels to the rendezvous without charge, making up the shortfall.[23] While Decatur regretted Jones's impending retirement, he believed the plan "cannot fail (barring accident) in my opinion of producing all the effects that this government could wish."[24] He was wrong. It was no accident that *President* was the first American warship taken attempting to leave port or that Jones's plan failed.[25] It was Decatur's fault.

Strategy and intelligence

In 1814 British forces on the American coast possessed good operational and strategic intelligence and, despite the distances involved, the Admiralty in London acted as the central clearing house for this information. When Commander in Chief Vice Admiral Sir Alexander Cochrane left the Atlantic theatre to attack New Orleans, the Admiralty took a larger role in directing operations. In overall command of the blockade of New York and Long Island Sound was Rear Admiral Henry Hotham (1777-1833), one of the brightest stars among British junior flag officers. Hotham had over 20 years of active service in the French wars behind him, had been commended by Nelson and had conducted blockades and captured enemy frigates. He directed the British blockade from Hardy's old anchorage off New London.

In late November the Admiralty warned him that as three American heavy frigates were ready for sea, he should be ready to meet a hostile squadron, keeping his ships in contact, ready to concentrate.[26] Orders issued the following day reflected an overriding concern for West Indian trade: Barbados and Jamaica were the rendezvous if the enemy escaped and their destination remained unknown. If no certain information could be obtained, news of an escape was to be communicated "by every possible exertion" as widely as possible.[27] Admiralty concern reached a crescendo in mid-December. A second battleship was sent out for Hotham, along with orders to "keep the force under your orders collected in such a manner as may ensure your being able to meet the enemy upon equal terms, should they make any attempt on our blockading squadrons." To avoid any repeat of the disasters of 1812, any ships at sea were to be fully manned, "equal to meet the enemy's vessels of the same force with themselves." To this end, damaged or slow ships should be paid off to complete the crew of his best units.[28]

Despite the Admiralty's anxiety, Hotham did not receive this urgent intelligence in time.[29] Fortunately the Admiralty could rely on senior officers to process locally gathered intelligence and respond with a consistency and clarity that reflected their experience and the existence of a strong unwritten doctrine. Hotham used locally gathered intelligence to capture the *President*. In mid-December Hotham returned to New London after refitting his 74-gun flagship, HMS *Superb*, at Halifax, Nova Scotia. He brought a large stock of victuals for the smaller vessels of his squadron – supplying the frigates *Pomone* and *Endymion* with beef, pork, bread, pease, rum, flour, lemon juice, vinegar and sugar on the 14th.[30] His well-established intelligence network provided copious information on American plans. On the 29th he reported, "The American Frigates *Constitution* and *Congress* sailed from Boston and Portsmouth about 17th inst." He detached three

British frigates under Captain Sir George Collier to search for them at Madeira and the Canaries before re-crossing the Atlantic to Barbados and returning to Halifax by way of the Carolinas.[31] Furthermore, Hotham expected *President, Peacock* and *Hornet* to sail from New York. At that time the blockading squadron off Sandy Hook comprised the razee *Majestic*, the 24-pdr. frigate *Forth*, the 18-pdr. frigate *Pomone* and the 18-gun brig *Nimrod*.

Confident he had penetrated the enemy's intentions Hotham reported "good information that the enemy's ships are to join, to intercept outward bound East India Fleet off Madeira, or Cape Verde Islands."[32] He stepped up target practice on the flagship, in case an American squadron came his way.[33] Although Hotham's intelligence reached London on 1 February, more than a month after the war had ended, it galvanized the Admiralty. Secretary John Wilson Croker hurriedly deciphered the text and minuted: "send extracts to the India House immediately." Croker knew that, peace or not, the American ships posed a serious threat to vital East India Company trade.[34]

Three days later Hotham reported to Cochrane, now far away off New Orleans. He knew *Congress* had not sailed from Boston, however:

> The *President, Peacock* and *Hornet* are waiting only for an opportunity to sail and the information I have received from New York, where they are, states that they are going to India, of which there is a probability if all I have heard of *President* being particularly victualled and stored for a very long voyage, with a large supply of anti-scorbutics, and of persons well acquainted with the Indian seas having been engaged to embark onboard her. I have a strong force off Sandy Hook at present and hope to be able to keep them in.

If the Americans escaped, he would inform the Admiral at the Cape of Good Hope, but he was short of suitable ships.[35] It was significant that while Hotham never secured reliable intelligence on the destination of *Constitution*, he was confident that *President* "is intended to go to India, if she can elude His Majesty ships which blockade her." His intelligence came from "different persons at New York," who reported her to have "a chosen crew, estimated in number at 550," and:

> that she is victualled and stored for a very long voyage, even to the extent of seven or eight months, with large supplies of Sour-Crout, and other anti-scorbutics; that charts of the East Indies have been bought up by her officers; and several masters of ships acquainted with those seas have been engaged to embark on board her.[36]

A week later Hotham had further intelligence. *President* and *Hornet* lay off Staten Island, ready for sea. *Peacock* was in the East River, waiting for bread. *President* had replaced her 32-pdr. carronades with 42-pdrs. belonging to *United States*. Her complement was 420 seamen and officers and 55 marines. Decatur had discharged a number of landsmen and boys and reduced the number of midshipmen to 12, so that the whole crew are nearly all seamen, 150 of whom were ex-*United States*, the remainder lately enlisted. Decatur had informed officers with families that the squadron would be absent 12 or 18 months, if war continued. This hint was supported by the fact that *President* had 7 months of provisions stowed – with more embarked on board a fast sailing brig.

> A gentleman was authorised by the Naval authority here, to call on Mr _____ who has passed all his life in the Indian Ocean; to know whether the island of Pulo Aor, at the entrance of the Straits of Malacca, would not be an eligible situation for a cruiser to intercept the Trade between Bombay and China, and whether it would be most prudent to pass thro' the Straits of Malacca, or those of Sunda.

He believed that Astor's brig *Macedonian* carried surplus provisions to the rendezvous. Furthermore, Decatur was reported to have said

> his instructions are not to man prizes, but to destroy every thing he meets. He has a number of ship masters volunteers; several of whom have been in the India Trade – Captain Rodgers from the command of a China ship, he has rated on the Books as Chaplain. Captain Robinson who went to Europe in the Dutch ship with Mr Rodgers, and lately returned in the *Jenny*, from Dartmouth, is appointed 2nd master the other names I could not obtain. The *Peacock* has 175 men at quarters, the *Hornet* 160.[37]

Hotham was remarkably well informed. There were few secrets ashore in New York or New London. Anticipating *President* would soon attempt to escape by way of Sandy Hook, Hotham dispatched HMS *Endymion* to join Hayes.[38]

HMS *Endymion*

HMS *Endymion* was an unusual ship. When the American 44s burst onto the stage, the British had not one frigate in commission carrying 24-pdrs. on her gun deck. Built between 1795 and 1797 on the lines and scantlings of a French prize, but fastened using British methods, *Endymion* proved fast and effective cruising in defence of trade.[39] After a brief period armed with 24-pdrs., *Endymion* was reduced to 18-pdrs. in 1800 –the heavier guns were unnecessary against

French ships and damaged the structure. Even so, the lightly built ship proved very costly to keep in service; her repairs cost double those of typical British ships. When the War of 1812 broke out, she was in the middle of a major repair. Anxious to respond to the demoralising loss of three frigates, the Admiralty began a crash programme to build *Endymion*-class ships. In May 1813 *Endymion* emerged from a year-long overhaul, re-fitted with 24-pdrs. and an extra pair of 32-pdr. carronades on the quarter deck in an attempt to match the American super-frigates.

Endymion was a superb sailing ship, combining astonishing speed, over 14 knots going large, and handiness in a near perfect combination, but she had not been designed or built to fight an American 44. The American ships were significantly larger, more heavily built and mounted four more main deck guns. *Endymion*'s lightly built hull had wider spacing between the main frame timbers than the American 44s. *President*'s frames were separated by a mere 2 inches, while 3 inches separated *Endymion*'s far lighter timbers. At close range this would be decisive: American shot would penetrate far more easily. But there was nothing else, so she would have to serve.[40] Re-arming with 24-pdrs. cost *Endymion* almost a knot, and she always required careful stowage and handling to bring out her best points, but her performance was never matched by a British frigate under canvas in the French and American wars. Even in the 1840s, she made over 13 knots and remained the benchmark against which all new frigates were tested.[41]

Henry Hope (1787-1863) was appointed to *Endymion* on 18 May 1813. Although a proven frigate captain, he owed this prize command to his cousin Admiral Sir William Johnstone Hope, a Naval Lord of the Admiralty. The son of a naval officer, Henry Hope was commissioned lieutenant in May 1804, obtained commander's rank in October 1806 and became a post captain in May 1808. He had commanded frigates in the Mediterranean for four years.

While on passage to America, Hope captured a fine privateer, the 220-ton schooner *Perry*, on 3 December 1813. Thereafter, his ship was under the command of Sir Thomas Hardy, blockading the USS *United States* and *Macedonian* in New London. Decatur issued a challenge to *Endymion* and the 18-pdr. frigate HMS *Statira*. Hope was willing but Hardy rejected the offer because *Endymion* was so obviously inferior in force to *United States*. After Decatur dismantled his ships, the blockading force dispersed. On 11 October the ship's boats attacked the privateer *Prince de Neufchatel* anchored off Nantucket but were beaten off losing 28 killed and 37 badly wounded.[42] In accordance with Hotham's instructions, these heavy losses were made up by a draft from other ships at Halifax. 50 came from the razee *Saturn*, then heading to Halifax for a refit.

Breakout

On 1 January 1815 *Endymion* lay at anchor off New London, close by Hotham's flagship, blocking the eastern exit from Long Island Sound. Another frigate squadron cruised off Sandy Hook, commanded by Captain John Hayes in the powerful 58-gun razee HMS *Majestic*. Armed with long 32-pdrs. and 42-pdr. carronades, the razees were old 74-gun battleships cut down by an entire deck to become super frigates. As a fighting ship *Majestic* was as superior to the *President* as the latter was to the *Macedonian*. In heavy weather and high winds the *Majestic* was a remarkably powerful ship, holding on to far more canvas than conventional frigates, and translating that power into speed through the water. In light airs her performance was poor, hampered by her bulky battleship hull, which had a far greater surface area below the waterline, adding significantly to frictional resistance. Having directed her conversion and commanded her ever since, Hayes knew his ship very well. Her battery of long 32-pdr. cannon would demolish an American 44, if he could get alongside. Contemporaries rightly considered "Magnificent" Hayes one of the finest seamen that ever strode the quarter deck.[43] His appointment reflected the Royal Navy's determination to deal with the American super-frigates.[44]

In early 1815 Hayes shared his vigil with the 24-pdr. frigate *Forth* and the 18-pdr. frigate *Pomone*. *Forth,* one of the new *Endymion*-class ships, was commanded by Nelson's nephew Captain Sir William Bolton. The French prize *Pomone*, Captain John Richard Lumley, had been a very successful cruiser, picking up numerous privateers and merchant vessels on the American coast. *Endymion* was sent to relieve *Forth*. Hayes needed a 24-pdr. frigate, as those armed with 18-pdrs. were completely outclassed by the big American 44. Under Admiralty standing orders, they could only engage such a ship in pairs. *Endymion's* arrival was no accident. Knowing an American squadron led by *President* was ready to sortie, Hotham relieved Hayes's 24-pdr. frigate with another vessel of the same class.

On 6 January Hope put to sea, in accordance with Hotham's orders, heading southwest to rendezvous with Hayes. Once under way, he exercised the ship's company at great guns, usually high-tempo loading drills followed by a few rounds of aimed fire, directed at a floating target. Hope paid careful attention to gunnery; his crew employed the same drills that Philip Broke had used on HMS *Shannon*.

Endymion made the rendezvous with *Majestic, Pomone* and *Forth* at midafternoon on the 8th, some 12 leagues off Sandy Hook. Hope quickly lowered a boat to visit *Majestic* where Hayes briefed him on the position of the enemy vessels, the specific problems of Sandy Hook and the standing orders and signals he should follow in the event of bad weather or battle. Then the squadron settled down to cruise in company, approximately ESE of the Hook, briefly disturbed by the need to catch and identify the 18-gun brig HMS *Nimrod*. On the 9th *Forth* set course for New London, and after a brief stop headed south for Bermuda.[45] Late on the 10th the 14-gun schooner HMS *Pictou* arrived with the latest intelligence from Hotham. That day Hotham detached HMS *Tenedos,* Captain Hyde Parker, from New London, bringing Hayes's squadron up to four frigates. Clearly Hotham was expecting trouble.[46] On the 13th *Endymion* exercised at great guns and small arms before running in closer to the American shore, to observe two American ships anchored close under the Hook waiting for their chance to escape. They were ideally placed for a break-out. Hayes expected they would try soon, exploiting a combination of westerly wind, high tide and a dark night. The sea was smooth and the weather overcast. Later that day the squadron chased the newly arrived *Tenedos.*

The following day the squadron was struck by a heavy snowstorm, *Endymion* losing sight of the other ships. When the snow squalls passed, the four frigates reassembled. Hayes kept close to the Hook. Both Hotham and Hayes had been able to get inside Decatur's head; they knew what to expect. And Decatur did not disappoint.

Confident the storm had dispersed the blockade Decatur decided to leave after dark on the 14th:

> I have come to the conclusion in consequence of the difficulties that exist in this ships pass[ing] the Bar, it being required that both Wind and Tide would be favourable; such a time when the Enemy are absent, may not occur again shortly, the other vessels are present at all times.

Unfortunately the transport *Tom Bowline* grounded on her way down from New York Navy Yard on the 13th, and he would not wait for her. He would sail that night with Astor's *Macedonian,* leaving poor *Tom* to accompany *Peacock* and *Hornet.*[47]

Decatur had judged his moment well. With a strong northwesterly wind blowing, intermittent snow showers and a dark night, he needed no second bidding. Dark and windy nights are fine for furtive movement, but less than ideal for careful navigation. The gunboats used to mark the navigable channel had been badly placed, and at 2000 hours the deeply laden *President* ran onto the bar. After labouring heavily for over an hour, breaking several pintles of her rudder, a combination of high water and strong westerly winds drove her into deep water, never again to grace an American harbour. Unable or more likely unwilling to re-

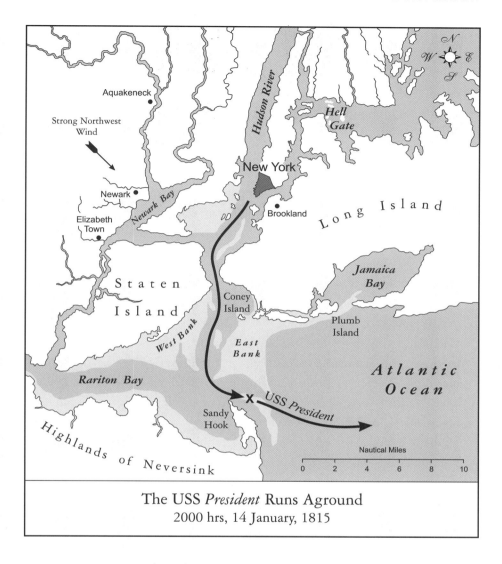

The USS *President* Runs Aground
2000 hrs, 14 January, 1815

turn to port, Decatur shaped a course along the south coast of Long Island, hoping to evade Hayes's ships in the dark. The "damage" that Decatur later blamed for his defeat is hard to assess. Had it been severe he had the option of anchoring and waiting for the tide to turn. Furthermore, the fact that a large, deep-laden ship was able to outdistance all but one of her pursuers the following day suggests the effect was limited. James Fenimore Cooper was uncertain whether to attribute the *President* being caught to damage received on the bar or "the manner in which she had been lightened."[48] Both men missed the simple explanation. Being caught by HMS *Endymion* at no more than one knot did not indicate impaired speed – *Endymion* was the Royal Navy's crack sailing frigate.

After running 50 miles northeast along the Long Island shore, Decatur steered

SE by S until 0500 hours when three ships were discovered ahead, one of them within gunshot (see map, page 113).[49] They were *Majestic, Pomone* and *Endymion*. Decatur quickly hauled up and passed 2 miles north of the enemy. He had met his match. "Magnificent" Hayes had anticipated his every move, making a near-perfect interception just before dawn. He had the whole day to catch and kill his quarry.

This was the pay-off for Hayes's hard work keeping station off Sandy Hook. Repeatedly blown off station by gales, his frigates always clawed their way back as soon as the wind eased, resuming their position on a point of bearing from the Hook that he judged likely, from existing circumstances, would be the enemy's route. His forethought and skill placed him in Decatur's path. Interception was highly probable. In his report Hayes observed that on the 12th

> the squadron was blown off again in a severe snow shower; on Saturday (13th) the wind and weather became favourable for the enemy, and I had no doubt but he would attempt his escape that night; it was impossible from the direction of the wind to get in with the Hook, and as before stated, (in preference to closing the Land to the southward,) stood away to the northward and eastward, till the squadron reached the supposed track of the enemy; and what is a little singular, at the very instant of arriving at that point, an hour before day light, Sandy Hook bearing WNW 15 leagues, we were made happy by the sight of a ship and a brig standing to the southward and eastward, and not more than two miles on the *Majestic*'s weather bow; the night signal for a general chace [sic] was made, and promptly obeyed by all the ships.[50]

Observing a strange ship to the NNW at 0530 hours Hayes let out the reefs in his topsails and courses and made all sail, lit a blue light and fired three rockets – the night signal for a general chase. He did not detach a ship to chase the brig, preferring to concentrate on the frigate, well aware that the lottery of wind and weather, sails and trim made it uncertain which of his ships would be best equipped to hunt down the flying enemy. The flying brig was irrelevant when set again the possibility of taking one of the mighty 44s.

Battle

At first light on 15 January Hope observed a frigate and a brig to the east; he assumed them to be the enemy because they did not answer Hayes's private signals. The wind was blowing hard, northwest by west. At Hayes's signal for the general chase, *Endymion* bore up and made all sail. With a strong breeze blowing, *Majestic* was the first to close on the enemy, and an hour after sighting his quarry Hayes was close enough to fire three shots at the flying *President;* they fell short. At dawn *Pomone* and *Endymion* were in company, but Hayes detached *Pomone* to investigate a strange ship seen off to the south. She proved to be the errant *Tenedos.* By 0800 hours the squadron was once more in company, in light airs, their quarry was 5 miles ahead bearing East ½ North. Anticipating Decatur might turn southeast, Hayes ordered *Pomone* to leeward, pinning *President* against the Long Island shore with only one course open. Through the morning the wind gradually fell away and the direction veered between NW by N and NNW, as the ships raced eastward, *Endymion* recorded various headings from North-East-North, to East ½ South.

By midday the American brig had escaped, showing a remarkable turn of speed for a laden transport, and *President* was now some 8 or 9 miles ahead of *Majestic.*[51] As the wind fell away, Hayes's ship dropped behind the lighter, finer-formed frigates. Once *Tenedos* had re-joined, Decatur found British ships on *President's* port and starboard quarters, and another pair astern. After midday the winds "became light and baffling." *President* had pulled away from *Majestic,* only to find another large ship coming up. As the wind dropped away, *Endymion* "passed ahead of our squadron fast" and by noon she was closing on two ships, which Hope took to be *President* and a brig of war.

Decatur responded by sending every spare man to lighten ship, starting the water, cutting away anchors, and jettisoning stores, boats, cables and anything else that could be spared. Neither Decatur and his officers nor the master had ever handled the ship at sea, making this a risky decision. It would be very easy to alter the trim and ruin the ship's sailing qualities.[52] Decatur also sent men aloft to wet the sails, from the royals down, which he hoped would improve their draw. At least he had the wind almost dead astern, although precious little of it. In light winds the fastest ship would have a major advantage. Predictably Henry Hope's elegant *Endymion* edged up on her quarry, making considerable distance through a fortunate shift in the wind in mid-afternoon.[53] The change did not favour *President.* As the wind dropped, all five frigates shook out the last reefs, set studding sails and skysails, adding extra canvas to the immense acreage already spread, desperately searching for every last ounce of effort.

Unable to keep pace, Hayes observed the contest of seamanship with an expert eye:

the chace [sic] became extremely interesting by the endeavours of the enemy to escape, and the exertions of the Captains to get their ships alongside of him; the former by cutting way his anchors, and throwing overboard every

moveable article, with a great quantity of provisions; and the latter by trim-
ming their ships in every way possible to effect the purpose; as the day advanced the wind declined, giving the *Endymion* the advantage in sailing.[54]

By 1300 *Endymion* was gaining on the enemy hand over fist and would soon
be within range. Hope ordered the crew to their quarters, a well-practised exercise conducted to the rat-a-tat-tat beat of the drums. The last partitions were
knocked down, the decks soaked and sanded to counter fire and blood, rammers
and sponges taken down, guns cast loose, magazines manned, charges measured
and shot arranged until finally the gun crews stood by their pieces – each man
in his station, and each aware of the vital role he played in the repetitive, lethal
business of loading, hauling out, aiming, firing, controlling the recoil, sponging,
ramming home the charge and being ready again. After the savage losses incurred
in the *Prince de Neufchatel* action, many gun crews had been reformed, and some
men must have wondered how the new team would stand up under fire. By the
time the two ships were within range, the gun deck was a silent study of tension,
from the lieutenant in command waiting for orders from the quarter deck to the
powder boys holding fresh ammunition charges in leather boxes, and the directing midshipmen stationed along the deck.

On the upper deck the picture was very different. With *Endymion* straining
every nerve to catch a flying enemy, most of the gun crews and marines were
hard at work attending to the set and trim of the sails. Most of the best seamen,
the topmen, were stationed at the upper deck carronades, which were only used
at close range, while the marines, who provided musketry and boarding parties, were the deck-level muscle power for tacking and wearing. The only upper
deck gun that was required in a pursuit action was the bow chaser, a long 18-pdr.
bronze gun mounted on the forecastle. So far nothing had happened that went
beyond the usual business of "exercise at the great guns." This much was routine
ordered drill. While everyone else went about their business, Hope "observed
the chace throw overboard boats, spars casks etc." at 1318 hours.[55] *Majestic* and
Pomone, now a clear second in the pursuit, also recorded this detail.[56] At the same
time *Tenedos* exchanged recognition signals with the brig HMS *Dispatch*, and
Hyde Parker cleared his ship for action. He was still 5 miles astern.

The action commenced at 1400 hours, Decatur opening fire with his stern
guns. Hope returned the compliment when his bow chaser bore on target. The
two captains had but one idea between them. They wanted to cripple the enemy's
rigging and slow them down, but Decatur had the better field of fire. Hope was
obliged to choose between closing the range and opening his arcs of fire. The

Americans hit home at 1439 hours, a shot cut through "the head of our lower studding sail, foot of the main sail, through the stern of the barge in the booms and going through the quarter deck lodged in the main deck without causing any other damage."[57] At 1500 the brig HMS *Dispatch* joined the chase, just as *Endymion* exploited a shift in the wind to close. Anxious to inform Hotham that he was in action, Hayes signalled *Dispatch*, but she did not respond until early evening.

Accurate long-range gunnery in a heavy swell was difficult. Hope was impressed by the rate and accuracy of the *President*'s fire, but *Endymion* suffered no significant damage. At 1610 *Endymion* shot away *President*'s jib halyards, followed by the fore topgallant staysail sheet 10 minutes later. With *Endymion* on his quarter, a blind spot for his guns, Decatur began to luff up into the wind, opening his stern arcs, desperately trying to cripple the flying *Endymion*. Confident he had the legs of his opponent, Hope pressed on. Standing at his exposed station on the starboard side of the quarter deck, Hope knew exactly what his opponent was trying to do, his observation that *President*'s fire was "passing over us" tells its own tale.

Battle tactics in the War of 1812
The British had been profoundly shocked by their defeats in 1812. For the previous 20 years they had hardly lost a single-ship action, easily defeating French and Spanish vessels that fought bravely but without tactical skill or seamanship. The inevitable result was a degree of complacency. Tactical thinking was reduced to coming to close quarters as soon as possible, firing into the enemy ship's hull and boarding if necessary. The American officers were better seamen than their continental contemporaries, much better. But their success flowed from the careful, cool-headed defensive tactics of their best commanders. When Decatur fought the *Macedonian,* he countered predictable British tactics and the superior sailing of his opponent's ship by using anti-rigging projectiles in the opening stages of the action. This reflected a conscious doctrinal choice: American ships carried a far higher percentage of such rounds than British ships – up to 20% of the total.[58] Star shot, bar shot and double-headed shot were used by American captains, along with langridge and grape shot, to dismantle the rigging and immobilize the enemy. By crippling the sails and masts of their opponents the American gained a "decisive advantage from the superior faculties of their long guns in distant cannonade," particularly against opponents anxious to close.[59] Having disabled the enemy, the Americans coolly exploited superior mobility to take up a commanding position for raking fire. Waiting on the defensive they made the best use of superior firepower.[60]

Decatur's tactics against *Endymion* were very similar to those he employed against *Macedonian*. He tried to cripple the enemy, before moving on to achieve his ultimate object, fight or flight. He did not adjust to the circumstances. Hope was a smart officer, he knew what to expect – the British had learnt their lesson – *and* he adapted his tactics as the action developed. It was no accident that Hope held a position that denied Decatur the chance to fire broadsides into his rigging throughout the chase. If Decatur would not stand and fight, he could be crippled and brought to bay. Had this been a single-ship action, Hope would have used a different combination of speed, position and firepower.

Closing for the kill

At 1500 hours the frigates were exchanging fire from bow and stern. "The object of each was to cripple the spars of the other"[61] as *Endymion* slowly closed the last 1,000 yards. Finally at 1700 she occupied a near-perfect position on *President's* starboard quarter, within half point blank range – little more than 100 yards – where the American stern and quarter guns could not bear (see upper tactical diagram, page 111). This was fine seamanship, combining tactical skill with wisdom. If he wanted to fire back, Decatur was obliged to luff up and open his stern arcs. Hope knew that every time Decatur did this, he would lose ground to the chasing squadron. Decatur knew that if he did not knock the British ship out of the chase, he would be defeated. He must have realized he was up against a real seaman, and a far abler tactician than blundering John Carden of the *Macedonian*: "The fire of the English ship now became exceedingly annoying, for she was materially within point blank range, and every shot cut away something aloft."[62]

The two ships were heading east by north, the wind now northwesterly, and racing along under every stitch of canvas that could be spread. This was very unusual; ships normally fought under head sails. Cooper reported that the poor performance of *President's* guns led many to think her powder defective.[63] For half an hour Decatur endured *Endymion's* galling fire, hoping that his wily opponent would come alongside, where his superior battery and much larger crew, (broadsides, 916 pounds to 676; crew, 480 to 346) would shift the balance of advantage in his favour. Anxious to end the combat quickly, he prepared his men to board, to exploit another major advantage.[64] Never a man for half measures, Decatur told the crew he planned to take *Endymion,* transfer the crew and scuttle *President* before fleeing from the rest of the British squadron.[65] It just might work, if he could get in a few effective broadsides. But Hope was not going to fall for such a ruse. Well aware of the disparity of force between the two ships and his increasing isolation, Hope constantly yawed across the American quarter, firing into her rigging: "every fire now cut some of our sails and rigging." There could only be one conclusion to this combat: *President* would be disabled and fall easy prey to the British. The French-built *Pomone* was slowly gaining ground; if she joined the action with *Endymion,* Decatur would be doomed. *Tenedos* was holding station in the middle distance, but *Majestic* had dropped a long way astern, despite Hayes's constant attention to the set and draw of his sails and the trim of the ship – he was desperately searching for a breeze.[66]

Up to this point Decatur must have believed his cagey opponent was just another standard British frigate, well handled, with well laid guns, but the only projectiles to hit the *President* were 18-pdr. shot. Having dropped the razee far astern, the other ships posed little threat. When he stepped up onto a shot box to get a better view over the bulwarks, Decatur was almost complacent. However, he had made a serious mistake. Hope had no intention of trading blows broadside to broadside with the powerful American ship. He knew his ship was significantly lighter than the American in scantling and frame, and he had no need for heroics. A skilful appreciation of the situation governed his every move. But he did shift his aim from the rigging to the hull, turning from catching to killing. Hope knew that to capture a resolute, professional opponent he would have to subdue her in a close-range gunnery duel. Having placed *Endymion* on *President*'s quarter, he poured a succession of broadsides at a range where every shot told, "galling him much." The first British broadside smashed into the American quarter deck. Decatur was flattened by a huge splinter that hit his chest and another cut his forehead. Although stunned and winded, he quickly resumed his position, but he had been lucky.[67] Among those standing beside him, First Lieutenant Fitz Henry Babbit had his right leg cut clean through and he was knocked down the wardroom hatch. He died two hours later. Another splinter fractured Lieutenant Howell's skull, with fatal consequences. This was no picnic. Many of those who stood in the "slaughter pen" beside the ship's wheel that day would die.

Between the sharp bark of his own guns Hope would have caught fragments of sound from his opponent, sounds of death and injury, of shattered timber and chaos. He must have been delighted by his ships accurate gunnery: "observed that our shot did considerable execution." The screams and cries of the wounded would have carried across the interval, along with the heavy stroke of iron balls on solid oak. Every minute he held his position he was levelling the odds; soon he could close for the kill. Struggling to regain his senses Decatur read the situation with equal facility: these were well matched combatants, experienced, confident and determined. Suddenly aware that the enemy was far more powerful than he had expected, Decatur had to act quickly. *President* could not endure such punishment for long. If she lost a mast the action was over; she had to get away. At 1730 Hope observed the *President* brail up her spanker and bear away on the wind, as if to cross *Endymion*'s bow and rake.[68] Decatur's only hope was to cripple *Endymion*'s rigging, and so he continued to fire high, searching for a killing blow, to smash a yard, cut a backstay or perhaps bring down one of *Endymion*'s masts. He had done it once before.

It was just after dusk when Decatur changed tactics. His sudden shift to a southerly course was a desperate attempt to knock *Endymion* out of the race with one or two raking broadsides and escape into the gathering gloom before the other English ships arrived. As he turned Decatur extinguished the battle lanterns.[69] One contemporary thought Decatur expected the *Endymion*'s masts, already damaged by his fire, would go by the board, leaving her disabled while he escaped.[70] This had been the main reason for the defeat of all three British frigates in 1812. Once again Henry Hope was ready.[71] Well aware that a raking broadside of 24- and 42-pdr. shot could cripple his ship, Hope immediately put the helm hard a'weather to counter the threat. No sooner had Decatur begun to change course than Hope followed suit, and *Endymion* was far handier than *President.*

This brought the two ships broadside on, giving Decatur one last chance to cripple his opponent and flee into the gathering gloom before the other British ships, now far behind the action, could get up (see lower tactical diagram, page 111). This time Hope did not avoid action, he considered the situation "favourable" and kept closing. His fire had slowed the American ship and inflicted serious damage. With the last daylight ebbing away, he had no choice but to engage – unless he was prepared to give up the greatest prize of the war. He would fight, on his own terms, and not risk a boarding action. At 1804 hours the *President* began firing musketry from her tops, which Hope countered with his marines. Frequent small adjustments, hauling up into the wind, allowed Hope to close the range without losing the favourable position he had taken for broadside fire, just abaft *President*'s beam. The two ships were now at half musket shot, 50 to 100 yards, and Decatur's fire was finally beginning to tell on *Endymion*'s sails and running rigging. But the real story was the terrible beating that *Endymion* inflicted on *President*'s gun deck. The American ship's fire visibly slackened as men fell and guns were disabled. Among the first casualties was Fourth Lieutenant Archibald Hamilton, commanding on the main deck. He was cut in half by a British 24-pound shot. Many of his men died at their guns, six of which were damaged or knocked off their carriages, while ten of the 15 starboard gun ports were hit.[72] The American crew took a beating from their opponents – strikingly similar to the one that *United States* had handed out to *Macedonian*'s badly led men.

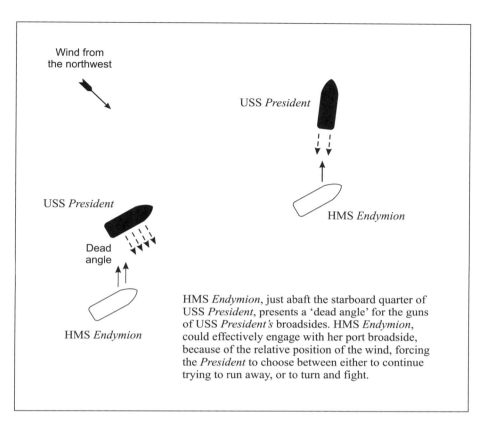

HMS *Endymion*, just abaft the starboard quarter of USS *President*, presents a 'dead angle' for the guns of USS *President's* broadsides. HMS *Endymion*, could effectively engage with her port broadside, because of the relative position of the wind, forcing the *President* to choose between either to continue trying to run away, or to turn and fight.

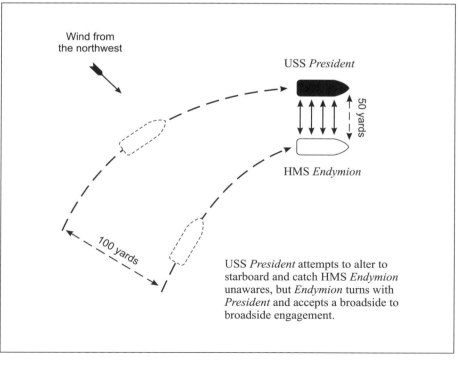

USS *President* attempts to alter to starboard and catch HMS *Endymion* unawares, but *Endymion* turns with *President* and accepts a broadside to broadside engagement.

With the two ships running south on parallel courses, Hope countered Decatur's attempts to close the range. Unable to grapple with this wary foe, Decatur continued to fire into her rigging, while Hope concentrated his fire on the American gun deck. At 1840 hours Hope observed the *President* haul up, which he presumed was done to avoid his fire, and immediately poured two raking broadsides into her stern, with devastating consequences, before resuming his preferred station on *President*'s starboard quarter. At 1915 *President* shot away *Endymion*'s boat from the starboard quarter and both the main and lower studding sails.

> At 1918 the enemy not returning our fire we ceased firing, at 1930 the enemy shot away the larboard maintopmast studding sail and the main brace, at 1932 the enemy hauled suddenly to the wind, trimmed sails & again obtained the advantage of giving him a raking broadside which he returned with one shot from his stern gun, the enemy much shattered.[73]

Between 1940 and 1950 hours *President* bore away from *Endymion* and ceased firing. At 1958 she ceased firing altogether, and displayed a light in her rigging, a recognized night signal of surrender.

After two and half hours of broadside fire, *Endymion* dropped back, her rigging cut to pieces and several sails shredded. Hope had sacrificed his rigging to win the battle. By adapting his tactics to the situation, he had defeated *President*. Later Decatur would imply – he scarcely had the gall to claim – that he had defeated *Endymion*. But this is simply incredible. Not only had the *President* deliberately fired into *Endymion*'s rigging, leaving her with light losses, but Hope was quite certain the American had surrendered.[74] Hope was far too astute to make a mistake on such a cardinal point, and no-one ever impugned his integrity. Rather than waste time closing on *President,* he called up all hands to bend on new sails and splice the rigging to ensure the prize did not escape. He would have sent a boat to take possession of *President*, but did not have one that could float, and at 2010 observed *Pomone* and *Tenedos* coming up.

Never one to miss an opportunity, Decatur kept under way and passed the nearly stationary *Endymion* at 2030 hours, resuming his course of E by N. Hope did not fire as his men were busy aloft and he believed the action was over. However, he had underestimated his enemy, misread the signals or been tricked. Decatur hoped to escape into the night under royal studding sails. Although it was quite dark when the *President* made her bid for freedom, the stars soon came out, revealing her to the pursuing *Pomone*, which had made up a considerable distance while the two frigates had been engaged.

At 2052 hours Hope completed shifting sails, having fitted a new main top-

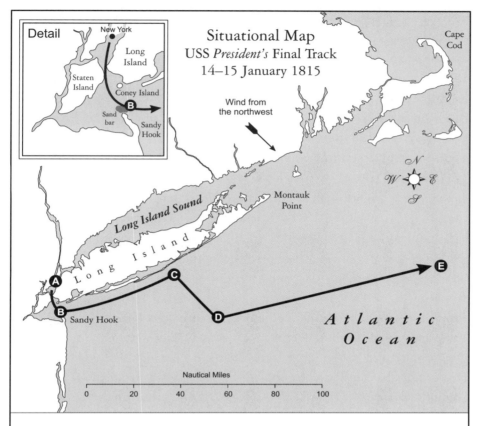

Situational Map
USS *President's* Final Track
14–15 January 1815

A. USS *President* departs New York the evening of 14 January 1815.

B. *President* runs aground on a sand bar off Sandy Hook at 2000 hrs. She stays aground for just over an hour, and then the NW wind and rising tide frees her.

C. *President* runs along the south coast of Long Island for 50 nautical miles before altering SE by S (146.5 degrees true).

D. *President* then sails SE by S until 0500 hrs, 15 January, at which point she encounters the British squadron consisting of HMS *Majestic*, HMS *Pomone* and HMS *Endymion*. *President* immediately alters course, and is chased for the rest of the day in a generally E by N direction.

E. *President* is finally captured at approximately 2105 hrs, 15 January 1815, south of Nantucket and ESE of Montauk Point.

sail, jib, foretopmast staysail and spanker. *Endymion* immediately trimmed sails and bore up to close on the enemy just after the other two ships passed. Hope could see *President* was heading east under a press of sail, and at 2105 he observed one of his squadron run up on her larboard beam and fire into her. *President* then shortened sail, luffed up and hoisted the light in her rigging higher. Decatur hastily conferred with his two surviving lieutenants, Shubrick and Gallagher, and merchant ship captain Robinson. While they concurred that they had no option but to surrender Robinson was far from happy with the decision.[75] Unaware that the enemy had surrendered, because *President* was actually trying to escape, *Pomone* delivered two close-range broadsides into *President's* larboard bow before Decatur hailed to make his situation clear.[76] These broadsides were ineffective, in part because they were not well aimed, and in part because Decatur had sent his men below to attend to their possessions.

In view of Decatur's subsequent claim that *President* had not surrendered to *Endymion,* it is curious that she made no effort to resist *Pomone.* Decatur's new opponent was a lightly built 18-pdr. frigate. At close range two or three broadsides would have crippled her if *President* were still in good fighting trim. In truth the latter was badly damaged. *Endymion* had fired steadily and accurately, smashing the gun deck and upper deck batteries of her opponent, killing and wounding officers and men, disabling guns, and damaging masts and spars. With several shot between wind and water and others below the waterline, the hold was filling up. *President* was in no state to fight or to flee. Two days later Decatur acknowledged "the crippled state of the *President's* spars" was a factor when he "deemed it my duty to surrender," hoisting the lantern in his rigging higher to signify the fact. He did not fire even the single shot at *Pomone* that was required for form and honour's sake.[77]

As *Pomone* came into action, *Tenedos* was closing fast, hastily shedding sail to avoid overrunning the concluding scene. At 2145 Hope hailed Hyde Parker to let him know that the enemy had surrendered, but he did not have a boat to take possession. *Tenedos* sent a boat, and Hyde Parker's men were the first Britons to board the *President. Endymion* finally caught up with her prize at 0045, but *Majestic* did not join until 0300. Hyde Parker recorded that the American ship had 480 men on board, from the captured muster book.[78] Having learned that he had taken the *President,* Hope ordered the guns secured and beat the retreat.[79]

The critical phase of the battle had been when Decatur changed course. Hope's decision to conform and accept broadside action gave the American a chance to exploit his superior firepower. He might have used the opportunity to knock out his opponent by decimating his crew, rather than continuing to fire into the rigging. He would have knocked *Endymion* out of action more quickly by beating her crew than by trying to ruin the rigging, which the combats of 1812 indicated would always be a long drawn-out business. By keeping up his attack on the rigging, Decatur gave Hope the opportunity he needed. While *President* crippled *Endymion's* rigging, *Endymion's* fire left *President* in a sinking state.

The *President* taken

After 24 hours of action, anxiety and bloodshed, including being knocked to the deck and cut about the head, Decatur must have been exhausted; his men were suffering, his officers falling at a frightening rate. He must have been demoralized by the knowledge that he had been comprehensively out-thought by Hayes and then out-sailed and out-manoeuvred by Hope. There was no reason to expect the other British ships to be any less dangerous. While Theodore Roosevelt considered Decatur had "acted rather tamely" in surrendering when he did, Henry Adams pointed to the root cause: "anxious to escape rather than to fight, Decatur in consequence failed either to escape or resist with effect."[80]

By contrast Hope had handled his ship very well, avoiding a close-quarters broadside action, which would have suited the more heavily built *President*, for as long as possible. A contemporary British verse celebrating the battle offers the following insight:

> "Be Silent, men!" was all his cry. "Bring all your guns to bear,
> And do not fire one shot in vain; both round and grape prepare."[81]

The battle had been settled by *Endymion*'s superior gunnery: her crew fought like professionals, calmly, deliberately and decisively. This was the system Philip Broke had used, and the state of the *President* after the battle bore mute witness to the results. Decatur confessed that *Endymion* fired accurately in the chase, while the American casualties suggest she did equally well in the broadside action. As Sir Howard Douglas, the father of modern naval gunnery, observed this was "some of the best gun-practice ever effected by British seamen."[82] William James, who had discussed the action with several of the British officers, declared, "Captain Hope, aware of the excellence of the Broke system, had long trained his men to the use of both great guns and small-arms, and many had

Captain Sir Philip Bowes Vere Broke, Royal Navy (1776-1841). During the long commission of the frigate HMS *Shannon* (1807-13) Captain Broke developed a strikingly modern, scientific approach to gunnery. He paid particular attention to gun drill and accuracy, insisting that his gun crews aim to hit precise targets. Broke also insisted that his crews operate in a highly disciplined manner, conducting their drills silently so that orders could be transmitted effectively even in the heat of battle. The dramatic success of Broke's gunnery system in the battle between HMS *Shannon* and USS *Chesapeake* laid the foundation for the eventual adoption of his approach in the Royal Navy, and ultimately all modern navies. (Lossing, *Pictorial Field-Book of the War of 1812*)

HMS *Endymion* and USS *President* exchange broadsides at close range in the later stages of their hard-fought battle. A third British ship firing at long range is shown in the distance, but this is artistic licence – *President* was only engaged by one British ship at a time. Broadside to broadside, *Endymion*'s superior gunnery proved decisive. (Naval Historical Center, NH 68511)

been the anxious look-out on board the *Endymion* for one of the American 44-gun frigates."[83]

Endymion's crew of 346 suffered 11 killed and 14 wounded. Her masts and yards were hit in several places and significant parts of the rigging and sails were cut, but she inflicted far more damage on *President*. On the starboard side *Endymion*'s 24-pdrs. had smashed through *President*'s hull all along the gun deck and the quarter deck, doing considerable structural damage and dismounting or damaging six guns in addition to the heavy human costs. Some shot went straight through the ship. During the chase and raking action, three shot went through the buttock area and one ended up in *President*'s magazine. Decatur admitted that *President*'s masts were "crippled." The damage inflicted by *Endymion*'s guns in the broadside action stood in stark contrast to that of the 18-pdrs. which simply bounced off the *Constitution*, earning her the name of "Old Ironsides." Size really did matter: 24-pdrs. were far more effective than 18-pdrs. Furthermore, they were used with "destructive precision" which, as the senior officer of the prize reported, left *President* with "six feet of water in the hold, when taken possession of."[84] Hayes reported *President* to be in a sinking state, and Hope attributed this to "the steady and well directed fire kept up by His Majesty's ship under my command."[85] *President* lost 3 lieutenants and 32 seamen and marines killed, and midshipman Richard Dale died at Bermuda in late February. Decatur, the master, 2 midshipmen and 66 seamen and marines were wounded for a total of 105 casualties, more than four times the number suffered by *Endymion*.[86] [Editors note: The contrast with similar actions set out in the previous chapter on page 77 is striking.] An anonymous officer reported that after the action Decatur declared to Hope, "You have out-sailed me, out-manoeuvred me, and fairly beaten me."[87] Similarly John Hayes had every reason to be pleased: he had outwitted the best officer in the United States Navy, and captured "the United States ship *President* – Commodore Decatur, on Sunday night, after an anxious chace of eighteen hours."[88]

The prize

The following day Hayes sent the rest of the squadron south towards Bermuda. Hope had his people up in the rigging, replacing the foretopmast and main topgallant yard, knotting and splicing the rigging, assisted by two shipwrights from *Tenedos*. That afternoon Decatur, his first lieutenant and 213 seamen were moved across to *Endymion* by boat. *Pomone* exchanged 140 Americans for 50 of her own crew; *Tenedos* took 160 and sent her first division across to man the prize. At the end of the day the easterly wind increased to gale force with occasional rain. The 17th was worse: *Endymion* lost her storm staysails and trysail, began labouring in the heavy seas and "a great quantity of water was taken in abaft" when the rudder head coat was washed away. Hope lightened ship, getting all moveable weight off the upper deck, heaving the quarter deck and forecastle guns overboard. Soon afterwards the mainmast fell, followed by the mizzen topmast. The foremast and bowsprit did not stand much longer.

President was also suffering. Shortly before *Endymion* lost her masts *Tenedos* observed the prize flying a distress signal, and ten minutes later the foretopmast went over, followed soon after by the main topgallant. As darkness fell *Tenedos* lost sight of the *President* amid the rain showers and gloom of an Atlantic night. With her lower masts carried away and several imperfectly plugged 24-pdr. shot holes between wind and water, she was in danger of foundering, the prize crew being exhausted at the pumps. Only the bowsprit kept the ship from lying to in the trough of the ocean. Prize Master William Morgan, First Lieutenant of

Endymion, veered out a sea-anchor made from two hawsers end-on from the forward ports, which quickly brought the ship's head to the sea. This enabled the pumps to keep on top of the water in the hold. The sea-anchor gave way after some 8 to 10 hours service, by which time the storm had abated. Jury masts were then rigged, and the ship navigated to Bermuda.

While *President* and her captors headed south, Hayes worked his way back to New London, arriving on the 22nd.[89] Having received a full report, Hotham wrote to the Admiralty and Admiral Cochrane the following day. His dispatch reached London on 17 February, being at the Admiralty Board the following day. Secretary Croker endorsed Hotham's dispatch "Copy for Gazette" and then "Own and impress their Lordships approbation of the judgment with which Sir H Hotham's dispositions were made and of the manner in which Captain Hayes carried them into execution, of the zeal and activity displayed by all the Captains and ships companies and the gallantry with which Captain Hope brought his ship into action and the distinguished bravery and good conduct maintained by him and all the other officers and ships company of the *Endymion* and as mark of their approbation they have promoted Lieutenant Morgan to be a Commander and Mr … Midshipman to be a Lieutenant and sent Capt. Hope the midshipman's blank commission to fill up for the midshipman. JWC 18.2.1815."[90]

A few days later Hotham had less agreeable news to report. *Hornet* and *Peacock* had escaped on the 20th "at which time HM ships stationed off Sandy Hook were unable to keep in with the land. I have no information of which I can rely as to their destination, but always understood they were intended to accompany the *President* and they may possibly proceed to a given rendezvous for meeting her." To make matters worse, Hotham had no ships he could detach in pursuit or even to warn other officers.[91]

Having got to sea, and unaware of the disaster that had befallen Decatur, the second squadron pushed on for the South Atlantic. *Hornet* arrived off Tristan da Cunha exactly one month later, where she met and captured HMS *Penguin,* despite Biddle having been told that the war was over.[92] *Peacock* went on to capture the East India Company sloop *Nautilus* on 30 June, despite her prize making it clear the war was over. William Jones had his Indiaman after all, a miserable little sloop.

Peace

Perhaps the battle had all been in vain. While the Treaty of Ghent had been signed on Christmas Eve 1814, it took months for the draft to reach the United States for ratification, so the fighting went on for another two months. Hotham only heard of the peace on 13 February.[93] Meanwhile *President, Endymion* and *Pomone* reached Bermuda, where the Vice Admiralty Court condemned the American flagship as a lawful prize to the four frigates and *Dispatch*.[94]

When the prize and her captors reached Bermuda, local political and commercial leaders were quick to celebrate a British triumph, presenting Hope with a public address and a piece of plate, together with an identical silver cup for the ship and all future *Endymions*. When delivered in 1821 the £500 vases featured the bow and stern of the successful ship and the motto "Old England Forever."[95] The American crew were well treated and given their liberty.[96] News of peace proved especially timely for some of *President*'s crew, who were recognized as British deserters.[97] No further action was taken.[98]

Another issue that spoiled the atmosphere at Bermuda was the allegation by a British officer of the prize crew that Decatur had concealed 68 armed men in the hold, with the intention of recovering the ship. The allegation appeared in the *Bermuda Royal Gazette* of 1 February. Such conduct by an officer who had already surrendered would have been utterly dishonourable. Consequently it met a strenuous denial from the Americans. On Decatur's honour the governor of the island compelled the editor, Mr. Ward, to retract. But Ward did so unwillingly and repeated the allegation on 16 March. His primary source was Lieutenant George James Perceval of the *Tenedos*, later Captain the Earl of Egmont.[99] Perceval may have misconstrued Decatur's decision to let his men attend to their personal belongings. This was common practice after surrendering. Certainly Perceval had seen the men with his own eyes, leaving the editor convinced the governor's demand for a retraction demonstrated undue delicacy.[100] For his pains he lost the government printing contract and received a beating in the street from American midshipman Robert Randolph. Randolph had good reason to be upset: first his brother had been captured on board the *Chesapeake*, now he followed suit on the *President*.[101] By this time Decatur was long gone, but there remained nagging doubts about his honour, his integrity and his conduct of the battle.

Similar disputes arose at Bermuda as to the precise nature of the action. Decatur was anxious, for obvious reasons, to claim that he had struck to the entire squadron, that he was outnumbered, and that he had defeated *Endymion*. He stressed that while the British ship bent on new sails and replaced some ropes she did not fire on *President*, despite having a fine view of her stern. Actually Hope was certain that a lantern raised in *President*'s mizzen rigging had signalled her surrender. While Decatur was desperate to embellish a defeat, his officers told a different story. Mr. Bowie, the chaplain, swore that a light had been hoisted before *Pomone* arrived. By contrast *Pomone*'s officers, considering the light merely a substitute ensign, fired two broadsides. *President* did not reply, and when challenged Decatur admitted that he had surrendered. He had no choice, for when *Pomone* arrived he had already ordered his men below decks to stow their bags, preparatory to capture. *Pomone*'s fire, which Decatur subsequently claimed killed and wounded many, was perfectly harmless, according to Mr. Bowie. Only one shot penetrated *President*'s larboard side, the side that *Pomone* fired into. All the other hits, on the starboard side, were from *Endymion*'s fire. Another explanation emerges in the testimony of Lieutenant John Gallagher and Marine Lieutenant Levi Twiggs: in the gloom of the night Decatur mistook the *Pomone* for the far more imposing *Majestic*.[102]

While the precise details of the capture of a ship, and the honesty of the officers concerned, might seem more than a little arcane 200 years after the event, such an impression would be misleading. Decatur, Hope, Hayes, Croker and James knew that this war had been waged in print, and that the impressions left by the printed word would long outlast the facts of the day. In this sense the War of 1812 was a very modern war.

Endymion and *President* were patched up with new masts but not refitted for service.[103] The British public needed to see the huge 44 for themselves, and the two ships sailed in early March. Victor and vanquished arrived at Spithead on the 28th, affording Englishmen "ocular demonstration of the 'equal force' by which their frigates had been captured."[104] The event was recorded in a lengthy verse by an author with considerable knowledge of the action.[105] Although the battered condition of the *President* prevented her being employed again, William James stressed that she served two vital purposes. Lying in Portsmouth Harbour she proved once and for all that an American 44 was not an "equal force" with a British 18-pdr. frigate and American claims to the contrary were unfounded.[106]

In addition to a well earned Naval Gold Medal, Henry Hope was gazetted a Companion of the Bath of 4 June. On August 16 *Endymion* reached Plymouth to pay off, the final entry in her log coming 20 days later. The empty hulk was laid up in the Hamoaze, in case the country had need of her again.[107] The 16th marked the end of Hope's sea service, although he became an ADC to King William IV and Queen Victoria, dying full of honours and years in 1863. Hotham also came home but was soon employed at sea. Shortly after taking the *President,* he captured an Emperor – his squadron arrested Napoleon after Waterloo. His later career included extensive service on the Board of Admiralty, and he died suddenly at Malta in 1833 while Commander in Chief in the Mediterranean. Fate had other plans for Stephen Decatur.

Honour and glory: The battle after the battle

No sooner had he arrived at Bermuda than Stephen Decatur realized that his actions were open to interpretations that did not add lustre to his reputation. Some argued that he had surrendered to a single, smaller ship, that he had not fought to the finish, and that he had attempted a dishonourable trick. Many years later Henry Adams observed: "Decatur was doubtless justified in striking when he did; but his apparent readiness to do so hardly accorded with the popular conception of his character."[108] He began to have doubts about the impact of his original report. Unusually for Decatur, this had been a long and detailed paper, not the brisk seven sentences it took to report his capture of an enemy frigate later that year.[109] It attributed the loss of the ship to her striking on the bar, emphasized the severe losses and damage suffered, and the heroism of the officers and men before he surrendered.[110] Perhaps this was not enough to satisfy his public.

The first report of the action only reached new Navy Secretary Benjamin Crowninshield in late January. It was a brief extract from Decatur's letter to his wife sent the day after the battle, sent in by his friend Captain Oliver Hazard Perry. In it Decatur claimed that he had only surrendered when two British ships were fairly alongside, and the other two close. When Perry, hero of Lake Erie, claimed, "The reputation of the Navy and his own stands yet pre-eminent,"[111] he fired the opening salvo in a desperate public relations battle. This soon overwhelmed such simple matters as what had actually occurred and who was really to blame.

The wounded hero arrived at New London on board HMS *Narcissus* on 22 February "sorely tried by concerns about his honor… Newspaper accounts … from Bermuda … suggested that he had surrendered to a single British warship after a less than stout defence." Perhaps "he had lost his lustre with the American public."[112] The adulation of the crowd, a celebratory dinner for the peace, attended by naval officers from both sides, and followed by a 21-gun salute from Hotham's flagship restored his spirits.[113] An admiring President would provide him with the perfect stage to re-gild his laurels and those of his country. The American government was anxious to restore the prestige of their heroic navy and its brand leader; they would be the bedrock of post-war diplomacy, the mark of the Republic's standing in the world.

The day after Decatur reached New London, President Madison sent a message to Congress calling for war with Algiers. With a powerful navy on a war footing, and few merchant ships yet at sea, it would be wise to settle the Algerine issue before it could flare up again. It would also be popular, and there was an election due. As the unquestioned hero of the last Barbary war the United States had need of Decatur, and he had need of his country. Just in case he had been forgotten, he sent in a report on the state of British naval preparations in the Gulf of Mexico.[114]

It was a mark of Decatur's concern that he sent a second, supplementary report on the loss of his ship on 6 March. He attributed a "considerable number" of his casualties to *Pomone*'s two broadsides, raised the red herring that 50 of *Saturn*'s crew were on board *Endymion* and stressed that he had surrendered to the squadron, as evidenced by his letter of parole. He also took care to blame the mischievous *Bermuda Gazette* for impugning his honour and his conduct.[115] This was a very different production to his original report, a carefully crafted exculpation based on dubious "evidence." *Pomone*'s broadsides had been wild, did little harm and caused no casualties. The men were already below decks. The 50 officers and men from *Saturn* were straight replacements for the savage losses *Endymion* had suffered in the engagement with the *Prince de Neufchatel*. As for surrendering to the entire British squadron, this may have been true, but only *Pomone* was in range when he did so, *Tenedos* was at least 3 miles distant and *Majestic* far further off. The capture was credited to the squadron because under British Prize Law every ship in sight of the capture was entitled to a share in the prize money. By his own admission Decatur had only ever engaged one ship, and he had surrendered before *Pomone* opened fire. With the war over British officers were prepared to be magnanimous; they had the prize they wanted. Decatur knew that by his own very high standards he had not done enough. His request for a Court to investigate was high risk, but essential.[116]

Crowninshield was in no mood to quibble. He was happy to call a Court, as soon as the other *President* officers reached America, but he believed the outcome was a foregone conclusion:

> from a conviction of the bravery and skill with which that ship has been defended, and a confidence in the result proving honorable to your high character as an officer, in sustaining a combat so vastly unequal and which terminated in a manner not derogatory to the credit of our Navy, or to the honor of our National Flag.

His letter closed with best wishes for a speedy recovery.[117] On the same day Crowninshield allowed Decatur to choose his command for the war with Algiers. He must have had the blessing of the President for this offer.[118] Replying by return Decatur took the first active command:

> It would be particularly gratifying to me at this moment to receive an active and conspicuous employment, in Europe it would be seen that my statement had been satisfactory to my Government.

He wanted to offer places to *President* veterans; this was just, and ensured they would not speak against him at the Court of Inquiry. Once the second division of the squadron arrived, under a higher-ranking officer, he would come home.[119] The concern for European opinion is revealing. Decatur knew his actions were open to question and grabbed the first opportunity to resume "honorable" high profile public service. That Crowninshield, and Madison, shared his anxiety was obvious: why else would the Secretary of the Navy venture to prejudge the outcome of a legal process required by the state. In fact the Administration was happy to send him off to war before the court met.

The Court of Inquiry was a whitewash and served only to further confuse the story – adding the *Dispatch* into the mix and removing *Endymion* from the final scene.[120] Commodore Alexander Murray, Captain Isaac Hull and Captain Samuel Evans, with Judge Advocate Cadwallader Colden, sat on board the USS *Constellation* at New York Navy Yard for three short days and about half an hour, beginning on Tuesday, 11 April. After reading Decatur's original report into the record, they interviewed Decatur, his two surviving lieutenants, four midshipmen, the master and a sick berth attendant. Decatur was allowed to question his juniors and used the opportunity to lead them on the key issues. The commissioned officers all backed their captain, stressing his cool, collected conduct and the inevitability of defeat. This was sensible for young men with careers to make and the prospect of a Mediterranean cruise. However, they overdid the praise and appear to have confused the Court, which produced a perfectly absurd judgment. Master James Rodgers was less willing to follow Decatur's lead, even suggesting that the ship might have been better handled. For his pains he found

his own honour under attack, forcing him to call a witness in his own defence. Captain Robinson, who had been so upset at the time of the surrender, was not called; he was a volunteer and therefore not required.[121]

In his Report Murray accepted all

Navy Secretary Benjamin Crowninshield saved Decatur's reputation in Court, and at sea. He knew the nation would need heroes after the war. Ironically it would be a fellow American, Captain James Barron USN, who killed Decatur in a duel. (Naval Historical Center) of Decatur's exculpatory claims, but added one of his own, that the *President* "had sustained but little injury" clearly contradicting Decatur's report of 18 January. Instead, Murray claimed "they did not give up their ship till she was surrounded and overpowered by a force so superior, that further resistance would have been unjustifiable, and a useless sacrifice of the lives of brave men."[122] Just in case anyone was minded to quibble that the whitewash had been applied too thick, Murray declared any suggestions Decatur might have done better were "without foundation, and may be the result of ignorance, or the dictates of a culpable ambition, or of envy." While this beggars belief, Murray went on to impute truly divine qualities to the performance of the ship and her captain: "In this unequal conflict the enemy gained a ship, but the victory was ours." Clearly Murray, aged 60, nearly deaf and "an amiable old gentleman" of limited naval pretensions[123] had allowed the brief to exonerate Decatur to run away with him. By accepting Decatur's argument that grounding on the bar was "the primary cause of the loss"[124] and Sandy Hook "an excuse" for the defeat, the Court prejudiced the U.S. Navy against New York as a naval base.[125]

Within five months of losing the *President,* Decatur had taken the Algerian frigate flagship *Meshuda,* completing one of the most remarkable reversals of fortune in naval history. He ended the war by forceful diplomacy and after coercing Tripoli and Tunis into new treaties cruised around the Mediterranean, garnering plaudits in Naples and Sardinia.[126] On his return to America he gave an afterdinner speech that was soon transformed into the stirring cry, "My Country, right or wrong!"[127]

Stephen Decatur was as clever as he was brave – he worked his fame with the same attention to detail as his ship. The British could see that such a high-profile celebrity was a dangerous instrument of national power. (After all Nelson was their national war god.) Consequently lawyer-turned-historian William James spent the remaining years of Decatur's life trying to cut him down to size.[128] The British counter-propaganda effort only ended when Decatur's pride and arrogance left him facing the wrong end of Captain James Barron's pistol. Barron shot Decatur in the groin, cutting the femoral artery, and he bled to death late on 22 March 1820.[129] With that, the taking of the *President* might have relapsed into the dust of history.

Judgments of the action vary: William James used it as a stick to lambaste Decatur's honour, integrity and courage, until James Barron rendered him a matter of mere historical curiosity.[130] James was convinced that had the two ships met in single combat *Endymion* "would ultimately have conquered; the dreadful precision of her fire, her quickness in working, and evident superiority in sailing,

A United States frigate fires on Barbary Coast corsairs in this battle scene. The appointment of Commodore Stephen Decatur to command a squadron in the Mediterranean shortly after he surrendered USS *President* enabled a "remarkable reversal of fortune" for this American naval officer." (Naval Historical Center, NH 2239)

added to the established bravery of her officers and crew, are strong grounds of belief."[131] While American naval historians have tended to dismiss James's judgments, his contemporaries did not disagree. Captain Edward Brenton, James's rival as the historian of Britain's naval wars between 1793 and 1815, disputed James's authority to write on naval subjects, because he was a Whig naval officer, rather than a Tory lawyer. Brenton outlived James and used his 1837 edition to have the last word. He did not share the polemical anxiety to degrade American officers; in fact he was an American, born in Newport, Rhode Island, before the Revolution. Even so, he was satisfied that in this case the Royal Navy had good reason to celebrate. "It would not be fair to the memory of that excellent man Commodore Decatur," Brenton observed:

> to say that this was an equal action. It might, perhaps, have ended in a drawn battle, had not the *Pomone* decided the contest; but no-one will contend that

the *Endymion* had not supported the honour of the British flag; nor that she would not, very probably, have achieved the conquest without assistance, if we may judge from the carnage on the decks of the enemy, and the damage sustained by him in the action.[132]

HMS *President:* Fighting for Canada?

Having taken one of the American super-frigates, the British were anxious to show the world that their defeats in 1812 had been unequal combats between standard British 18-pdr. frigates and American opponents one-third larger, more heavily armed and manned. When the *President* arrived at Portsmouth in triumph, an existing HMS *President*, a French prize, was quickly renamed to ensure the American *President* would never be forgotten. Initially the Admiralty planned to return her to service, but she was in poor shape, seriously afflicted by old age and dry rot. The survey did not discover any signs of "damage" caused by grounding on the bar at Sandy Hook. Yet *President* was far too important a trophy to abandon, so the Admiralty ordered the construction of a precise copy to preserve the name and the image. *Endymion* was also rebuilt, a tribute both to her fame and her superior sailing qualities. Whatever the merits of the action, it is highly significant that the United States Navy has never revived the name. There have been no more USS *Presidents*, and although many presidents of the United States have had ships named for them, not one ship has been named for the office they hold.[133] If the defeat had been as glorious as contemporary American accounts claimed, surely such a proud name would have been revived.

This action mattered because after 1815 Canada was essentially undefended and indefensible. The Americans had the manpower and the position to invade and conquer the two provinces. British strategy depended on naval mastery to deliver a superior army to Canada and drive the Americans back whence they came. Advised by the Duke of Wellington, the British Government wisely restricted the fortification of the frontier, built an inland canal system to link Montreal with Kingston, the key to the Great Lakes, and based Canadian strategy on retaining Quebec and Halifax in the first year of war, as the basis for an eventual recovery of the country. By reminding Washington who was the master of the oceans, the British could exert diplomatic leverage and strategic deterrence. The reminder would take a naval form and the more potent the image, the more likely it was to be understood. *President*, and *Chesapeake*, *Endymion* and *Shannon*, Philip Broke's sons and other relics of British glory would be the key to maintaining the peace between Britain and America. Whether the United States ever sought to conquer Canada after 1815, the frontier, from Maine to Vancouver, remained highly contentious, with frequent outbursts of alarming rhetoric. Naval power mattered: it was the symbol and the strength of the British Empire at its zenith.

Endymion returned to service in 1832, the standard against which a new generation of sailing frigates would be judged, "the fastest frigate of the late war, which took the *President*."[134] She went to war again, this time with China, sailing up the Yangtze River as far as Nanking in 1842. Hulked in 1860, the old hero was finally broken up in 1868, giving her name to fresh *Endymions*.[135]

When Anglo-American relations reached a particularly low point in 1833, the First Lord of the Admiralty despatched HMS *President* to carry the flag of Admiral Sir George Cockburn on the North American Station. It was a clear and simple reminder of what had happened during the War of 1812. Cockburn was the man who burned the White House.[136] Early in Queen Victoria's reign the action was pictured for the third and last time, in a print that may have been inspired by the heightened tensions of the early 1840s. The unfortunate *President* is shown with *Pomone* firing into her, *Tenedos* directly astern and *Majestic* closing fast.[137] HMS *President* served as a flagship during the Crimean War before becoming the headquarters of the new London Naval Reserve Unit in 1862. The old ship lasted through to 1903, being replaced by another hulk that bore her prestigious name. In the late 20th century the Naval Reserve Headquarters came ashore, but the new building still bears the proud name HMS *President*. Names matter. They reflect an honest pride in the achievements of the past.

The decision of the United States Navy to leave the name *President* to lie fallow since 1815 suggests there was something dishonourable in her loss, and thus the flagship of President James Madison's Fleet, once graced with the figure of the founding father of the country, has left her name to another navy.

Appendix

Endymion, killed 11

John Read, 29, Middlesex, quarter master
Stephen Murphy, 24, London, captain of the
    maintop
James Fair, 22, Scotland, ordinary seaman
William Ash, 20, Sheerness, Kent, landsman
M. Norton, 33, Dublin, Ireland, able seaman
H. Jenkins, able seaman
William Mitchell, 21 Portsea, able seaman
Robert Annard, 35, Aberdeen, 2nd quarter
    gunner
Peter Connell, 30, Westmeath, Ireland, recently
    re-rated able seaman from landsman
William Hope, 21, Lancashire, landsman
J Smith, serjeant of marines

Endymion, wounded 14

3 of whom died of their wounds, while 2 more
    men lost their lives before the *President* was
    safe in Bermuda.
Thomas Duff, 42, Irish, captain of the after-
    guard, died on the 16th of his wounds.
Peter Newland, 26, London, ordinary seamen,
    fell overboard from the prize 19 January 1815.
Benjamin Neil, Port Glasgow, died from a fall 24
    January 1815.
William Lane, 25, Berkshire, trumpeter died of
    his wounds 24 January 1815.
Robert Lyons, Glasgow, landsman, died of his
    wounds 27 March 1815.[138]

President, killed

Lieutenants
Fitz Henry Babbitt
E F Howell
Archibald Hamilton

Crew
Henry Hill
Sam Gaines
Sam de Coster
Michael Barton
John Weary
John Briggs
Char Conway
Wm Smith 3rd
Wm Keeler
Ja Chapman
Geo W Swift
Francis Dea
Edw. James
Amos Peasly
Wm. Barrett
Cahal Pratt
Thos. Kelly
And. Sestrom
Wm. Moore
Aaron Lynn

For the *President*'s wounded see RG 45 CL 1815
    vol. 1 no. 50. Log of USS *President*.

Courses and tracks

ADM 52/3904 Masters Log HMS *Endymion*.
During the day ran 78 miles on course steadily
    shifted from East by North at 13.00 to ENE at
    5 and SE at 18.00, and east again at 22.00.
Winds varied from NW to NNW
At the close Montauk Point lay North 37 miles,
    West 95 miles.

PRELUDE

The Renewed Wolfpack Attacks of September 1943

Tᴏ HE NEXT THREE CHAPTERS all deal with naval actions in the latter part of the Second World War. By the summer of 1943 the growing strength of Allied navies made the task of the *Kriegsmarine,* the German navy, increasingly difficult. Greater Allied strength was certainly necessary to combat the *Kriegsmarine,* but – as all three of these studies will clearly show – it was not sufficient for victory. The simplest way to explain this, and which will also be shown in many ways in the next chapters, is that the German crews were too skilled, determined and coura-geous to be easily beaten. The Allies had to find better ways to fight even as they gradually increased the number of escorts and aircraft devoted to suppressing and sinking their opponents.

Growing Allied strength and skill had made the north Atlantic convoy lanes a deadly place for U-boats in May 1943, resulting in the temporary suspension of wolfpack operations there late that month. Both sides were well aware that this suspension could not last long unless the Germans were willing to concede defeat in their effort to sink Allied merchant shipping. Throughout the summer of 1943 the *Kriegsmarine* strove to introduce new weapons and sensors, along with new tactics, that would allow them to once again sink greater numbers of ships.

There were clear limits to German options, however. With the U-boats avail-able, the only viable option was to mass large numbers of submarines around a convoy and then hope that enough could break through the ring of escorts to ensure destruction of many merchant ships. More effective U-boats might have enabled smaller numbers to inflict significant damage, and radically new U-boats were ordered that summer. But these new submarines took time to build and faced serious teething problems before they could be used operationally. It would be many months before they could be put into service.[1] Better torpedoes might also make a difference, but even the innovative new weapons loaded aboard U-boats in August 1943 were not all that much more effective than those already in use. In the end, quantity would have to continue to be used, as not enough qualitative improvements could be achieved soon enough to matter. Wolfpack tactics remained the only way that significant numbers of U-boats could be gathered around a convoy.

Allied electronic sensors that enabled escorts and aircraft to more and more often detect U-boats in poor or limited visibility proved important, and were poorly understood by the German forces. U-boats had relied on the small visual profile they presented when operating on the surface – especially at night – in the early part of the war for much of their success. By mid-1943 Allied radar and radio detection systems meant that German submariners could be found much more often when they were on the surface. Nonetheless the requirement for speed meant that older U-boats needed to operate on the surface for wolfpacks to be effective. The Germans understood that their submarines were becoming more vulnerable on the surface, but desperately sought to find ways to keep older tactics alive, adopting hazardous tactics and hoping that more daring and more anti-aircraft guns would hold off the growing number of Allied planes found deep in the Atlantic.

In many ways the German navy was seeking to renew battle with only minor improvements, hoping the skill and courage of their sailors would compensate for the advantages Allied forces now had. They also hoped to achieve surprise, choosing to employ new weapons and different operational patterns in an effort to catch the Allies off guard.

The results may seem predictable in hindsight, for though there is no doubting the courage and determination of the German crews, the Allied crews proved equally courageous and determined. When two sides are equal in determination, technical and organizational factors loom large, and the Allies displayed important advantages in these dimensions as the war ground on.

Surprise did play an important part in the battle, although in the event both sides were startled by unforeseen events. The Allies had to adapt to an unexpected method of use for a new German weapon, even though they had anticipated the weapon would soon appear. On the other hand the Germans had to counter an Allied decision to manoeuvre convoys in an unusual way. By the end of the battle the Allies were well on their way to countering the new German weapon, while the Germans were still confused by the actions of the Allied convoys, and failed to grasp the true results of the battle.

Though hardly the deciding factor in the battle, this difference in perception is in many ways symptomatic of the overall results of the shipping war – the Allies more often successfully adapted to and neutralized the initiatives of their opponents, while the *Kriegsmarine* proved less effective in understanding the measures taken by the Allies, and were consequently less effective in adapting to operational realities. The battles that swirled around convoys ON202 and ONS18 briefly seemed to offer the German navy hope that their wolfpacks could once again inflict serious losses on Allied shipping. The true reality was that Allied technological and tactical developments along with increasing numbers were fast making wolfpacks convenient gatherings of targets for antisubmarine forces. The Germans would have to find a new way to stalk shipping if their U-boats were going to operate effectively in the Atlantic.

SEPTEMBER 1943

Hollow Victory

GRUPPE LEUTHEN'S ATTACKS
ON CONVOYS ON202 AND ONS18[1]

Douglas M. McLean

HMS *Lagan* raced at 20 knots toward a radar contact 3,200 yards away. Minutes later, at 0300 hours 20 September 1943, the contact disappeared, as *U-270* submerged. This boat, along with 19 others – *U-229, U-238, U-260, U-275, U-305, U-338, U-341, U-377, U-378, U-386, U-402, U-422, U-584, U-641, U-645, U-666, U-731, U-758, U-952*[2] – was part of the German navy's renewed attacks on convoys in the North Atlantic, ending a strategic withdrawal that had lasted all summer. *Lagan*, now some 10 miles off the port beam of convoy ON202, stood on for another minute, before slowing to 15 knots and altering 20 degrees to south, actions designed to bring her within range of the submarine at a speed that optimized asdic (sonar) detection. Two minutes later, when *Lagan* was only 1,200 yards from where the U-boat had dived, an explosion shattered the frigate's stern, killing 29 of her crew.[3]

The frigate represented a state-of-the-art Allied antisubmarine warfare vessel, in commission for less than a year. She had enough speed to overtake U-boats on the surface, unlike the smaller corvettes rushed into production at the outset of war. Her sensors included Type 271 radar, whose centimetric wavelength could not be detected by U-boat radar detectors, as well as modern – for the period – asdic. Armed with 4-inch guns, depth charges and the new forward-launched, contact detonation antisubmarine weapon known as Hedgehog, she was well equipped to locate and destroy her prey.

Lieutenant-Commander A. Ayre, RNR (Rtd),[4] the captain of HMS *Lagan*, had done everything right up to the time of the explosion. Acting on orders from the Senior Officer of Escort Group C2 (EGC2), Commander P.W. Burnett, RN, *Lagan* had headed southwest to find the U-boat detected by High Frequency

Direction Finding (HF/DF) operators aboard the rescue ship *Rathlin*. They estimated a U-boat was sending contact reports of the convoy to its headquarters in
Paris (*Befehlshaber der U-boote*, or *BdU*). The boat's transmissions were detected
shortly before 0200 hours and reported to Burnett two minutes after the hour.
A little over 20 minutes later, Burnett ordered *Lagan* to hunt along a bearing of
210° from the convoy to a range of 10 miles. At 0225 the frigate began her fateful
search, first detecting the U-boat at 0244 at a range of 4,400 yards. Less than 20
minutes later *Lagan* was dead in the water.[5]

The presence of *Lagan's* opponent in this engagement, *U-270*, marked a determined effort by the German navy to reverse the success of Allied antisubmarine forces in May 1943. A deadly combination of growing numbers of Allied
escorts and aircraft, using increasingly effective sensors, weapons and tactics, and
directed by superior staffs and operational intelligence ashore, inflicted crippling
U-boat losses. In his memoirs Grand Admiral Doenitz referred to the U-boat
losses in that terrible month as "frightful."[6] If Germany could not attack convoys
effectively, then the western Allies could rapidly build up their strength in the
United Kingdom, paving the way for an assault on the Continent. Determined
to renew the attack on Allied shipping, Grand Admiral Doenitz ordered new
initiatives to regain U-boat effectiveness in the summer of 1943. He also directed
that new submarine designs be rushed into production, but these designs were
months if not years away – the battle would have to be continued with the existing fleet, with modest improvements in weapons and sensors.

Lagan's opponent, *U-270*, represented a typical U-boat of the period, a Type
VIIC. In many ways an aging but effective design, the small vessel – only 220 feet
long and about 761 tons on the surface – was the workhorse of the U-boat fleet.[7]
Her submerged speed was a maximum of just over 7 knots, but proceeding at
that pace would exhaust the boat's battery, its sole means of underwater propulsion, in a little over an hour. Nor was there power to replenish the boat's air supply once she submerged, as nuclear submarines do today. Consequently, U-boats
needed to stay on the surface as much as possible if they were to locate and intercept their prey effectively. This posed a clear dilemma to U-boat commanders, as
their small vessels were very vulnerable to enemy warships and aircraft when not
hidden beneath the waves.

Type VII U-boats used passive sonar to assist in locating Allied ships – convoys could usually be heard before becoming visible. The eyes of the boat's commander through the periscope, or those of lookouts when the boat was surfaced,
were perhaps the most vital sensors, however, often providing first detection of
their opponents. Very few U-boats were equipped with radar at this point in the

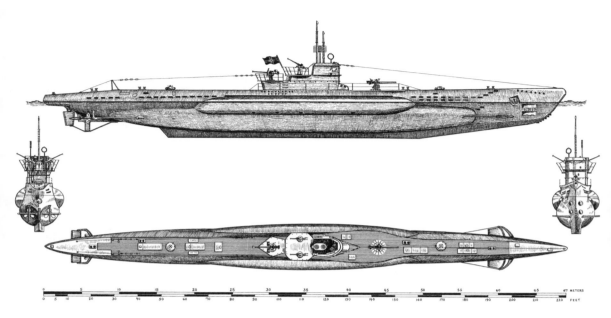

Workhorse of the wolfpacks. The Type VIIC formed the backbone of the U-boat arm during the war, with well over 500 of this main type being built. It was more a submersible than a submarine as designers had optimized the hull form for surface rather than submerged operations, which is why it looks so different from submarines found in the navies of today. By mid-1943 the number of anti-aircraft guns fitted was growing, as can be seen here. The large deck gun forward of the conning tower, still shown on this boat and once used in surface actions to sink merchant ships, had become so seldom used that it had been removed from all U-boats involved in this battle. (Drawing from *Tin Hats, Oilskins & Seaboots* by L.B. Jenson, courtesy of Alma Jenson)

war; German ship-borne radar was significantly less sophisticated than that of the Allies. German scientists were still reluctant to accept that the Allies could produce centimetric band radar waves in a set compact enough to be installed in a small warship or aircraft, despite the evidence of one found in a crashed British bomber earlier in 1943. Growing evidence of the Allied ability to locate U-boats in poor visibility and at night required the German navy to try to deploy equipment to at least warn their crews that they had been found, but German radar warning devices remained a step behind Allied radars. In August 1943 a new radar warning set, the FuMB 9 *Zypern* ("Cyprus," also known as the Wanze G1 and Hagenuk) was rushed into production. The FuMB 9 covered wavelengths of 1.3 to 1.9 metres, and therefore could not detect the shorter centimetric waves used by the most modern Allied radars.[8]

The German navy did not grasp how badly their boats were falling behind in the war of electronic measure and counter-measure. Without a good understanding of their opponent's capabilities – the German navy had woefully little knowledge of Allied weapons and sensors throughout most of the Battle of the Atlantic – German naval staffs largely planned in a vacuum. *BdU* staff better understood their own weapons requirements, and therefore more efforts were focused on providing more and better weapons to counter Allied warships and antisubmarine aircraft.

At the outset of the war most U-boats were equipped with a single 20mm cannon for air defence, with a standing order to dive on those rare occasions when antisubmarine aircraft were in the area. The growing prevalence of Allied aircraft in ever-expanding areas of the Atlantic made immediate diving much less desirable, as it slowed boats rushing to intercept a convoy. *BdU* decided to massively increase anti-aircraft armament and advised U-boats to fight off aircraft whenever possible. For the attacks on convoys ON202/ONS18, *BdU* ensured all boats were fitted with no fewer than eight 20mm cannon, in two twin and one quadruple mounting. Tactical direction that called for all U-boats to concentrate around a convoy and then surface to fight off aircraft was also devised. Somewhat paradoxically, U-boats were not to gather in groups of more than two – rather they were to be scattered around the perimeter of the convoy, with the hope of dispersing defenders.

This questionable guidance reflected desperation as opposed to careful study. There were two critical weaknesses. The first was that U-boats were small warships without the radios or command and control facilities to work closely together. The second weakness also related to size – the tossing deck of a U-boat in all but the calmest sea made it a poor gun platform. Allied warships and aircraft worked together as teams on a regular basis, and had developed equipment and doctrine to facilitate coordination This meant that they were more likely to synchronize their efforts effectively in the face of a mass U-boat attack. Aircraft could also use their greater speed to concentrate attacks on one U-boat at a time, a tactic that would work especially well for carrier based aircraft that could rapidly call up reinforcements. Allied operational research scientists soon appreciated the weaknesses of the new German tactics, observing that the longer a U-boat remained on the surface, the more vulnerable it was.[9] The German navy had no operational research process to review tactical concepts. Instead, U-boat sailors determined the value of new doctrine the old fashioned way – trial by fire in combat.

The first test of the new German anti-aircraft policy occurred the day before *Lagan* encountered *U-270*, as U-boats of group *Leuthen*[10] moved toward the positions they had been assigned in a patrol line, not all of them proceeding as stealthily as *BdU* had directed. The need for speed to reach its patrol position may have resulted in one of the boats coming to the attention of the most flexible antisubmarine weapon in the Allied inventory, Very Long Range (VLR) Liberator aircraft. There would be 23 flights by these well-equipped planes in support of the defence of the two convoys during the course of the battle. The first three patrols were by Liberators from Iceland, two of them Canadian aircraft en route back to Newfoundland, on 19 September. Directed to sweep ahead of westbound convoy ONS18 (still a bit ahead of the faster, and therefore overtaking, ON202), Liberator A/10 of number 10 Squadron RCAF detected *U-341* racing along on the surface about 100 miles northwest of the first convoy (see inset map, page 143). This U-boat followed *BdU*'s direction to fight aircraft on the surface if surprised. By staying on top of the waves, *U-341* gave this Liberator time to conduct a determined attack. The first pass proved too high to drop depth charges accurately, so Flight-Lieutenant J.F. Fisher came around for a second attack just over the waves. Six depth charges straddled the gamely fighting *U-341*, blowing the boat's bow out of the water. Fisher came around for another pass, dropping his last four depth charges on the sinking conning tower of the U-boat, destroying it.[11] This was not an auspicious start to *BdU*'s anti-aircraft policy for this battle.

A key reason for VLR Liberators patrolling in the vicinity of the threatened convoys was the work of Allied intelligence staffs. The Royal Navy's Operational Intelligence Centre (OIC) became aware in mid-September that a wolfpack was forming but proved unable to determine the precise location of its intended patrol line. A main source of OIC information derived from the decryption of German radio messages encoded on their Enigma machine. The resulting intelligence information, classified Ultra, provided important insights into German operations and intentions. Remarkably, the Germans failed to appreciate Allied code-breaking capability throughout the war. Despite the overall value of this source, decryption was not always timely or complete, and there had been long periods earlier in the war when decryption was not possible to any useful degree at all.[12] The limitations of Ultra would have important consequences in this battle, despite the growing Allied mastery of German codes. The OIC proved able to gain a general understanding of the German intention to once again undertake wolfpack operations against convoys in the North Atlantic, but mistakenly placed the initial position of the *Leuthen* patrol line about 150 miles south and 125 miles west of the actual location (see map, facing). German tactical intentions were also not understood until the battle was actually in progress, providing U-boats with a small but important degree of surprise in their operations. The problems experienced by the OIC in penetrating German plans and intentions may seem small in retrospect, but had a significant impact on this battle that has not been well understood by historians.

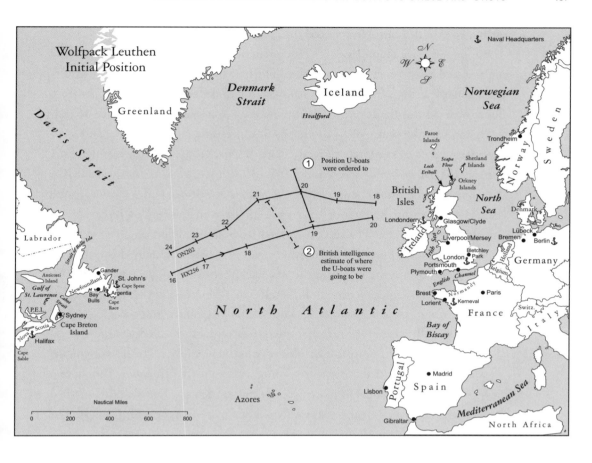

The OIC's mistaken assessment of *Leuthen's* initial patrol position partially resulted from a successful deception effort on the part of *BdU*. The German navy regularly changed geographical reference positions, and often issued these reference positions to boats before they proceeded to sea, precluding any chance that their exact location could be determined through decryption of radio messages. Patrol assignments relative to these reference positions were then radioed after the boats left port. Allied intelligence officers found determining the location of reference points challenging immediately after each change, as their locations had to be determined by inference over time. (The Allies, however, remained thankful that *BdU* did not provide more details to boats by hand before they sailed, a simple procedure that would have undermined decryption significantly.)

Careful plotting is required to understand the impact of *BdU's* success in disguising the intended patrol position of Leuthen. *BdU* ordered Leuthen to establish a patrol line from "AK 3441 to AL 4762" (*KTB* 14 September 1943). This equates to a line starting at approximately 59° 30' N / 028° 10' W in the north running south-southeast some 340 miles to 54° 15' N / 025° 15' W.[13] The OIC, however, estimated that the patrol line would run from 56° 30' N / 032° 00' W to 51° 40' N / 027° 00' W[14] (see map, page 137). The small difference in longitude and latitude contributed to two significant mistakes. The first was to identify the most threatened convoy as the eastbound HX256, which had departed New York City on 9 September bound for Liverpool.[15] The intended track of this convoy passed directly through the centre of *Leuthen*'s position as estimated by the OIC, and measures were taken to reinforce the escort. The support group EG9, consisting of the frigate HMS *Itchen* (SO Commander C.E. Bridgeman, RNR), destroyer HMCS *St. Croix* and the three corvettes HMCS *Chambly*, *Morden* and *Sackville*, found itself diverted from its intended patrol destination in the Bay of Biscay to the central North Atlantic. The second mistake was that the effort to divert convoys ONS18 and ON202 around the northern end of *Leuthen* actually drove the two convoys almost into the centre of the wolfpack, because a modest course change of only about 20 degrees was used. If the OIC had discerned the actual location of *Leuthen*'s intended patrol positions, a course change of closer to 60 degrees might have successfully steered the merchantmen clear of the U-boat concentration.

The diversion of EG9 into the North Atlantic to support the wrong convoy serendipitously placed these warships much closer to the threatened merchant vessels than they would have been in the Bay of Biscay. The Royal Canadian Navy had great hopes for these ships, the second support group for Canada. The ships left Plymouth harbour on 15 September for their first operational patrol. The next day the five ships in EG9 received orders to continue westward toward the central North Atlantic, placing themselves under the orders of Commander in Chief Western Approaches (CinCWA). The group's suspicions that they were now intended to augment convoy escorts were confirmed late on the 16th, when Western Approaches ordered them to rendezvous with the eastbound HX256. On 18 September Western Approaches ordered EG9 to proceed to the assistance of ONS18 if no submarine threat to HX256 was perceived. Shortly after EG9 received the signal, however, HX256 was indeed attacked by *U-260*, which happened upon the convoy while en route to her *Leuthen* patrol position.[16] The attack failed to inflict significant physical damage, the anti-torpedo net defences of the Liberty ship SS *William Pepperell* proving sufficiently effective that the ship remained in convoy.[17] Nonetheless, the attack, coming where and when it did, must have reassured British intelligence experts that they had indeed accurately estimated the patrol position of *Leuthen*, and therefore psychologically the attack must be considered a success. EG9 reached HX256 early on 19 September, but remained with that convoy for only about six hours.[18] Finding that no U-boats lurked in the area, the group then proceeded northeast toward ONS18, as evi-

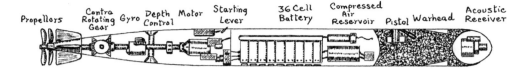

An electric torpedo with a difference: the German navy had high hopes for the new acoustic system deployed for the first time in this battle. Designed to adjust course toward the broadband noise produced by propellors, this torpedo provided a much better chance of hitting the small and difficult targets presented by Allied escorts. Teething problems and swift Allied counter-measures resulted in these weapons having only a modest impact on the Battle of the Atlantic. (Drawing from *Tin Hats, Oilskins & Seaboots* by L.B. Jenson, courtesy of Alma Jenson)

dence mounted that the U-boats were concentrated further north and east than where OIC originally estimated.

The inadequate course diversion of the two westbound convoys had more serious consequences. Prior to the arrival of the diversion message late on 18 September, the two westbound convoys had a mean course of 268°.[19] The diversion north to a base course of 287° was comparatively small – a larger diversion was certainly possible, and major diversions were done on other occasions with other convoys. If these two convoys had been successfully diverted, then there might have been a delay of several days or a week before the U-boats succeeded in locating a convoy. This delay would have provided the Allies an opportunity to react to another German surprise – their plan for using their new torpedo – and might have significantly reduced the impact of the new weapon.

BdU had pinned great hopes on this torpedo, one that homed on the propeller noise produced by Allied escorts. The Germans referred to this weapon as the Type V torpedo, codenamed *Zaunkoenig*, a type of bird that roughly translates as "hedge king." The Allies referred to the German Naval Acoustic Torpedo as the GNAT, an appropriate acronym as this weapon proved troublesome for the remainder of the war, but not nearly as deadly as the Germans hoped.

The battles around convoys ON202 and ONS18 would be defined by the threat of the Type V. *BdU* hoped that the surprise of this new weapon would result in substantial losses before the Allies could develop effective counter-measures. The German's planned to aim Type V torpedoes at the convoy escorts during the initial part of the attack, so weakening and disorganizing them that the merchant ships would become easy prey for conventional torpedoes. Part of the reason for the focus on warships was the small supply of the new weapon; only 80 were ready by August 1943, and this was only achieved by substantial pressure from Grand Admiral Doenitz to accelerate production. This meant that most boats in *Leuthen* had only four of the new weapons, and the three boats that sailed from

Norway did not carry any. With only two Type V torpedoes available for the bow tubes and two for the stern, there were clearly not enough for general use.[20]

Worse for the Germans, Allied intelligence staffs anticipated that an acoustic homing torpedo would be introduced at some point, although the decision to deploy the Type V torpedo for the convoy attacks in September was not specifically discerned by Allied intelligence in advance. Unfortunately, the Enigma message sent by *BdU* detailing the intention to concentrate initial U-boat attacks on escorts was only deciphered by the OIC on 20 September, too late to give warning to the escorts.[21] Nonetheless, the possibility of a torpedo fitted with an acoustic homing system had been expected for some time – the Allies already had such a torpedo in use against U-boats – and Allied operational research studies had estimated the potential impact. These reports, one of which had been written several months earlier in June 1943, expected that the main use of such a new torpedo would be against merchant ships. The prospect of their being used primarily against escorts was identified as the third of three possibilities, and not assessed to be likely.[22] The Allies general anticipation of a new German acoustic homing torpedo limited the degree of surprise that the Germans could achieve. However, general knowledge of an opponent's capabilities does not mean that his intentions are understood, and the Allied intelligence failure to anticipate the use of the Type V primarily against escorts resulted in tactical surprise being achieved by the U-boats.

U-270's attack on HMS *Lagan* therefore was no accident. Before the introduction of the Type V, head-on torpedo shots at warships racing directly towards a U-boat were rarely successful, which was one of the reasons that warships charged towards a submarine as soon as they detected it. With the new weapon, *Lagan*'s intercept course toward the U-boat provided an ideal target for *U-270*. *Kapitänleutnant* Paul-Friedrich Otto, the boat's commander, used his experience – he had commanded the boat since early September of 1942, and this was her second war patrol – to good effect, resulting in a devastating attack.[23] Arguably, this proved the most successful Type V attack of the entire battle. *Lagan*'s damage ensured it could no longer defend ON202, and in fact now required a screen of her own if she was to return to port safely. Commander Burnett assigned the trawler HMT *Lancer* to this task, and the deep ocean tug, HMS *Destiny*, was tasked with towing *Lagan* back to port. With one stroke, ON202 lost two defenders and the tug intended for towing damaged merchant ships to port. This left EGC2 with five ships: HMCS *Gatineau*, an E-class destroyer handed over to the Canadians in June 1943, the I-class destroyer HMS *Icarus* and three corvettes: HMCS *Drumheller*, HMCS *Kamloops* and HMS *Polyanthus*.

Unfortunately for the merchant ships of ON202, *U-270* was not the only

The corvette HMCS *Chambly* slices through the fogbound North Atlantic in a picture that illustrates well the conditions that persisted around the merged convoys for much of 21 and 22 September. Fog was a mixed blessing. Escorts with their radar and better radio-interception equipment could often find their foes before being seen, but limited visibility in the vicinity of a convoy increased the risk of collision. (Library and Archives Canada, PA 115352)

U-boat closing on the convoy. Just over one hour after *Lagan* lost her stern, HMS *Polyanthus*, stationed ahead of the convoy, detected a radar contact 3,400 yards away. She closed the contact and sighted her prey at 0428 hours after firing star-shell.[24] The U-boat dived shortly after this, and *Polyanthus* closed, obtained asdic contact and delivered two depth charge patterns on her contact.[25]

Most U-boats would have been driven off by this attention. After the battle Commander Evans would mistakenly comment in his Report on Proceedings that "It was satisfactory to find that the Hun had become no more enterprising during his summer vacation and that the old rule of 'A submarine detected is an attack averted' still held true."[26] Unfortunately this was not the case for *Polyanthus*'s target, *U-238*, commanded by *Kapitänleutnant* Horst Hepp.[27] Hepp doggedly followed the corvette back toward the convoy after the attacks, finding himself well placed ahead of the port columns of ON202 at dawn.[28] He conducted a periscope attack, hitting the lead ships of the two port columns, SS *Frederick*

Douglass (11) and *Theodore Dwight Weld* (21), on their port sides.[29] The damage to the *Douglass* was serious but not immediately fatal, while the *Weld* broke in two with the after part sinking promptly.[30] Rescue efforts appear to have gathered up all 70 of those aboard the *Douglass*, but not all of the *Weld* crew survived.[31]

This dawn attack highlighted the vulnerability of ON202 to the gathering pack. If ON202 and its escorts had been the only ships in that lonely stretch of the North Atlantic, a significant German success would have been likely. *BdU* had strived to achieve surprise for this first operation against North Atlantic convoys since May, and had taken great pains to assemble the *Leuthen* pack covertly.[32] These measures had, unknown to *BdU*, only been partially successful, and unfortunately for German U-boat crews there were far more than five escorts in close proximity to the area of ocean they were headed for.

Two additional escort groups were within a few hours steaming of ON202 on the forenoon of 20 September 1943. One of these groups, B3,was escorting slow convoy ONS18, approximately 45 miles southwest of ON202 when *U-238* attacked.[33] The close proximity of these two convoys in the midst of the *Leuthen* U-boats resulted from a combination of circumstance, chance and the OIC's imperfect intelligence estimate of where the wolfpack would be established. The circumstance resulted from the nature of the convoy cycle at this point of the war. Ships returning from the United Kingdom to North America were grouped into two different classes of vessels, slow and fast. Fast merchant ships had to be capable of maintaining at least 10 knots, while slow ships need only achieve a steady 7 knots.[34] ONS18, a slow convoy of 27 ships, including the Merchant Aircraft Carrier (or MAC ship) *Empire Macalpine*, had departed Liverpool for North America on 13 September, two days ahead of ON202. Both convoys were routed along roughly parallel great circle routes, and both had been diverted to the northwest on 18 September. The faster speed of ON202 meant that it would, under the circumstances, overtake the slower convoy during the course of 20 September. The actual location of *Leuthen* was, fortunately for the Germans, directly in the path of both convoys, and, fortunately for the Allies, where the two convoys would be closest to each other on their passage across the ocean.

The other EG reasonably close to ON202 was EG9. The earlier diversion of this group from the Bay of Biscay to the middle of the Atlantic proved to be one of the OIC's most important contributions to the battle. Allied intelligence, despite some errors such as the misplotting of *Leuthen* and failure to anticipate the primary use of Type Vs against escorts, proved significantly superior to the German intelligence perspective at *BdU*. The German code-breaking organization, *B Dienst*, had achieved significant success up to the middle of 1943 and could

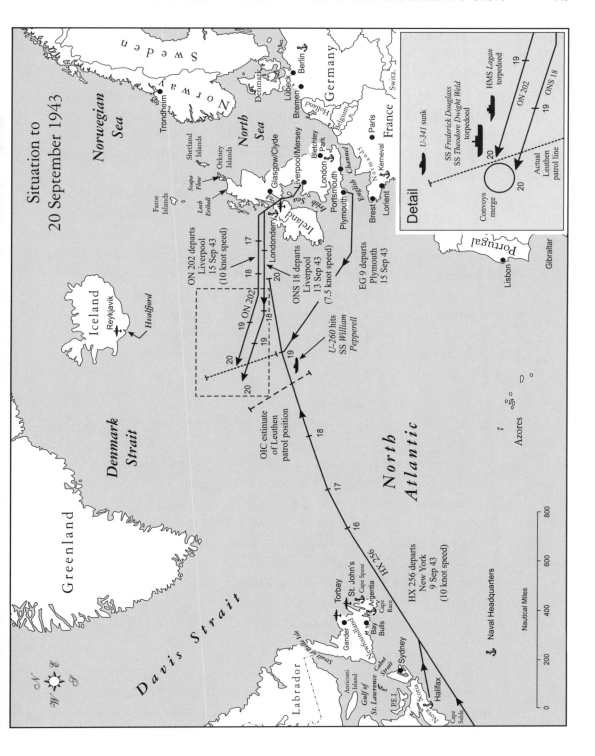

still provide some insight into Allied convoy activity. However, a realization on the part of the Allies that some of their codes were being broken by *B Dienst* resulted in significant changes to encypherment procedures starting in June 1943.[35] Small amounts of information continued to be gained by German efforts, but overall an incomplete and often confusing picture proved the best that could be provided to operational staffs at *BdU*. Limited German knowledge of their opponent's capabilities and intentions contrasted markedly with the general Allied knowledge of German capabilities and broad intentions.

The growing number of U-boats at large in the North Atlantic in early September 1943 only gradually became known to the OIC. On 13 September 1943 Commander Roger Winn, RNVR, estimated that "there are thought to be about 20 U-boats in the general area of the Azores or further north, and it would not be surprising if a marked renewal of activity occurred in the next week or 10 days."[36] Winn, whose long-time service as head of the OIC and remarkable analysis talents gave him an almost uncanny sense of *BdU* intentions, had long expected a return of the U-boats to the North Atlantic. Six weeks earlier, at the end of July 1943, he had produced an estimate that anticipated events of late September 1943 with remarkable prescience:

> It is common knowledge both to ourselves and the enemy that the only vital issue in the U-boat war is whether or not we are able to bring to England such supplies of food, oil and raw material and other necessaries, as will enable us, (a) to survive and (b) to mount a military offensive adequate to crush enemy land resistance. Knowing this is so, the enemy must have intended an ultimate return to this area, so soon as he might be able, by conceiving new measures and devising new techniques, to resist the offensive which we might be able to bring to bear upon him there … but it might be the last dying struggle of a caged tiger for the enemy to send back in September or October into the North Western Approaches his main U-boat forces…. Even if heavy losses of merchant shipping and escort forces on the North Atlantic convoy routes were to be suffered … no fear need to be felt as to the ultimate outcome.[37]

EG9's shift at mid-day on 19 September to support the slow, rather than the fast, westbound convoy likely resulted from two factors. First, ONS18 was still a little ahead of ON202 in its trek westward across the Atlantic, and therefore more likely to encounter U-boats as the convoy entered the region of minimal air coverage. Allied planners were well aware that U-boats preferred to operate where antisubmarine aircraft were less likely to be encountered. By this point in the war the air gap was far from the absolute hole that had been the case in earlier years.

Refuelling at sea proved very important to escorts, most of which were not designed for the great distances involved in accompanying convoys all the way across the ocean. The procedure was relatively primitive, with the ship supplying the oil – usually a specially fitted merchant ship – trailing a hose astern that was picked up by the escort, such as the Town-class destroyer (possibly HMCS *St. Croix*) shown here. Three ships, SS *Beaconstreet,* SS *Thorhild* and SS *Mexico City,* provided almost 600 tons of oil to the escorts for the two convoys. (Library and Archives Canada, PA-116335)

Nonetheless, the central North Atlantic remained farthest from the airfields of land-based aircraft, minimizing the time these dangerous U-boat hunters could remain in the vicinity of convoys. Second, slow convoys were inherently more vulnerable to attack than fast convoys. Two or three knots may seem like a minor difference in speed, but in relative terms the difference is significant as a fast convoy was approximately 33% faster than a slow one. U-boats on the surface could travel more than twice as fast as a slow convoy, with a speed advantage of approximately 10 knots. This enabled U-boats closing on a slow convoy or repositioning for attack to get in position much more quickly than might be the case for a faster-moving convoy. In terms of relative motion a slow convoy was much more vulnerable than a fast one.

As evening fell on 19 September 1943, three Allied groups of ships – ONS18 with its escort EGB3, ON202 with its escort EGC2, and EG9 – were rapidly converging on a small part of the North Atlantic, where the 19 remaining U-boats of *Leuthen* were also moving toward their patrol positions. The escorts with ONS18 and ON202 both began detecting growing numbers of radio transmissions from U-boats, alerting them to the danger ahead. Commander M.J. Evans, RN, the Senior Officer (SO) of ONS18's escort embarked in HMS *Keppel*, believed that

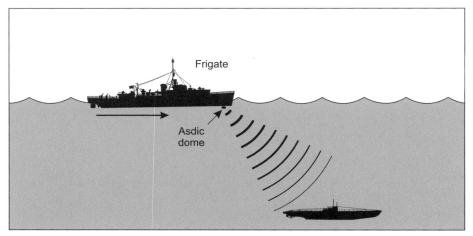

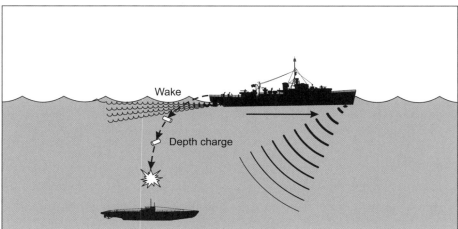

Depth charges were the most common antisubmarine weapon of the war. The depth charge was very basic weapon, little more than a drum filled with explosive dropped by a ship as it passed ahead of a U-boat, and a successful attack required skill and anticipation. The escort would have a good idea of the location of the U-boat as it closed, and – depending on the sonar fitted – may have had some indication of the target's depth. However, contact would be lost as the ship passed over top, and there would also be a significant delay while the charges descended to the depth at which their hydrostatic fuses had been set. This delay provided U-boats with an opportunity for evasion that reduced the chance of a successful attack. (Drawing by Christopher Johnson)

most of the transmissions came from astern and were therefore more likely to threaten ON202.[38] EG9, closing from the southwest on ONS18, also considered that "it seemed probable that the group had also been seen by the enemy" and planned to join the convoy by an evasive course that would place them on the quarter of ONS18, hopefully drawing any U-boats that might be tracking their progress away from the merchant ships.

Despite Commander Evans's estimate that the greater threat loomed for

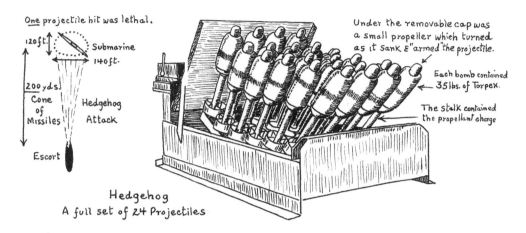

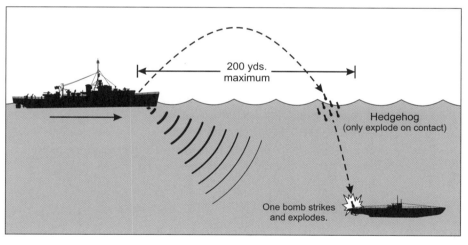

Hedgehog was the first ahead-thrown antisubmarine weapon and an attack differed significantly from one employing depth charges. An escort using hedgehog never lost contact with its target, which reduced the possibility of successful evasion by the U-boat. Hedgehog attacks also caused much less disturbance in the water, as the bombs only exploded if they struck their target. Despite the potential advantages provided by the new weapon, the Royal Navy experienced problems in getting the level of expected success for a long time after the new weapon was introduced. (Upper: From *Tin Hats, Oilskins & Seaboots* by L.B. Jenson, courtesy of Alma Jenson. Lower: Drawing by Christopher Johnson)

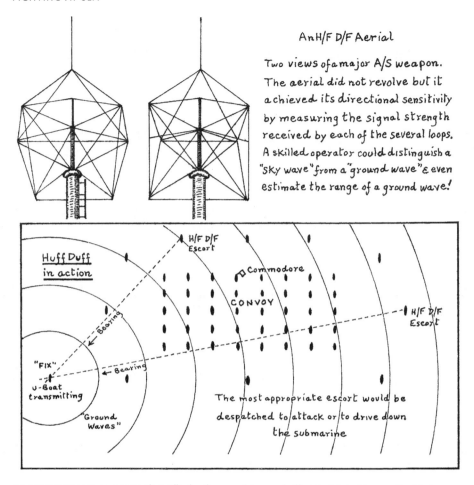

An H/F D/F Aerial

Two views of a major A/S weapon. The aerial did not revolve but it achieved its directional sensitivity by measuring the signal strength received by each of the several loops. A skilled operator could distinguish a "sky wave" from a "ground wave" & even estimate the range of a ground wave!

An important secret weapon that Allied sailors used to good effect in this battle was the High Frequency Direction Finding (HF/DF) set. German sailors knew that their radio transmissions could be intercepted, but their scientists were convinced that the equipment needed for this purpose was so large that it could not be fitted in a small escort. Allied scientists succeeded in producing compact HF/DF sets, which proved very helpful in locating U-boats as they closed to attack. (Drawing from *Tin Hats, Oilskins & Seaboots* by L.B. Jenson, courtesy of Alma Jenson)

ON202, the first U-boat probe was toward ONS18 at 2130 hours. The destroyer HMS *Escapade* detected the approach of what turned out to be two U-boats. The first U-boat was forced down by a quick depth charge attack. *Escapade* then turned to the second U-boat, forcing it down as well, and then began a more deliberate series of attacks. Good asdic contact and careful manoeuvring seemed to offer the possibility that a warship would also destroy a U-boat before any Allied losses were suffered when fate intervened. Running in to deliver her fourth Hedgehog attack, *Escapade* suffered a terrible blow from her own weapon, when a number of the Hedgehog rounds detonated while being fired. There were 21 fatal casualties and considerable damage to the forward end of the ship, effectively putting the destroyer out of the battle.[39]

Hedgehog consisted of 24 spigot mortar projectiles fired from the bow of a warship, designed to land in a circular pattern approximately 130 feet in diameter about 200 yards ahead of the ship.[40] Each of these small bombs weighed about 65 pounds, half of which was the high explosive Torpex, and was designed to explode on contact. In theory a far more deadly weapon than depth charges, which a warship had to drop astern as it passed over a U-boat with the depth of explosion pre-set, Hedgehog had yet to achieve a kill percentage anywhere close to what had been predicted. Reports prepared by Allied scientists indicated that in the fall of 1943 Hedgehog was only a little more deadly than depth charges, and Allied staffs struggled to determine how to improve this unsatisfactory situation. Clearly, devastating early explosions such as the one that occurred aboard *Escapade* would not enhance the reputation of the new antisubmarine weapon. RN ASW staffs hastily put together an article for the issue of the *ASW Review* immediately following the event, outlining the measures being taken to prevent a recurrence of a premature Hedgehog explosion.[41]

The experience of Allied crews with Hedgehog points to a common challenge associated with new weapons. A variety of issues, often summed up in the catch phrase "teething pains," almost inevitably preclude a new weapon achieving optimal effectiveness when first introduced. In the case of Hedgehog these teething pains included a lack of manuals and well-understood doctrine for the use of the weapon. The nature of the weapon did not sit well with some sailors. The theoretical advantages of a weapon that only exploded on contact were certainly valid: explosions in the water often produced disturbances sufficient to cause a warship to lose asdic contact with a U-boat, either for a short period or even permanently. Nonetheless Allied sailors found the explosive detonations from depth charges viscerally satisfying: even if the explosions did not destroy the target, surely they must instill caution in their opponent, perhaps even fear. The RN initiated special trials and studies in January and February 1944, leading to improvements in doctrine, training and maintenance. These changes led, finally, to results closer to predicted levels: in the second half of 1944 Hedgehog attacks were successful about 35% of the time, a rate over four times higher than for the same period a year earlier.[42]

That all lay in the future, but as HMS *Escapade* limped away in the darkness, returning to the UK for repairs and thereby reducing the escort of ONS18 to seven ships, EG9 detected a submarine by radar just before midnight on the 19th. HMCS *Chambly* confirmed the contact was a U-boat when she detected the distinctive hydrophone effect of high-speed screws as the range closed to

3,000 yards. The U-boat appears to have been surprised by the appearance of warships coming from an unexpected direction; the range closed rapidly until the U-boat suddenly turned to starboard when the warships were less than a mile away. The escorts fired starshells, illuminating the boat's conning tower, forcing it to dive when they were 1,500 yards away. A depth charge attack was delivered on what *Chambly* considered good sensor information, but with no apparent result. *Chambly* and *Itchen* circled the area where the U-boat submerged for about two hours, while the remaining three ships of EG9 continued to close with ONS18 and had just altered to head for the convoy themselves when a submarine was again detected at 0142 hours in the morning of 20 September. Whether this was a different U-boat, or the same one seeking to escape, remains unknown, but attacks failed to inflict any serious damage, and shortly afterwards *Itchen* and *Chambly* broke off to head for the convoy.[43]

These last two escorts to join ONS18 were still en route when they were advised of a torpedo striking *Lagan*, the first successful U-boat attack of the battle. The converging groups of warships and growing number of U-boats in the same area of ocean clearly multiplied opportunities for engagements, and incidents increased sharply as 19 September came to a close and the day planned for the initiation of *Leuthen*, 20 September, began. In a general sense *BdU* could be quite satisfied with their intelligence, which had estimated that a westbound convoy would reach the area of *Leuthen* around or soon after 20 September. However, the limitations of German intelligence became very clear as the operation unfolded. German estimates of Allied movements were based on the convoy following the great circle route, which required a course close to due west at the point where *Leuthen* intercepted ONS18 and ON202. More critically, *BdU* never grasped that *Leuthen* had located not one but two convoys. Contact was lost with ON202 shortly after *U-238*'s successful attack, and significant difficulty in regaining contact is evident in *BdU* records. Perhaps more telling is the comment that "there were great variations in fixes from the reports received, which made it difficult for the boats to find the convoy…."[44] The variation in fixes would have been easier for *BdU* to understand if they had been aware that there were two convoys and three escort groups engaging *Leuthen* as the operation began. *BdU*'s failure to grasp the situation would contribute to flawed assessments of events as the operation unfolded.

The U-boats' loss of contact with both convoys before noon on 20 September reflected the efforts of not only the three escort groups now working against *Leuthen*, but also the air support provided by five Liberator aircraft from the RAF's 120 Squadron flying out of Iceland. These aircraft had a busy day, conducting eight attacks while engaging four different U-boats.[45] The immediate assessment of these attacks was that they did no more than shake up the U-boat crews, without inflicting any damage, even upon the two U-boats that were engaged by escort and aircraft combinations. In the first of these combined ASW efforts, HMS *Icarus* headed north to join Liberator X/120 about 15 miles off the starboard beam of ON202 just before 1000 hours. The Liberator conducted two attacks while the frigate closed the position, *Icarus* sighting the U-boat at 1002 hours, 9 minutes after she saw the aircraft circling above. *Icarus* closed the diving position, gained contact and conducted what are believed to have been an unsuccessful depth charge and then a Hedgehog attack before returning to the convoy.[46]

The uncertain identity of the U-boat in this incident – it may have been *U-238*[47] – is an unfortunately common aspect of this battle. The uneven nature of surviving records, number of incidents occurring in a small period of time and space, and natural difficulty of identifying opponents in ASW all combine to make untangling the historical record very difficult in many of the encounters over the next few days. The next combined ASW action also included an unidentified U-boat, even though this submarine came relatively close to ON202. Just after 1400 hours HMCS *Drumheller*, stationed on the port bow of the convoy, reported a U-boat approximately 6 miles off the port beam of ON202. Liberator X/120 had also seen this threat but could only monitor the action as she had already expended all her depth charges. *Drumheller* charged toward the U-boat, opening fire at 14,000 yards – an immense distance for the rather limited fire control capabilities of a corvette. Thirty rounds were fired, some of them landing quite close according to the spotting by the Liberator, but the U-boat stayed on the surface for almost a half hour before submerging.[48] *Drumheller* detected a contact by asdic after closing the submerged position, but classified its only contact as non-submarine. However, shortly afterwards an "unaccountable explosion" was heard and a disturbance seen in the water around the ship, presumably from the detonation of a magnetically fused torpedo below the ship. Six single depth charges were then dropped at short intervals astern, "in the hope of discouraging further attention" as the escort returned to her station.[49] Again, it is not clear which U-boat was involved.

At approximately the same time *U-338* gave the signal for all U-boats to remain surfaced to repel aircraft. "Unfortunately the short signal came at a time when there were still too few boats in the vicinity of the convoy so that the intended effect, i.e. that of dispersing the escort, was not successful...."[50] One source indicates *U-338* succumbed to the attacks of HMCS *Drumheller*, another credits Liberator N/120 later on the 20th, but the authoritative review of U-boat destructions conducted by the British Ministry of Defence disagrees and indicates there is at present no known cause of that boat's loss.[51]

HMCS *St. Croix*, the ex-USS *McCook,* entered the RCN in late 1940. One of the 50 destroyers exchanged by the U.S. for eight British bases, the flush-decked four-stacker was a Great War design that spent most of the inter-war period in mothballs. The impact on her hull of numerous patrols in the rough seas of the North Atlantic is noticeable even here in the quiet waters of a harbour. (DND Canada DHH, PA104474)

Precise information may never be determined in these confused incidents. What remains known is that for almost 12 hours U-boats lost contact with both ONS18 and ON202, from before noon on the 20th until late that evening. As a result a critical Allied decision remained unknown to the Germans throughout the battle, and indeed throughout the remainder of the war. For just before noon on 20 September CinCWA ordered that ONS18 and ON202 merge.[52]

The decision to join the convoys ensured that there would be a healthy number of escorts available for defence. The removal of *Lagan* and *Lancer* from the battle had already caused Commander Burnett concern. *Escapade*'s removal was of less concern to Evans because he had a larger escort to start with, and, more importantly, had the five warships of EG9 about to join his convoy. Before the signal to merge the convoys arrived, Evans and Burnett had already been discussing the possibility of splitting EG9 so as to reinforce ON202. Nonetheless, the decision to merge the convoys posed hazards and difficulties for the Allies that mitigated the benefits of concentrating escorts. To begin, adding a slow convoy to a fast one instantly reduced the maximum speed of both convoys to about 7 knots, a 25% reduction in the speed of advance for ON202. This increased the time that would be spent in the dangerous mid-Atlantic region and eased the task of U-boats racing to intercept or manoeuvre around the convoy. The second complication, and one that would bedevil the escort for the remainder of the attack, was that two separate convoys are much more difficult to control than one, a condition exacerbated by their difference in speeds.

Ideally, a convoy presented a broad front but short columns. This enabled all merchant ships to pass by a point in the ocean relatively quickly, as compared to a formation of long columns with a narrow front. This rapid passing complicated the efforts of U-boats, as they strove to intercept the moving point in the ocean where ships would pass within torpedo range. On the other hand, the greater the number of columns abeam one another, the more difficult it became to alter the course of the convoy. The difficulties in stationing and manoeuvring resulting from placing convoys beside one another led to the Convoy Commodore of ON202 – the senior of the two commodores, and hence placed in charge of both formations – desiring to station ONS18, the slower convoy, astern of ON202. Commander Evans, the most senior of the three escort group commanders, and therefore placed in overall command of the escorts and defence of the convoys once the junction occurred, strongly desired that the convoys be stationed abeam one another. This difference of views and the tyranny of circumstance would complicate the efforts to defend the two convoys throughout the period when U-boats strove to attack the merchant ships, resulting in only brief periods when the convoys travelled abeam one another.

Other complications became evident as ONS18 proceeded northeast – away, that is, from its destination – to rendezvous with ON202. The signal ordering the convoys to merge arrived at ON202 in a garbled form, resulting in that convoy steering a course of 230° when ONS18 expected a course of approximately 279°. Commander Evans summed up the resulting confusion by dryly noting that the "convoy junction was not a success."[53] His further observation that the "two convoys gyrated majestically around the ocean, never appearing to get much closer and watched appreciatively by a growing swarm of U-boats" colourfully described his assessment of events. However, he failed to appreciate how successfully the escorts were keeping U-boats from actually observing merchant ships.

"Junction battle"

Commander Evans's assessment was undoubtedly affected by an incident just before the two convoys came within sight of each other. At 1629 on 20 September HMS *Keppel* detected a U-boat and ran in to conduct an attack. Doubt as to the quality of the contact disappeared when a periscope was sighted close aboard *Keppel's* starboard beam, but this definite proof came too late to ensure a good attack. Instead of 10 depth charges set shallow, only five left the ship set for the initial estimate of medium depth. The attack surprised the U-boat as much as the sighting surprised *Keppel*, as the periscope of *U-386* had been trained toward HMS *Itchen* and HMCS *St. Croix*. These two ships from EG9 had been engaging a contact in the vicinity of the planned convoy rendezvous as ONS18 approached.[54]

The engagement with *U-386* continued as ONS18 and ON202 struggled to come together, the official time of merger being 1715 hours according to the Commodore of ON202. Evans decided that *Keppel* should continue to hunt this confirmed submarine contact for at least a few hours, despite his concern to ensure that the convoys and their defence be organized as quickly as possible. *Itchen* and *St. Croix* also remained hunting *U-386*. The importance of maintaining the offensive spirit aboard escorts was a common theme for the key Allied escort leaders in this battle, and Commander Evans judged that it was important, for a short while at least, to allow the crew of *Keppel* the opportunity to conduct attacks against a seldom-seen foe.[55]

The issue of morale in this battle is important but difficult to assess with precision. The commanders at sea throughout the battle repeatedly stressed the importance of maintaining an aggressive response to U-boat detections, as had become standard by this point in the battle of the Atlantic.[56] Shore authorities were beginning to develop doubts as to whether the standard aggressive attacks should still be doctrine as evidence mounted of new German tactics and weapons. Shortly after *Lagan* lost her stern to a Type V, Ultra provided definitive confirmation that this attack was no anomaly, but resulted from a dramatic change in U-boat tactics. This signal, sent by *BdU* on 18 September and decrypted early on 20 September, directed:

> The destruction of the escort must be the first objective. The destruction of a few destroyers will have a considerable moral effect upon the enemy and greatly facilitate the attack on ships of the convoy in addition. When engaged in getting ahead of the convoy on the surface, do not form any group that consists of more than two boats. Your aim must be to spread yourselves evenly round the convoy in order to split up the defence. I expect of all commanding officers that each chance of a shot at a destroyer will be utilized. From now on, the U-boat is the attacker – fire first and then submerge.[57]

RN shore authorities, finally aware of *BdU*'s true intentions, responded swiftly. A signal from CinCWA arrived mid-afternoon on 20 September, as the convoys closed toward each other, ordering escorts to avoid positions directly astern of U-boats, avoid steady courses, and to maintain maximum speeds while keeping U-boats 30 degrees on the opposite bow as much as possible.[58] Providing new tactical direction this promptly indicates how quickly the Allies were able to adapt to changed situations. Unfortunately, the tactics were vague and ambiguous – escort commanders were not enlightened as to how they could close the range on a U-boat while keeping it 30 degrees off the bow, an almost impossible task. Escorts needed at least to force U-boats away from convoys in order to defend merchant ships effectively, and this required heading toward radar and asdic contacts. The inadequate quality of this initial tactical response reflects the surprise achieved by the Germans in the way they used their new torpedo.

The problem posed by the acoustic torpedo was that it enabled U-boats to engage escorts effectively from any direction, as opposed to those previously rare occasions when a U-boat found itself just forward of the beam of an escort at close range, where a straight-running torpedo had a reasonable chance of hitting a minor warship. Sea-going commanders acknowledged that the new weapon increased the threat, but deemed the importance of taking the offensive against U-boats worth placing their vessels at risk. Shore staffs struggled to develop effective methods to counter acoustic torpedoes, but some of their proposals required turning away from submarine contacts. How to resolve the tension between sea-going commanders determined to maintain an offensive posture against U-boats while avoiding significant losses to escorts would become an enduring challenge for the Allies for the remainder of the war.[59]

The engagement between *Keppel, Itchen, St. Croix* and *U-386* was one of several encounters as the rendezvous occurred. Liberators "N" and "J" of 120 Squadron conducted attacks on two U-boats well out to the port side of the conjoining convoys, while EGC2's *Kamloops* and *Gatineau* attacked an asdic contact ahead of the port wing of ON202.[60] The growing number of U-boats engaged by sea and air escorts of the convoys indicated clearly that the boats of *Leuthen* were converging on their target, encouraged by a signal from *BdU* during the morning of the 20th exhorting them to close with the convoy and reminding them to "think of the chances of a surprise blow at the beginning of the operation."[61] *Keppel* remained with her target long enough to deliver four depth charge attacks, damaging *U-386* sufficiently to force her return to base, and summon HMS *Narcissus* from B3 to join and assist the prosecution before detaching at 2000 hours to return to the convoy. Commander Evans could then finally assess for himself how well the convoy formation had come together since the junction.

Shortly before *Keppel* departed, *St. Croix* had set off to investigate yet another U-boat reported surfaced astern of the convoy by a Liberator. At 1956 hours, as *St. Croix* closed the position of *U-305*, a Type V struck her stern, leaving her adrift.[62] *Itchen* headed toward the crippled Canadian warship, circling the settling vessel whilst searching for the attacker. HMS *Polyanthus*, escorting the rescue ship *Rathlin* back to the convoys, was diverted to assist *Itchen*, screening the latter as she took on *St. Croix* survivors. This rescue effort was aborted before it could get started, however, when *U-305* conducted another attack, the first of two torpedoes in this salvo hitting *St. Croix* again at approximately 2052 hours, and the second, a Type V, aimed at *Itchen* exploding about 30 feet astern of her. This time the Canadian ship went down quickly.

Commander Bridgeman aboard *Itchen* suspended rescue efforts until daylight after this further evidence U-boats were aggressively targeting escorts with their new torpedoes. Commander Evans, who had ordered *Narcissus* to return to the convoys, now reversed his decision and sent that ship back to support *Itchen* as he assessed that "at least four offensive submarines were operating in the area."[63] This decision came none too soon, for at 2236 hours HMS *Polyanthus*, investigating yet another radar contact, was hit by a Type V, the little ship sinking rapidly. *Itchen* and *Narcissus* remained busy for the remainder of the night, engaging or being fired upon at by as many as half a dozen, rather than four, U-boats.[64]

The loss of *St. Croix* was a heavy blow to the Canadian Navy. As the only destroyer in EG9, albeit of ancient vintage, she represented the only fast RCN warship in this Canadian support group,[65] only the second such group in the entire RCN. The more modern HMS *Itchen*, a River-class frigate, reflected the more successful modernization program of the Royal Navy. While the RCN had dedicated enormous effort to the Battle of the Atlantic, the results had not proved as successful, particularly in terms of U-boats destroyed, as desired both inside the RCN and among the public and politicians. Canadian naval leaders hoped to improve the RCN's record by fielding support groups, with their greater opportunities for offensive action against U-boats. The dispatch of EG9 to counter the first wolfpack in months provided exactly the opportunity that had been sought. The deadly impact of the new German torpedo on *St. Croix* not only destroyed that ship, but drastically reduced the chance that the RCN would destroy any U-boats in the battle.

The human cost of the battles swirling around ON202 and ONS18 would also be heavy. *St. Croix*'s captain, Lieutenant Commander A.H. Dobson, RCNR, DSC, survived this night, but would not survive the passage.[66] Commander Dobson's background is representative of the manner in which experienced mariners were found to command the large number of escorts needed for the vast convoy system. Born in northern England in 1901, he went to sea as a crewman aboard a merchant ship when he was 17. Determination and hard work resulted in his achieving a Foreign Master's Certificate when he was 25, qualifying him to stand watch on the bridge of a sea-going merchant ship. While moving from job to job, seeking a good position at sea, he also joined the RN as a sub-lieutenant in the Royal Naval Reserve. His prospects, as for many others, dimmed badly when the Depression struck, and he and his wife immigrated to Canada, settling in Halifax, Nova Scotia. Work remained difficult to find there as well, but he achieved a position aboard the Hudson's Bay SS *Nascopie* in 1935, plying the waters of Davis Strait regularly for several years. The birth of his second child the next year probably contributed to a decision to join the Royal Canadian Naval Reserve, as the added income would help make ends meet.[67]

When the war started Dobson was called up almost immediately, his personal qualities of determination and leadership as well as his significant sea-going knowledge needed by the small RCN. He received command of a small ex-RCMP vessel, the *Fleur-De-Lis*, taken over by the navy to conduct local patrols off the Nova Scotia coast. Armed with only an ancient Lewis machine gun, it is doubtful the ship could have done much in the event of a serious encounter, and its use indicates the weak state of Canada's navy when war began. Dobson assessed his contribution as minor, and even succeeded in returning to SS *Nascopie* for one last run through the northern outports of Labrador in the spring and summer of 1940. However, as newly constructed ships began to leave their slips, Dobson took his Escort Commanders course and was given command of HMCS *Napanee*. He gained experience with wartime convoys aboard this corvette, earning a reputation as a reliable commander and excellent seaman, before assuming command of the destroyer *St. Croix* on 6 January 1942.[68] Returning to escorting North Atlantic convoys, Dobson received his Distinguished Service Order for *St Croix's* destruction of *U-90* on 24 July 1942 while defending ON113.[69]

St. Croix spent the remainder of 1942 on the North Atlantic run, where Dobson gained experience as the Senior Officer Escorts for several convoys. At the outset of 1943 *St. Croix* went alongside for several weeks for maintenance, and then proceeded to Tobermory, the RN's renowned ASW training establishment. In late February she started the United Kingdom–Gibraltar route, assisting HMCS *Shediac* in destroying *U-87* on 4 March.[70] Soon after, *St. Croix* returned to the North Atlantic, but by June a longer refit was required, returning her to Halifax. She was fitted with a Canadian-designed centimetric radar, Hedgehog and a High Frequency Direction Finding set, making *St. Croix* relevant again in ASW.[71] After briefly escorting North Atlantic convoys as part of her usual close escort group, *St. Croix* received orders to join EG9 in early August. Dobson and

St. Croix clearly had had a busy war before their fateful rendezvous with ON202 and ONS18.

Dobson and about half the crew of 148 survived *U-305*'s attack. He and roughly 60 others spent the night packed into *St. Croix*'s whaler, while the remainder clung to carley floats.[72] The next morning *Itchen*, which had avoided more torpedo attacks overnight, recovered 71 of *St. Croix*'s crew alive (10 others succumbed to the sea before being rescued) as well as *Polyanthus*'s sole survivor.[73]

The fierce fighting between escorts and U-boats astern of the convoy had a profound impact on the entire battle, shaping perceptions on both sides of their opponent's intentions and capabilities. *BdU* received many reports of escort sinkings by U-boats. The German staff concluded that on the 20th U-boats accounted for three merchant ships and no less than seven destroyers, with three more destroyers probably sunk.[74] The next day these achievements were labelled as a "great success." Reports of numerous escorts still remaining around the convoy were explained away by *BdU* as reinforcements sent out to try to prevent heavy merchant ship losses.[75] That escort losses were not nearly as heavy as claimed was given only modest consideration by German naval authorities.

Reality was grim enough. One frigate had been knocked out of the battle by Type V damage, while a destroyer and a corvette had been sunk. U-boat claims for merchant ships were much closer to actual, with two sunk as opposed to the three claimed. In fact, the merchant claims were reasonable, in that *U-238* had first hit and damaged the ship that *U-645* later finished off, the *Theodore Dwight Weld*. The great discrepancy in escort claims arose primarily from the tendency of Type V's to explode in general proximity to their target, as opposed to hitting a warship. The requirement that U-boats dive after firing so as to avoid becoming targets of their own homing torpedoes clearly made observing results difficult. Finally, multiple U-boats attacking a small number of escorts resulted in several claims for the same ship destroyed when a Type V did strike home.

The Allies could take cold comfort from the confused picture drawn by German shore authorities. The loss of *St. Croix* and *Polyanthus* confirmed that U-boats had a new anti-escort weapon and that U-boats were targeting escorts as a priority. Allied shore staffs, both operational and scientific, focused swiftly on measures that would reduce the effectiveness of the Type V. A new message sent out by Western Approaches late in the evening of 21 September – the day after the three warships were torpedoed – reiterated the original guidance of 20 September and noted that "steps are being taken to produce counter-measures to acoustic torpedoes on the highest priority," but, most significantly, ended by directing that "for the moment escorts are to adopt less offensive tactics confining the object to immediate defence of the convoy."[76] A diary entry by an RN officer not involved in the battle but aware of intelligence indicating that the German navy might soon deploy an acoustic homing torpedo penned his personal assessment of the news in his diary, noting that "everyone is in a flat spin about the acoustic torpedo, which the U-boats are using and which has accounted for no less than 3 escorts."[77] The reactions of RN shore authorities to the new acoustic homing torpedoes indicate clearly that *BdU* had achieved a measure of technical/ tactical surprise in this battle.

The junction battle made escort commanders on the scene very aware that U-boats were using a new weapon. However, their reactions varied. The senior escort officers apparently remained convinced throughout the battle that escorts needed to aggressively hunt down any U-boat detected, but some escort commanders took the latest Admiralty guidance more seriously, leading to tension as the battle progressed.

First night of the merged convoys

The convoy plodded on as the junction battle unfolded and shore authorities struggled to determine how best to react. The distraction provided by the battle at the junction, the accidentally evasive convoy course to the southwest and the poor intelligence picture at *BdU* all combined to minimize the number of U-boats that even came close to the convoy on the night of 20/21 September.[78] Three U-boats were detected and turned away before reaching the convoy, allowing the merchant ships to proceed unmolested this night.[79] This was fortunate, for Commander Evans found the two convoys stationed astern of one another when *Keppel* caught up. He convinced the commodores to align the two convoys abeam of one another on the correct course at daylight, but deteriorating weather prevented manoeuvring. Thick fog enveloped the area from the morning until mid-afternoon of the 21st. When visibility cleared Commander Evans found to his surprise that ONS18 was stationed almost exactly on the starboard beam of ON202 (see figure 1). This desirable state of affairs apparently occurred without order, moving Evans to ascribe the arrangement to "a higher power."[80]

Poor visibility on the 21st complicated the efforts of U-boats to attack the convoy, but also provided them with cover from aircraft, and therefore allowed them to remain nearby in favourable positions for attack once the weather cleared. Finding the exact position of the convoy proved challenging for the U-boats during the day and through the night of September 21st/22nd. Most of the reports of the convoy to *BdU* were based on hydrophone bearings, as U-boats listened to the massive noise coming from the ranks of ships, with a few reports of radio traffic

Convoys ON 202 and ONS 18
21–22 September 1943
Order of Battle

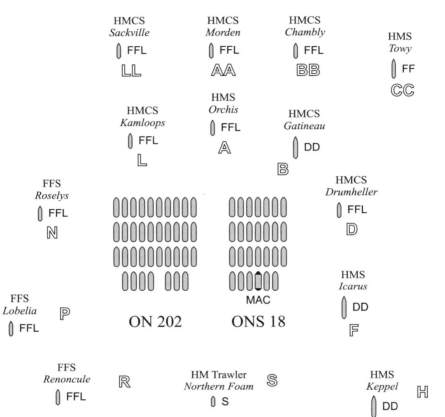

NOT TO SCALE

The two convoys were in good formation abeam each other on this night, giving the escorts a better chance of defending them.

Single Hollow Letter Designations (from standard RN screening diagrams) indicate escort station in the inner screen; generally 6000 yards outside the convoy.

Double Hollow Letter Designations indicate escort station outer screen position; generally 12000 yards.

Distance between columns in a convoy was 1000 yards, and distance between ships in a column was supposed to be 800 yards. These distances were often honoured in the breech as manoeuvring was often a challenge for the merchant ships.

LEGEND

DD – Destroyer

FF – Frigate

FFL – Corvette

S – Trawler

Merchant ship

MAC – Merchant aircraft carrier

Black silhouette indicates sunk.

from merchant ships in the convoy. During the night escorts were sighted by U-boats, but not a single merchant ship.[81] Lacking electronic sensors as effective as those carried by Allied warships, especially radar with its ability to give a reasonably clear tactical picture, U-boats could only fumble blindly in the poor visibility, unable to close and attack. This part of the battle could be assessed as a draw.

The weather ensured that aircraft played no effective part in the battle during this period, but strong efforts to provide air cover were made. Five VLR Liberator aircraft – four RAF and one RCAF – flew escort missions during the course of the day, three of which succeeded in locating the convoys.[82] They achieved little – *BdU* remained unaware that aircraft were even operating on this day[83] – but this steadfast effort would continue, aircrew risking their lives in marginal conditions so that if the weather cleared they would be on station rather than having to travel hundreds of miles first. ONS18's modest aircraft capability also played a part during the day, the *Empire Macalpine* launching a Swordfish biplane shortly after weather cleared at 1430 to patrol around the convoy. The return of fog soon after the launch resulted in the Swordfish being stranded in the air with visibility of only about 50 yards. Consummate flying brought the biplane back aboard the merchant aircraft carrier, avoiding the loss of either aircraft or aircrew. Commander Evans recorded the event with evident relief and admiration for the skill involved.[84]

Second night of the merged convoys

The poor weather and better station-keeping of the convoys resulted in ON202/ONS18 being reasonably prepared for the tentative attack efforts of the U-boats that night. The convoys were a little spread out, but their position abeam one another provided a compact perimeter for the escorts to defend. The escorts had also been augmented by the frigate HMS *Towy*, delayed from sailing with B3 because of mechanical defects but catching up before dark. Fourteen escorts encircled the convoy, and the number would rise to 16 when HMS *Itchen* and *Narcissus* rejoined after rescuing *St. Croix's* and *Polyanthus's* survivors. In the early years of the war, poor visibility had favoured the U-boats, which took advantage of their low visual profile to charge in toward convoys at night, anticipating that encounters with escorts would be rare. By the fall of 1943, Allied advances in electronic sensors, especially radar and HF/DF, had reversed the advantage.

The losses and damage suffered by the U-boats on the 19th and 20th meant that there were now 17 submarines still hunting the convoys,[85] resulting in near parity in numbers between attackers and escorts. The strong escort and poor visibility presented a difficult challenge for U-boat commanders, who maintained contact by listening on sonar, occasionally augmented by detection of some medium frequency radio transmissions. These cues provided only the general bearing of the convoy – no reliable ranges could be determined. Nor could a U-boat readily isolate the sound of the escorts against the background of the merchant ships, rendering them vulnerable to surprise. That at least half a dozen U-boats attempted to close and attack the convoy through the night of 21-22 September therefore indicates both the high level of individual training of these skippers, and their determination despite the heavy losses suffered by U-boats up to this point in the war.[86]

Night battle 21/22 September: Advantage escorts

Commander Evans followed the efforts of the U-boats to close on the convoy through HF/DF interceptions, which suggested that attack could be expected from any direction. The first engagement developed at 2100 hours on the port bow of the convoy, where Free French Ship (FFS) *Roselys* reported two surfaced submarines and one submerged, although Evans assessed that this was, in reality, only one U-boat trying to get through to the convoy.[87] *Roselys* conducted two depth charge attacks; there were no further indications of U-boats approaching from the port bow that night.

A little more than two hours later, at 2330, HMS *Icarus* detected a surfaced U-boat on the starboard beam of the convoy and gave chase briefly, forcing the U-boat to dive. *Icarus* resumed station after less than 20 minutes, at 2349 hours, re-establishing the integrity of the screen. Under an hour later, at 0039, HMCS *Chambly*, in the extended screen about 12,000 yards on the starboard bow of the convoy, engaged a U-boat with its main gun. At least one hit was claimed, although no U-boat reported shell damage this night, before the submarine dived and was depth charged. The frustrated attacks resulted in a high volume of HF traffic from the U-boats, which Commander Evans noted as being so intense that in the early hours of 22 September the boats were "all trying to talk and jamming each other most effectively."[88]

A two-hour lull was broken at 0440 when the trawler *Northern Foam* surprised *U-305*, *St. Croix*'s nemesis, trying to break into the convoy from astern.[89] The trawler reported hitting abaft the submarine's conning tower with its 4-inch gun after narrowly missing while attempt to ram, after which the U-boat dived. Commander Evans considered that the shell hit must have damaged the boat as it "had not been underwater five minutes when surfaced and made off to the south at high speed."[90] In this case, however, the trawler's shot had only passed close by the submarine's tower.

In addition to the U-boats detected near the convoy by escorts, other boats had approached unseen, and even fired torpedoes at ships in the convoy, although without success. *U-377* launched a Type V against a merchant at 0341 hours but it malfunctioned and detonated after running just 26 seconds without hitting anything or even being recognized as an attack by the ships in the convoy.[91] This would prove to be one of the few times during the battle when this weapon was not aimed at an escort.

The escorts were aware that they had driven off at least four U-boats, although they proved unable to take sufficient advantage of their better sensors to inflict crippling damage on any of these submarines. Commander Evans considered, however, that the screen provided by his escorts was sufficiently robust as dawn approached that he could make an offensive effort. At 0542 hours HF/DF intercepts provided evidence that a U-boat engaged in keeping contact on the convoy had been detected. Evans directed *Keppel,* stationed on the starboard quarter of the convoy, to sweep out on the intercepted bearing. Thirty-eight minutes later, *Keppel* detected a U-boat on radar 6,000 yards away. Racing toward its unseen enemy in the fog, the British destroyer closed rapidly on the crew of *U-229,* which only reacted when their attacker was about 800 yards away, far too late to take effective evasive action. The boat may have been swinging to try and bring its stern to bear, as it is quite possible that the two Type V torpedoes embarked in her bow tubes had already been expended.[92] Commander Evans account of what followed cannot be improved upon, and is quoted in full here:

> A few seconds afterwards she was sighted fine on the port bow, her creaming wash against a setting of dense fog and heaving swell making a picture that will not easily be forgotten. Handling the ship with great skill, Lieutenant-Commander Byron [Captain of *Keppel*] fired a couple of rounds into the submarine and then drove *Keppel's* forefoot into her pressure hull just abaft the conning tower. Although the U-boat was trying to dive, the conning tower, up to the last minute, showed as a black pit surrounded by the white froth of her wash. As we passed over a ten charge pattern set at 50 feet added insult to injury.... A most satisfactory episode, in which HF/DF, radar, plotting, asdic, depth charges and ship handling all worked faultlessly to the required conclusion.[93]

The destruction of *U-229* brought a night highly satisfying to the escorts and extremely frustrating for the U-boats to a close. Clearly, new torpedoes were not enough to overcome the growing sensor advantage of the Allies in low visibility. Escorts appearing suddenly at ranges of less than 1,000 yards gave U-boats little chance of using their best anti-escort weapon. It does not appear that German staff fully appreciated the Allied advantage in low-visibility sensors, but when *BdU* ordered the pack to continue the operation, it noted, "It is to be hoped that visibility will become better on the 22nd and will allow a massed attack in the night of 22nd-23rd."[94]

Day 22 September 1943: Out of the fog

As dawn broke, fog continued as thick as ever. Low visibility, despite the relative advantages it conferred on the escorts, was something of a mixed blessing for the Allies. The merchant ships in the convoy could neither stay in station easily nor could the course of the unwieldy formation be altered. As a result a course of 210°, farther south than the most direct route toward Halifax and New York, had to be maintained. Until visibility cleared there was nothing Commander Evans could do to affect the situation of the convoy itself, so he focused on active efforts to keep the U-boats at bay. HF/DF intercepts continued to pour in throughout the forenoon, mostly astern and on the quarters, and ships were dispatched to search for the originators of these radio signals. Commander Evans recorded his disappointment that none of these chases resulted in a sighting.[95] Since U-boats were keeping in contact with the convoy by listening to the sound of the massed ships, this suggests that most German hydrophone operators were very carefully listening for the sound of high speed screws approaching, allowing boats to evade the searching escorts.[96]

The three corvettes of EG9 – *Sackville, Chambly* and *Morden* – had been ordered to sweep the convoy's quarters and then patrol astern as dawn broke, but initially failed to find any boats. The first contact occurred somewhat by accident in mid-morning when the rescue ship *Rathlin* sighted a U-boat somewhere astern of one of the convoys at the point blank range of 100 yards. *Rathlin*'s effort at ramming failed but forced the U-boat to dive. There was little chance of an escort finding and attacking this U-boat, as *Rathlin* – not fitted with radar – had little idea of her actual position and was apparently astern of the wrong convoy, having drifted out of her assigned station in the poor visibility. *Gatineau* was nonetheless sent to see if she could find the rescue ship. Commander Evans described how the Canadian destroyer "steamed repeatedly through the invisible columns [of ships in the convoy] at high speed, and in hand steering in a most gallant manner which made my hair stand on end."[97]

Two encounters between escorts and U-boats did occur during the fogbound morning and early afternoon of 22 September. The first took place just before noon, when at 1142 hours 10 miles astern of the convoy *Sackville* first shelled a U-boat into diving, and then depth-charged the boat for good measure. Though

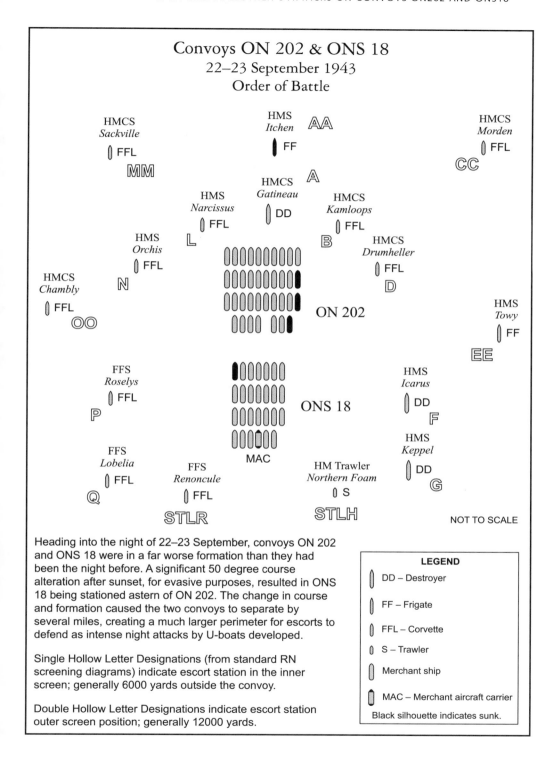

Convoys ON 202 & ONS 18
22–23 September 1943
Order of Battle

HMCS *Sackville* FFL MM

HMS *Itchen* FF AA

HMCS *Morden* FFL CC

A

HMS *Narcissus* FFL

HMCS *Gatineau* DD

HMCS *Kamloops* FFL

L

HMS *Orchis* FFL

B

HMCS *Drumheller* FFL D

HMCS *Chambly* FFL N OO

ON 202

HMS *Towy* FF EE

FFS *Roselys* FFL P

ONS 18

HMS *Icarus* DD F

FFS *Lobelia* FFL Q

FFS *Renoncule* FFL STLR

MAC

HM Trawler *Northern Foam* S STLH

HMS *Keppel* DD G

NOT TO SCALE

Heading into the night of 22–23 September, convoys ON 202 and ONS 18 were in a far worse formation than they had been the night before. A significant 50 degree course alteration after sunset, for evasive purposes, resulted in ONS 18 being stationed astern of ON 202. The change in course and formation caused the two convoys to separate by several miles, creating a much larger perimeter for escorts to defend as intense night attacks by U-boats developed.

Single Hollow Letter Designations (from standard RN screening diagrams) indicate escort station in the inner screen; generally 6000 yards outside the convoy.

Double Hollow Letter Designations indicate escort station outer screen position; generally 12000 yards.

LEGEND

DD – Destroyer

FF – Frigate

FFL – Corvette

S – Trawler

Merchant ship

MAC – Merchant aircraft carrier

Black silhouette indicates sunk.

"After living under a blanket [of fog] for so long, it was very nice to come into the open air and find it filled with Liberators." The Very Long Range (VLR) Liberator shown here was the source of inspiration for this quotation from Commander M.B. Evans, RN, in his RoP for 22 September 1943. The growing number of VLR Liberators from early 1943 on allowed the air gap in the North Atlantic to be closed, significantly worsening the odds of survival for U-boats. Allied sailors clearly appreciated the steadfast efforts of aircrews. (Library and Archives Canada, PA-107907)

Sackville reported the smell of oil, subsequent analysis indicates these attacks did no serious damage.[98] A few hours later at 1508, *Roselys*, on the port bow of the convoy, reported a surfaced U-boat detected by radar and forced it to dive.[99]

Just after this, at 1520 hours, visibility cleared. Commander Evans registered his relief that "After living under a blanket of fog for so long, it was very nice to come into the open air and find it filled with Liberators."[100] These aircraft did not immediately sight U-boats, despite the radio traffic that had been detected numerous times around the convoy. Commander Evans judged this lack of sighting ominous, as it suggested that the U-boats were already in good positions ahead of the convoy, awaiting nightfall to attack. Good visibility also made evident a gap between the two convoys, with ONS18 lagging about 4 miles astern of ON202.[101] The commodore of ON202 immediately ordered the slow convoy to move up to its ordered station abeam of his convoy, but quick orders could not overcome the reality that the slower and older ships of ONS18 could only catch up gradually. However, by 1800 hours, both convoys were reasonably in station, on a course of 210°.

Shortly before 1800 hours the two RCAF Liberators on station near the convoys both became engaged with U-boats at almost the same time. Liberator L/10, flown by Warrant Officer J. Billings, found *U-270* – *Lagan's* nemesis – ahead of the convoy on its starboard bow and attacked at 1740 hours, the submarine remaining on the surface to fight it out. Billings radioed for assistance from another plane from his squadron circling the convoy and discovered that Liberator X/10 was engaged in battle with another U-boat on the port quarter of the convoy.[102] The contest between *U-270* and Liberator L/10 resulted in damage to both parties, the Liberator losing an engine to determined flak and *U-270* suffering multiple hits, most seriously a hole in her pressure hull, necessitating a return to port in France.[103] WO Billings tried to use all his possible armament, but could not employ his last weapons because *U-270* refused to dive. Liberator L/10's bomb bay held two Mk XXIV mines, the code name for the Allied acoustic homing torpedo that had entered service in May 1943. Unknown to the Germans, this weapon homed on the sound of a submerging submarine's screws. It was used only as the U-boat's dived for two reasons: first, to keep the new weapon a secret – apparently with success, as German crews do not appear to have discovered this threat until after the war – and secondly, because this torpedo had a low speed, making it unlikely to be effective against a surfaced submarine.[104] This engagement removed *U-270* from the battle, but fighting it out on the surface perhaps saved that boat from destruction by an acoustic homing torpedo.

Or perhaps not, as the Allied acoustic homing torpedo – like its German counterpart – was still a "temperamental weapon" at this stage in its development and sometimes failed to find its mark.[105] Liberator X/10's final attacks on *U-377* with two homing torpedoes missed. Before this, however, the two opponents had engaged each other heavily, the U-boat seeking to follow *BdU* anti-aircraft doctrine as well. Liberator X/10, flown by Flight Lieutenant J.R. Martin, first used four depth charges, which missed, as well as its machine guns, which struck the commanding officer of *U-377* in both arms.[106] *U-402*, about 7 miles away from this engagement, also drew the attention of Flight Lieutenant Martin, and the two exchanged machine gun and cannon fire until *U-402* disappeared into a fog bank.[107] Martin and his crew patrolled between the last positions of the U-boats and the convoy until fuel concerns forced them to head for Newfoundland.

Escorts in the close screen also detected U-boats attempting to penetrate to the merchant ships as night approached. At 1930 hours FFS *Lobelia*, on the port beam of the convoy, detected a submerged U-boat. Conducting a number of depth charge attacks, the corvette brought wreckage to the surface. The incident is a little mysterious, as the wreckage rose some distance from the attacks, which Commander Evans commented "do not appear to have been too good."[108] Nor can a U-boat be associated with these attacks. Nonetheless, the event contributed to the growing sense that U-boats were indeed closing for attack.

Empire Macalpine, the MAC ship with three Swordfish biplanes aboard, launched her first patrol of the day an hour before the Canadian Liberators began to engage their opponents. Probing ahead of the convoy, Swordfish "B", armed with two depth charges, sighted a U-boat 8 miles ahead of the convoy at 1818 hours just as the sun was setting. Her radioed reports resulted in the launch of Swordfish "C", armed with four rocket projectiles, five minutes later. The two aircraft, unable to communicate because of radio problems, conducted an uncoordinated and ineffective attack in the twilight at about 1840 hours. The U-boat remained on the surface throughout the engagement, manoeuvring at high speed, sending up what the Swordfish pilots described as intense flak. The rocket projectiles were fired from long range, while Swordfish "B" climbed to 5,000 feet before attempting to attack with depth charges that, predictably from that high altitude, missed. The biplanes' report of a U-boat resulted in two ships – HMS *Keppel* and HMS *Narcissus* – attempting to search for a dived U-boat ahead of the port bow of the convoy. Word that the U-boat was staying surfaced never reached Commander Evans. Confusion worsened when Evans, believing that the Swordfish had been recalled because of nightfall, saw anti-aircraft fire arcing into the sky 4 or 5 miles ahead of the area he was probing with asdic. He assessed this as a ruse on the part of another U-boat designed to draw attention away from the boat he was then searching for. The true situation, that the Swordfish were attacking the U-boat he was searching for, would not become apparent until well after the event.[109]

The final Liberator, N/10 flying out of Newfoundland, experienced difficulty homing on the convoys, only finding the ships after the sun had set and with only about an hour's flying time left. Patrolling ahead of the convoy, the aircraft sighted a wake and encountered flak, possibly from *U-275*.[110] Denied permission to use flares for a night attack by the convoy's escort commander, Liberator N/10 had to be satisfied with forcing the U-boat to submerge before turning for home.

The growing number of engagements between aircraft and U-boats as evening approached on 22 September indicated that *BdU*'s orders to close the convoy for a night attack were being followed. At least five U-boats had been attacked,[111] two of them suffering damage that either reduced their effectiveness or required their return to base. Nonetheless, Commander Evans could draw only limited comfort from the efforts of the supporting aircraft as night fell. He had clear indications that probably a dozen or so U-boats were operating in the vicinity of the convoy. The improved visibility meant that the U-boats could attack much more effectively than had been the case the night before.

A final complication occurred as Evans sought to turn the convoys onto a course more suited to reaching their destination. His initial intent of altering the convoy course 40 degrees to starboard after night fell had to be changed when a U-boat was reported 40 miles along this intended course. He decided to increase the course alteration a further 10 degrees, so that the convoy would steer 260° once it turned. This major course change, necessitated by the train of circumstances set off by a garbled message two days previously and poor visibility that precluded altering the convoy course earlier, resulted in a setback that undermined the defensive posture of the escorts that night. Convoy ONS18 had been struggling to resume station on the starboard beam of ON202 since visibility cleared in mid-afternoon. However, on learning of the requirement to alter course 50 degrees to starboard after dark, the senior convoy commodore insisted that ONS18 take station astern of ON202 once again, on the grounds that attempting to wheel a combined convoy of 18 columns that far in the dark would be impossible. Commander Evans unhappily agreed to this new arrangement, a decision he would regret afterward.

A long and deadly night

Commander Evans had a good opportunity to view the results of his convoy formation decision (see figure 2). HMS *Keppel*'s night screen station was on the starboard quarter of the formation, which meant that the destroyer had to proceed past all the merchant ships as it travelled from its day station on the port bow. He observed that the formation change, placing ONS18 astern of ON202 and the large alteration of course to 260°, had dispersed the entire formation greatly. The slower ONS18 had fallen 3 to 4 miles astern of ON202, and the interval between the ships in column inside ONS18 "had opened out alarmingly."[112] The result was that the merchant ships were spread over an area of 30 to 35 square miles, producing a perimeter approximately twice as large as had been the case the night before when the two convoys were abeam one another.

BdU's hopes for better weather were answered with good visibility as darkness fell. The aggressive behaviour of boats fighting off the air attacks during the day of the 22nd also raised German hopes, *BdU* staff commenting that: "In

The name on this rescue ship is not *Rathlin*, but the scene shown certainly parallels one that must have been similar aboard the rescue ship that accompanied ON202. Rescue ships performed magnificently during the Battle of the Atlantic. In addition to their primary duty of pulling sailors from the sea, rescue ships were often fitted with HF/DF equipment, which provided invaluable information to escort commanders. (Library and Archives Canada, PA-153052)

warding off attacks the boats remained on the surface which resulted in aircraft not achieving their purpose, namely that of shaking off the boats, and moreover, contact with the convoy was maintained."[113] Nevertheless, expectations of causing major merchant ship losses were fading somewhat, as boats continued to report a strong escort around the convoy despite the apparent losses.

That there were a number of U-boats still striving to attack the convoy, and hopeful of success in the better visibility, became evident early in the night. Around 2100 hours the trawler *Northern Foam*, in station astern of the formation, sighted a U-boat at 1,500 yards and began illuminating with starshell. The boat dived to avoid the trawler, which then conducted two depth charge attacks to discourage this submarine from further pursuit. Shortly afterward, HMS *Itch-*

en, in station 12,000 yards ahead of the convoys, detected a U-boat and put her deep with an attack. *Itchen* resumed her station at 2200 hours.

A long and confusing melee began about 40 minutes later, *Itchen* and HMCS *Morden* – in station 12,000 yards ahead of the convoy's starboard bow – each reporting a surfaced U-boat attempting to close on the convoy from ahead. *Itchen* reported chasing her boat toward the port bow of the convoy, eventually forcing it to dive. At 2347 hours she reported another surfaced U-boat 2 miles to the north of her, apparently attempting to get to the front of the convoy as well. By this time HMCS *Morden* had also returned from chasing her radar contact away from the convoy and detected an asdic contact ahead of the convoy at 2345 hours, attacking with depth charges nine minutes later. As *Morden* attempted to turn for another attack, she was forced to alter course by *Itchen* charging toward a surfaced U-boat that *Morden* does not appear to have detected.

To further complicate an already very confused situation that was now drawing near to the main body of the convoy, HMCS *Gatineau*, part of the close screen directly ahead of the convoys, had also detected the U-boat. *Gatineau* had already been steering to starboard of the convoy course as a result of her normal patrol zigzag, and therefore began a long (about 270°) turn to starboard to cut off this U-boat before it could reach the merchant ships. The U-boat in question was most likely *U-666*, which had sighted *Morden* (even though the corvette failed to see her) before *Itchen* or *Gatineau* detected her. *U-666* had launched a Type V torpedo at *Morden* while continuing to approach the convoy. The sudden arrival of *Itchen* forced *Morden* to alter course (possibly saving her from the first Type V) and evidently took *U-666* by surprise, but *Kapitänleutnant* Herbert Engel, her commander, coolly fired a second Type V at the newly arrived British frigate from close range. HMS *Itchen* turned on her signal lamps shortly before midnight, illuminating *U-666*, and opened fire at the boat. Seconds later, at 2359, the second GNAT from *U-666* struck *Itchen*, apparently finding a magazine, as the ship blew up in a thunderous blast. Debris from the frigate rained down on the conning tower of *U-666*, which then submerged. The first Type V failed to find a mark, exploding harmlessly at the end of its run.[114]

The loss of *Itchen* was triply devastating as she carried the survivors of HMCS *St. Croix* and HMS *Polyanthus*. Most of three warship companies were lost, almost 400 men, as only three sailors were picked up when a merchant ship, probably the SS *Wisla*, courageously stopped in the midst of all the confusion to pluck them from the water. The first two, Petty Officer Clark and Able Seaman Flood, were both from HMS *Itchen*, while Able Seaman W.A. Fisher, a "stoker" or member of the engineering department, became the sole survivor of HMCS

St. Croix. No one remained from HMS *Polyanthus*.[115] The loss of *Itchen* also went unnoticed in the extremely confusing melee, as the escorts closest to the blast initially assumed that a U-boat had blown up.

Less than an hour later, at 0047 hours, HF/DF indicated another U-boat probe from the starboard bow. Evans's warning sent *Gatineau* searching in that direction, and by 0105 hours the Canadian destroyer was in contact, pushing this attacker away from the convoy. Ten minutes later the Free French corvette *Lobelia* reported a U-boat trying to close in from the port quarter, another attack beaten off by the escorts this night. An hour later, at 0215 HMS *Icarus*, on the starboard beam of the convoys, reported that her hydrophones had detected torpedoes passing down her side. Her radar then detected a U-boat at 2,400 yards, almost certainly *U-238*. Much to Commander Evans's disgust, *Icarus* pursued the contact cautiously.[116] *U-238* eventually submerged some 12 miles from the convoy, by which time the five torpedoes launched by the boat had struck three ships on the starboard side of ON202 – SS *Skjelbred* (102), SS *Oregon Express* (103) and SS *Fort Jemseg* (94). Two of the ships were hit right aft, causing Commander Evans to wonder if perhaps these were more of the new acoustic homing torpedoes. Historical evidence confirms none were, although two were pattern-running weapons that altered course automatically at a set distance in order to maximize the chance of hitting ships in convoy.[117] Added to the two other merchant ships hit on 20 September, also from ON202, this attack made *U-238* by far *Leuthen*'s most successful boat. There is some irony in this, as *U-238* did not have any Type V torpedoes embarked. Clearly, skill and daring still counted for much.

Two of the ships immediately began to sink, but SS *Skjelbred*'s damage, though serious, did not immediately threaten that vessel. Indeed, the convoy commodore considered she could have been saved had the rescue tug *Destiny*, en route to Liverpool with HMS *Lagan* in tow, still been in company with ON202.[118] In the end the Norwegian captain of *Skjelbred* had to be ordered to transfer himself and all of his crew, a total of 43, to the trawler *Northern Foam*, as no warship could be spared to stand by the stricken vessels while temporary repairs were completed. The busy trawler also picked up 33 of *Fort Jemseg*'s crew, another 21 from that ship being rescued by the merchant ship *Romulus*. The crew of the last of these three merchant ships, the Motor Vessel *Oregon Express,* proved less fortunate. At least 8 died, while a number of the 36 to 39 survivors – records are inconsistent – were injured, some badly. Most of the survivors were picked up by the merchant ship *Kingman*, with one more picked up by the *Romulus*.[119]

The U-boats were clearly making a very determined effort on this night, and the successful attack by *U-238* caused a few of the merchant ships in the convoy to fire illumination rockets. The dim glow of the resulting flares, their light diffused by the low clouds overhead, probably aided U-boats more than the escorts. The resulting light silhouetted HMCS *Chambly* on the distant port beam of the convoy as it pursued yet another boat, *U-260*. This one had been detected by radar just as *U-238*'s torpedoes were finding their mark on the starboard side of the convoy, and then a dim red light was seen as the corvette closed to 4,000 yards. *U-260* dived at 0231 hours when the Canadian warship had closed to 1,700 yards, firing a Type V that exploded astern of the fortunate corvette. *Chambly* delivered one depth charge attack, also inflicting no damage on the departing U-boat before its asdic broke down.[120] The drawn engagement deflected another effort to get at the merchant ships.

HMS *Icarus* reacted to yet another probe on the starboard beam of the convoy at 0250. In a familiar pattern, the U-boat was chased until it dived, depth charged without success, and then *Icarus* resumed station. The lull this time lasted four hours, during which Commander Evans began to suspect that HMS *Itchen* may have come to harm – in the confusion during the midnight attack he had not yet learned of the fate of that frigate. At 0445 hours Evans tasked HMCS *Sackville* to investigate whether *Itchen* was present, and then amended the order to have the Canadian corvette search the area where the explosion had occurred.

These concerns were temporarily put on hold when the last attack on the convoy developed on the port beam as night drew to a close. In the morning twilight *U-952* fired four older torpedoes, two of which found marks at 0650 hours. One, a dud, struck the liberty ship SS *James Gordon Bennett* of ON202, causing no apparent damage. The second, apparently fitted with a proximity fuse, detonated just in front of a second American ship, the SS *Steel Voyager*, causing modest but survivable damage to the bow area of the merchant ship. The crew of *Steel Voyager* showed little interest in keeping their ship afloat, however. FFS *Renoncule* induced the captain and some of *Steel Voyager*'s crew to re-embark after they hastily abandoned ship, but this success proved temporary and only resulted in time wasted. The merchant ship's captain insisted the damage was too significant to continue, a judgment later questioned by Evans once he was fully briefed on events.[121] At the time, with *Steel Voyager* drifting ever further astern of the convoy, the decision was made to abandon the ship and take her crew aboard other ships.

HMCS *Chambly* sighted *U-952* about half an hour later, a lookout first sighting a dim red light on the port quarter. Training the radar to the bearing of the sighting produced a contact at 5,000 yards. *Chambly* shaped an intercept course, apparently coming to the U-boat's attention as the range closed, as the red light was extinguished. A conning tower became visible at 4,000 yards against the eastern twilight, just as the U-boat's superior speed began to draw it away from the corvette. Lieutenant A.F. Pickard noted with satisfaction later that one salvo was sufficient to make him submerge.[122] By this point *U-952* was well clear of the convoy, and so *Chambly* was recalled.

The end of a battle

The growing light signalled an end to the night's attacks, and would indeed prove the substantive end of the battle. RCAF Liberators – the first of five being P/10 at 0842 – again met the convoy, and there were a few further engagements, but the merchant ships in ON202 and ONS18 would not be threatened again on this passage. One of the major reasons for this was not just the successful efforts of Commander Evans to reorganize the convoys into a more defensible formation,[123] but the decision by *BdU* to call off *Leuthen*. Weather, especially fog, proved a major consideration, as German shore staff believed the convoy still to be wreathed in fog and knew that the endemic fog off the coasts of Newfoundland would soon be reached.[124] The second significant reason for the German decision was the assessment that after three days of intense effort U-boat crews must be exhausted, and prolonging the attack with the poor visibility anticipated risked high losses with little chance of success.

The ocean escorts handed ON202 and ONS18 over to two groups of the Western Local Escort Force, W.1 and W.2, at 1100 hours on 25 September, before heading into St John's, Newfoundland. With the battle ended, both sides strove to identify lessons and ready themselves for the next round of the grueling campaign. The Allies set in motion a well-practised process that started immediately after a battle and unfolded for several months as various reports and analyses proceeded.

The initial Allied assessment, one that would be mildly adjusted but not substantially altered as time went on, was that the enemy should have been disappointed by the results of his effort to renew attacks on North Atlantic convoys.[125] Evans assessed that "In a series of attacks spread over nearly five days only three submarines succeeded in firing at the convoys."[126] Post-war analysis suggests that there may have been five attacks, but two were such complete failures that no ships in the convoy or escort were even aware an attack was attempted.[127] The Allies understood, as a result of both intelligence and battle experience, that new weapons and tactics had been adopted by the U-boats, and focused on the acoustic homing torpedo as the most threatening new development. Increased flak armament and a greater tendency to stay on the surface to use these guns was apparent, but analysts assessed that the increased threat to aircraft would be more

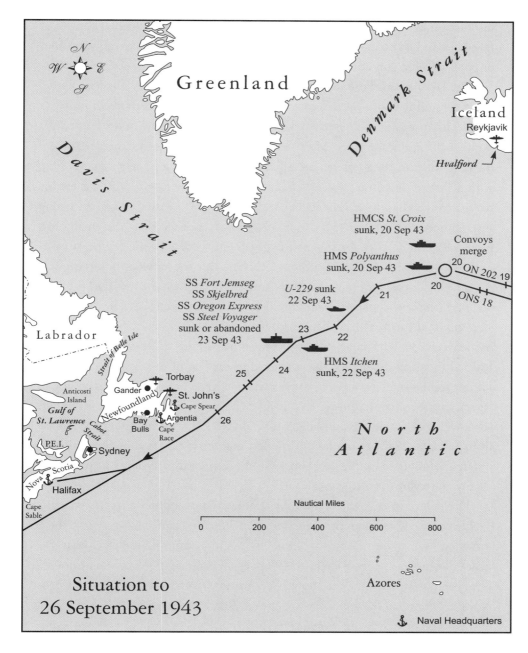

Situation to
26 September 1943

than offset by the greater opportunity now available to attack U-boats before they dived to safety. The possibility that U-boats were using red lights to attract escorts and then attack them, and that U-boats were operating as coordinated pairs to trap escorts, were both raised in light of incidents observed by escorts during the battle. However, Allied analysts apparently never attached significant weight to these possible tactics, which appear to have been products of circumstances insofar as they occurred at all.[128] Finally, escort commanders remained convinced that U-boats had no radar, as they were frequently surprised and notably cautious in poor visibility conditions.

The German staff produced their own analysis very quickly after the battle. The final summary of *Leuthen's* convoy battle appeared on 24 September in *BdU's KTB*. This assessment differed markedly from the Allied view of the battle, in a number of significant ways. Most remarkably, *BdU* remained convinced that a dreadful toll had been exacted on Allied escorts, claiming no less than 12 sunk, with a further three probables.[129] *BdU* rationalized the fact that Allied escorts to this convoy had been very strong and, despite these losses, remained strong throughout the battle for two reasons. First, *BdU* assessed that "the enemy had provided extra defence for this first engagement" – that is, the return of wolf-packs to the North Atlantic – although this did not trigger what in hindsight is the obvious question of how the Allies knew to strengthen the escort of the convoy that *Leuthen* happened to run into before the boats attacked. The continuing strength of the escort after their apparently heavy losses in the first part of the battle received attention as well, and *BdU* indicated that "The enemy must have realized the great danger for the convoy after the loss of so many escort vessels. He apparently then did all in his power with the great advantage of 2 days' fog to surround the convoy as quickly as possible with a new escort force." It is difficult to view this assessment as anything but muddled, given the distance of the battle from Allied ports. The simpler explanation that losses reported by U-boats were inflated, possibly reflecting problems with the new acoustic homing torpedo, was clearly given short shrift.

The failure to consider significant alternative possibilities suggests systemic problems with learning in the *BdU* staff. The difficult technical history of U-boat torpedoes during the war was well known. The disastrous failure of U-boat torpedoes to function effectively in the Norwegian campaign of 1940 had caused a furor in the German navy and should have made German naval officers conscious of the possibility of flaws in a new weapon.[130] *BdU* even noted in their own record of the development of the *Zaunkoenig* that development had been rushed, commenting that this was "a great risk owing to lack of the normal course of tests."[131] Despite this knowledge, the *BdU* analysis of the battle accepted claims that were so clearly at odds with events that reasons for the continued presence of a large escort had to be contrived, a process that hardly contributed to a rigorous analysis.

German assessments were better when addressing the level of success against merchant ships, where attacks were predominantly made with older torpedoes. Nine of these were claimed sunk, with two more torpedoed, and it was recognized that "Hardly one boat sighted the convoy itself." The real tally of six ships lost, and one further torpedoed without damage, is reasonably consistent with what *BdU* gathered from the U-boats involved. In comparison with the real toll of escorts – three sunk and one damaged against the claim of twelve and three – the merchant assessment is noteworthy for its comparative accuracy.

German assessments of losses to U-boats were more accurate, as might be expected. The loss of *U-341* was not included as part of the convoy action, allowing *BdU* to claim only two boats lost, *U-338* and *U-229*. Only two boats were recorded as receiving significant damage, *U-386* and *U-270*. The wounding of the commanding officer of *U-377* did not rate a mention, as *BdU* preferred to stress:

> The fact that the boats remained surfaced and warded off attacks with all guns led to success even though one boat was probably lost. Contact with the convoy was, however, maintained and the boats remained in the vicinity. It is essential that as many boats as possible of the group are always up to the convoy so that convoy escort – aircraft and escort vessels – are split up as much as possible and just a few boats off the convoy do not bear the brunt. If aircraft only engage a few boats many others can approach the convoy.

In short, *BdU* desperately wanted to believe that the new doctrine of surfaced boats fighting off aircraft on the surface was working.

It is hardly surprising, once a general understanding of the analysis is gained, that *BdU*'s final assessment of the attack on ON202 and ONS18 was that "This was therefore a very satisfactory result which might have been considerably better if the weather had been favourable."

The reality is that the convoy battle proved a limited German tactical success. The surprise achieved by new weapons and, especially, new tactics, had sunk and damaged ships and aircraft, but the advantages gained were at the cost of alerting a tenacious and adaptive enemy to the existence of these new methods and tools. The surprise gained had been limited to the manner in which the new torpedo had been used, not that a weapon of this type existed. The Germans failed to understand that only tactical and not operational surprise had been gained. The reason for the large number of escorts protecting ON202 and ONS18 was that Allied intelligence had gained sufficient knowledge of the deployment of U-boats into the North Atlantic to shift a support group from its intended patrol area off the Bay of Biscay to the central convoy route in mid-ocean, not the panicked reaction after the first attacks hypothesized by *BdU*. German success in disguising the initial location of the *Leuthen* patrol line proved significant because it resulted in a battle occurring quickly, before the Allies could react to the deciphered message on 20 September revealing that U-boats intended to target escorts as a priority. As a result, Allied reaction to the deliberate targeting of escorts by the Type V was mildly confused in the first few days, as shore staff and commanders at sea struggled to determine the best response to this surprising development.

By the end of the battle, however, reasonably effective Allied tactical and technical counter-measures were in development. The Type V could never be completely neutralized when fired from ambush – an acoustic homing torpedo is a significant improvement in a submarine's arsenal – but escorts alert to the presence of a U-boat had effective tactics available that made a hit by a Type V unlikely, at the cost of delaying the approach to the vicinity of a U-boat somewhat. By the end of September escorts were being fitted with a towed noisemaker that could be deployed to seduce a Type V away from a ship's screws. This technical counter-measure had limitations as well, as the noise from the decoy significantly hindered asdic effectiveness. There were also concerns that a single noisemaker might not be effective, and it took months for defence scientists and intelligence analysts from the Allied nations to agree, finally, that a single noisemaker towed about 200 yards astern of a warship provided reasonable protection from a Type V. Even after all this, Type V's sank warships until the closing days of the war, as the problems associated with decoys and avoiding tactics meant that a surprise attack by a U-boat still had a good chance of success.[132]

The Allied staff superiority in responding to enemy initiatives and adapting new technology and tactics to the ongoing campaign became evident even as the engagements swirled around ON202 and ONS18. Allied staffs quickly developed awareness of new enemy weapons and tactics and began to develop counter-measures immediately. German staffs, on the other hand, failed to grasp the teething problems with their new torpedo. The anti-aircraft policy adopted by *BdU* was never properly analyzed before being adopted, and results from the field were consistently viewed through the most optimistic prism. *U-341*, which *BdU* became aware was possibly lost on 19 September and declared lost on 23 September, was not considered sunk as a result of the new tactic, although the possibility that radio signals were sent by the boat in accordance with the new stay-surfaced-and-fight-aircraft direction was mentioned in passing. More significantly *BdU* decided not to include the loss of *U-341* in its summary of the attack on ON202/ONS18, despite the boat being designated as part of *Leuthen* and likely being sunk in the general vicinity of the convoys. The wounding of the commanding officer of *U-377*, also a result of the new anti-aircraft tactic, was also omitted by *BdU* in its final summary of the battle, where staying surfaced and fighting aircraft was judged a success. The Allies came to a completely different conclusion: "A curious aspect of the U/B countermeasures is their tendency to remain on the surface and fight it out. Actually, the longer they remain on the surface, the less chance they have."[133]

Evidence that *BdU* had failed to grasp the true implications of the battle for ON202/ONS18 arrived swiftly. The surviving U-boats from *Leuthen* and additional boats fresh from base were formed into a new pack, *Rossbach*.[134] The 21 boats of this group struggled to find Allied convoys from 27 September 1943 on, starting with the next westbound convoys from the UK.[135] Using scattered radio decrypts of Allied convoy straggler routes, *BdU* placed *Rossbach* to the southwest of Iceland in pursuit of ONS19, ON203 and ON204.[136] Equipped with a better understanding of the new German grid references, Allied intelligence diverted these convoys safely clear of the patrol line. In the meantime Allied patrol aircraft were vectored into the area of the pack, using the cover of an HF/DF intercept of a weather report from *U-631*.[137] Giving up on westbound convoys, *BdU* redeployed *Rossbach* against HX259 and SC143. Allied intelligence again decrypted this intention and diverted HX259 north around the intended patrol line. SC143, on the other hand, was reinforced with EG10, a British support group of four destroyers, and headed directly at *Rossbach*. Heavy air support augmented these surface ship reinforcements. Once again most U-boats failed to sight any merchant ships, despite Luftwaffe assistance in the form of a long-range reconnaissance flight on 8 October that sighted and sent homing signals while circling SC143.[138] The Polish destroyer ORP *Orkan* sank quickly when hit by torpedoes from *U-378* during the early hours of 8 October, and *U-645* succeeded in torpedoing the American merchant ship *Yorkmar* about 24 hours later. These two successes were more than offset by the heavy losses suffered by *Rossbach*, three U-boats going down on 4 October,[139] another three on 8 October,[140] all the result of air attacks. Four other boats were forced to return to port with damage caused by air or surface attacks. South of the main battle, the milch cow* *U-460* and another boat, *U-422*, succumbed to attacks from carrier aircraft from the USS *Card* on 4 October. Both these latter boats were victims of the Allied acoustic homing torpedo.[141]

BdU still persisted in massed attacks on convoys after this setback, but results continued heavily against the U-boats. The limited ability of German naval commanders and staffs to analyze events and adapt to Allied reactions contributed to what was clearly a losing proposition for German submariners sent out to battle. This proved fatal for many U-boat crews, while losses to the Allies remained, in

---

* "Milch Cow" was the nickname the Germans gave to the large Type XIV U-boats designed to replenish other U-boats with fuel, stores and sometimes torpedoes so that they could remain on patrol longer.

comparison, moderate. By the end of the war 12 of the commanders of the 20 U-boats associated with the ON202/ONS18 battles had been killed in action, including the fearless *Kapitänleutnant* Horst Hepp of *U-238*, and another was a prisoner of war. Only 7 survived the war, one of these being *Kapitänleutnant* Paul-Friedrich Otto, whose devastating torpedo hit on *Lagan* marked the first success of the German acoustic homing torpedo. By the end of the war, almost three quarters of those who served in the U-boat arm of the *Kriegsmarine* were lost at sea. In terms of percentage, far fewer Allied crewmen perished, although Allied ships were lost until the final days of the war. In the end, Allied adaptability proved superior not just in the autumn of 1943 but for the remainder of the campaign. The modest results achieved against ON202 and ONS18 proved the last significant success of the wolfpacks, a hollow victory indeed.

The last corvette. Alongside in her final berth near the Maritime Museum in Halifax, Nova Scotia, HMCS *Sackville* is the last surviving Flower-class corvette out of 269 built, of which 123 were built in Canada. This veteran ship, so active in battles such as that described in this chapter, is a fitting memorial to the brave escort crews. The Government of Canada officially designated HMCS *Sackville* as Canada's National Naval Memorial on 26 April 1985. She is restored with wartime equipment, and her present location allows easy comparison with modern warships of the Canadian navy, making her small size readily apparent. (Courtesy of Canadian Naval Memorial Trust)

PRELUDE

Clearing the
English Channel

T HERE WERE A NUMBER OF OBSTACLES that Allied navies had to overcome to
land and sustain forces in northwest Europe. Thwarting the major U-boat
threat to North Atlantic convoys was certainly one of them, and, as we saw in the
last chapter, had been generally achieved by the fall of 1943. Yet the German navy
tenaciously struggled to remain a threat. The surface fleet of the *Kriegsmarine* was
not large, but the skill and determination of German sailors ensured that this
small naval force continued to pose a problem for Allied planners as preparations
for D-Day unfolded.[1]

Growing numbers of Allied warships were gathered in an effort to gain control of the narrow seas. Once again, however, numbers alone were insufficient.
The German surface forces sought to maximize their chances by operating at
night as much as possible, to reduce the threat from Allied air power, and developing highly effective tactical reactions to Allied surface attacks. The Allies
needed to achieve better levels of training as well as develop new tactics to prevail,
and over the winter of 1943-44 this was slowly accomplished.

The battles that were fought in the English Channel during the war were usually comparatively brief engagements. The German navy, well aware of its overall
inferiority in numbers, strove to engage only when the odds favoured success.
Yet even when faced with daunting numbers of opponents or caught completely
by surprise, German sailors proved highly effective at launching stinging attacks,
often inflicting more damage than they received. Governed by the principle of
calculated risk and immediate counter-attack, followed by a swift departure from
the scene of action toward one of the many protected ports along the coast of occupied France, German surface tactics were often frustratingly effective.

The next chapter, by Michael Whitby, traces the evolution of night fighting
effectiveness of the Royal Navy in the English Channel from early in the war
until the climactic battles coincident with the landings in Normandy. Several
battles earlier in the war are covered, albeit briefly. These engagements make clear
that success required considerable change on the part of the Royal Navy.

Overcoming the highly effective German forces required the development of new tactics that optimized new technology, better team training and a better set of command arrangements. All of these measures were challenging to implement, and the slow progress and occasional setbacks at the hands of the *Kriegsmarine* caused tensions between staff officers at the Admiralty and those closer to the sharp end in operational commands or fighting at sea.

In smaller battles between courageous and determined foes, personality often proves a critical factor. Superb sketches are drawn of many of the key actors in these battles. German and Allied commanders are revealed clearly, with their many sterling qualities and occasional human flaws.

Even as the elements of Allied success began to come together in the spring of 1944, problems continued to undermine the chances of success. The chaos inherent in night actions proved difficult to overcome. In the final action that broke the back of German surface forces in the Channel, the Allies did have a significant numerical advantage, but confusion – arising from a number of factors that are examined in detail – resulted in this quantitative advantage being essentially squandered. That the remaining Allied warships achieved an important success indicates very clearly how important the other elements of training and tactical development had worked to level the battlefield between the *Kreigsmarine* and the Allied navies.

"Shoot, Shoot, Shoot!"

DESTROYER NIGHT FIGHTING
AND THE BATTLE OF ÎLE DE BATZ[1]

Michael Whitby

Sailors have never been comfortable fighting at night. Quite simply, too much can go wrong. Command and control can be confused, the risk of engaging friendly forces high, collision a constant worry, and the chance of surprise from an unexpected quarter an ever-present danger. The renowned British fighting admiral Sir Andrew Cunningham, a successful destroyer captain in the First World War and victor in the dramatic night battle off Cape Matapan in 1941, summed up the hazards well, observing that "in no other circumstances than in a night action at sea does the fog of war so completely descend to blind one of the true realization of what is happening."[2] Whereas in earlier maritime conflicts naval commanders could usually avoid night action if they chose, by the Second World War it had become an inescapable element of naval warfare. Nowhere was this truer than in the English Channel – the Narrow Seas – where the German *Kriegsmarine* and the Royal Navy and its allies clashed almost nightly.

The western entrance to the English Channel – the "Chops of the Channel" of maritime lore, where the Narrow Seas meet the Atlantic Ocean – was the scene of much of this fighting. At no time during the Second World War was control of these waters so critical as during the early stages of Operation NEPTUNE, the Allied invasion of Northwest Europe in June 1944. From the eve of the invasion onward, scores of transports lumbered across the Channel bearing Allied troops and materiel to Normandy. The *Kriegsmarine* knew they had to interdict that shipping to have any chance of defeating or even hindering the invasion, and although the German surface fleet was in many ways a spent force, it still possessed a small number of destroyers, including some of the most powerful in the European theatre, which, if allowed, could ravage thin-skinned transports. Due to the Allies' overwhelming air superiority, German destroyers had to attack at night to have any real chance of success, but throughout the war they had demonstrated prowess in that setting. This posed a threat the Allies could not ignore, and in early 1944 the RN formed its own strike force, the 10th Destroyer Flotilla, to counter German destroyers on the west flank of the Channel. On the night of 8/9 June 1944, in an action that can be dubbed the Battle of Île de Batz, eight destroyers from that unit – four British, two Canadian and two Polish – faced off against four German counterparts in a long running battle off the coast of northern Brittany.

Typical of most battles, this action has a complex anatomy. To appreciate its nuances as well as its historic consequence requires an understanding of how the RN worked to master the art of night fighting throughout the Second World War, which entails analysis of several earlier actions. The tension between technology and tactics also has to be considered, especially in the development of seagoing radar. Ship design and naval weaponry are also factors, and since people do the actual fighting at sea, aspects such as leadership, training and experience come to bear, including within the *Kriegsmarine* destroyer force, whose skill and professionalism is too often overlooked. Finally, Normandy was truly an Allied victory, and at sea, as well as on land and in the air, Canada's contribution was indispensable. Two destroyers of the Royal Canadian Navy (RCN) played an instrumental role in the action, and the reasons why they were there and how they faced up to battle are important ingredients of Canada's naval heritage.

Of the elements that shaped the Battle of Île de Batz, the necessity of seizing the initiative in the first moments of action arises as perhaps the most critical. "The tactical maxim of all naval battles," naval theorist Wayne Hughes emphasized, "is *Attack effectively first*. This means that the first objective in battle is to bring the enemy under concentrated firepower while forestalling his response."[3] This was well understood going into the Second World War and was highlighted in the RN's tactical doctrine. However, the rapid development of radar, or radio direction finding (RDF) as the British referred to it officially until July 1943,[4] had a tremendous impact on how that maxim could best be achieved so that decisive results could follow. Radar extended the range of detection between units fighting at night, which shaped thinking about how one achieved position to attack effectively first, and altered notions of tactics and weaponry. The process by which sea-going and staff officers of the RN grappled to take best tactical advantage of radar – and how the *Kriegsmarine* counteracted – is a major component of the story ahead.

Early destroyer contests

From the outset of the Second World War, the *Kriegsmarine* could not match the Allies in sheer naval strength and of necessity relied upon the cloak of darkness to overcome its desperate shortage of destroyers. It had just 21 in commission at the end of September 1939 and could expect no reinforcements until late 1940, whereas the RN boasted some 176 built or building, with 96 based in home waters.[5] Nonetheless, during the first winter of the war the Germans made superb use of a slim resource. In a daring campaign between November 1939 and February 1940, German destroyers laid some 2,000 magnetic and contact mines along the British east coast convoy routes and in focal points such as the Thames, the Humber and the Tyne. These were carried out with great boldness. *Korvettenkapitän* Theodor von Bechtolsheim, the commanding officer of the destroyer *Karl Galster* who would later lead the German force in the Battle of Île de Batz, recalled on one occasion his ship was still in the Thames Estuary at dawn and had to make its way home through scores of British fishing boats in growing daylight. Pulling down their ensign and waving to all in sight, they managed to convince fishermen they were "Tommies" and slipped away with their identity preserved.[6] Overall the mining campaign claimed tens of thousands of tons of shipping, and such was the German success in mounting these sorties without detection, let alone interception, that it was not until after the war that the British realized that most of the mines had been laid by destroyers and not U-boats and aircraft as suspected.[7] Although they seldom made contact with British surface forces during this period, confidence among German destroyermen skyrocketed.

That feast was followed by an abrupt famine. The *Kriegsmarine* destroyer force was routed in the Norwegian campaign of April 1940, losing ten ships in the two battles of Narvik. That spread the force thinner still, and it was not until late 1940 that they were able to exploit the capture of French ports to mount a mini-offensive against shipping along England's southwest coast. Operating under the doctrine of the strategic defensive, destroyer officers were under strict instructions to engage only targets of opportunity and to withdraw if they encountered any serious opposition – "Shoot and scoot" one British officer had derisively dubbed the doctrine during the First World War.[8] In no way should this be construed as a lack of boldness; the Germans simply lacked the strength in numbers to risk using destroyers as expendable assets. The RN's Plymouth Command tried to intercept the raiders operating on their doorstep, and after several close brushes, in the early hours of 29 November 1940, five ships from the 5th Destroyer Flotilla (5 DF) met three German counterparts in a typical early Second World War destroyer night action.

That night, Captain Lord Louis Mountbatten, RN, commander of the 5th DF, or "D5", had been ordered to patrol an east-west line about 25 miles south-west of Plymouth to cut off any German destroyers attacking shipping in Falmouth Bay. The mission suited Mountbatten just fine. An ambitious, aggressive officer, who had earned public acclaim – much of it self-generated – when he coaxed his heavily damaged destroyer HMS *Kelly* safely home after she had been torpedoed by an E-boat, Mountbatten yearned for applause for victory in battle, not for saving a stricken ship.[9] With *Kelly* under repair, Mountbatten embarked in HMS *Javelin*, commanded by Commander Anthony Pugsley, RN, and was accompanied by *Jupiter*, *Kashmir*, *Jackal* and *Jersey*, modern destroyers of the identical "J" and "K" classes commanded by skilled naval professionals.

Beginning his patrol at about 1900, Mountbatten elected to deploy his ships on a line of bearing.[10] Line ahead, where ships follow directly behind each other at about 1,000 yards apart, was the standard night formation, but Mountbatten thought that his captains would have difficulty remaining closed up in that formation at the high speeds of 27-30 knots he chose to use. Instead he deployed in a line of bearing, where ships were offset from the leader, and throughout the night he practised variations in speed and bearing to increase his flotilla's efficiency.

Gun flashes on the horizon and sketchy reports of attacks from Plymouth indicated that enemy destroyers were about, and at 0521 hours Mountbatten sighted a searchlight off the starboard bow. Six minutes later lookouts observed the glow of a burning ship on the same bearing, and Mountbatten altered course to 340°. Then at 0537, as he described in his report, "a turn to 310° was made, to make sure of keeping between the enemy and his base." On *Javelin*'s bridge, a dozen pairs of eyes probed the darkness. The destroyer was one of the first British ships equipped with radar, but the unreliable, primitive Type 286 – the aerial was fixed directly ahead and had no real search capability – only worked sporadically.[11] Finally, "At 0540, two darkened vessels were sighted fine on the starboard bow, crossing ahead of our line from starboard to port, and I immediately ordered 'Alarm Bearing 310' to be passed by W/T. The range was estimated at 5,000 to 6,000 yards and it was closing rapidly. The R.D/F set of the JAVELIN, which had unfortunately broken down at 0450, came into action again at this time, and reported three ships at a range of 3,500 yards. This report, unfortunately, did not reach the Principal Control Officer [PCO] of the JAVELIN, who thought that the R.D/F was still out of action."[12]

As he described in his postwar memoirs, Pugsley, standing beside Mountbatten on *Javelin*'s bridge, "set in motion the often practised routine for getting the guns on to a target at night":

This dramatic image depicts the close range of the fighting on 29 November 1940. In the background to right a German destroyer, probably *Hans Lody,* is illuminated by the blast from HMS *Javelin*'s main armament. Not visible are the two German torpedoes that would devastate *Javelin* in a matter of seconds. (National Archives (U.K.), ADM 199/530)

"Enemy in sight. Follow captain's sight," and I swing the sight that sits astride the gyro compass binnacle on to the enemy ship. An electric pointer at the Gun Director [atop the Bridge] flickers round a dial and is pursued by a companion pointer as the Director follows round. The Director steadies, pointed in the same direction as the captain's sight on the bridge and the target comes into the Director's telescope. The two pointers come to rest lined up with each other. At the guns a similar process takes place as they in turn follow the Director. Range and deflection are passed by the Gunnery Control Officer and quickly applied to the Director sight. "Director Target," shouts the Gunner's Mate at the Director to the Gunnery Control Officer, who reports "Ready to open fire!" to the captain.[13]

Before Pugsley could order "Open fire" and the gunnery control officer could issue the final "Shoot, shoot, shoot!" the situation changed.

Mountbatten had a choice of continuing straight for the enemy to force an immediate engagement or turning parallel to cut them off from their base. The latter move would cause a delay in opening fire as the Director Control Towers that guided the three twin-4.7-inch turrets would have to slowly swing around to

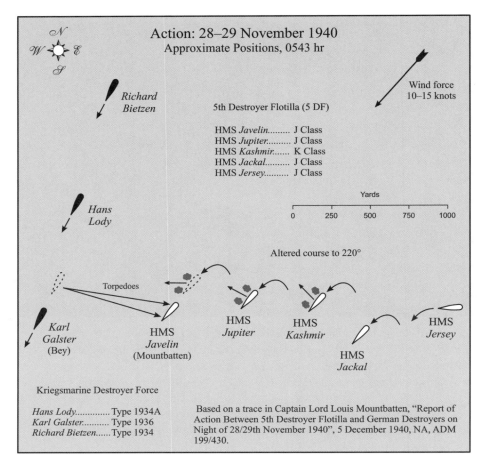

Action: 28–29 November 1940
Approximate Positions, 0543 hr

N
W   E
S

Richard
Bietzen

Wind force
10–15 knots

5th Destroyer Flotilla (5 DF)

HMS *Javelin*......... J Class
HMS *Jupiter*.......... J Class
HMS *Kashmir*....... K Class
HMS *Jackal*.......... J Class
HMS *Jersey*.......... J Class

Yards

0   250   500   750   1000

Hans
Lody

Altered course to 220°

Torpedoes

Karl
Galster
(Bey)

HMS
Javelin
(Mountbatten)

HMS
Jupiter

HMS
Kashmir

HMS
Jackal

HMS
Jersey

Kriegsmarine Destroyer Force

Hans Lody..............Type 1934A
Karl Galster...........Type 1936
Richard Bietzen......Type 1934

Based on a trace in Captain Lord Louis Mountbatten, "Report of
Action Between 5th Destroyer Flotilla and German Destroyers on
Night of 28/29th November 1940", 5 December 1940, NA, ADM
199/430.

the new bearing. Pugsley turned to Mountbatten: "Straight on at 'em, I presume,
Sir?" D5 demurred: "No, no. We must turn to a parallel course at once or they
will get away from us."[14] Mountbatten later explained he wanted "to try to get as
close to the enemy as possible, whilst remaining between him and his Base, and
so I held on until 0543, when I altered the course of the Flotilla together to 220°
by W/T and fire was opened by all ships except JACKAL who, through being out
of station had her range fouled by the KASHMIR."[15]

Following Mountbatten's orders, Pugsley brought *Javelin* around to port and
gave the order to open fire (see map above). However, the director lost bearing on
the target during the turn and a delay of about a minute passed before the first
salvo rang out. Due to an over-estimation of range – a common occurrence at
night – the first rounds were fired well over at 5,000 yards. Range was immediately reduced to 2,000 yards, but before any shells could hit home two torpedoes
exploded against *Javelin's* starboard side, bringing the destroyer to a dead stop.[16]
The confusion and disarray that so often accompanies night action now settled

over the British formation. In *Jupiter*, the ship following *Javelin*, the turn to port absorbed the attention of Lieutenant Commander N.V. Thew, RN, junior CO in the flotilla, who lost sight of his target. Thew swung out on *Javelin's* engaged quarter to maintain his line-of-bearing but then had to veer sharply to port to avoid collision when *Javelin* was torpedoed. The torpedo explosions temporarily blinded all on *Jupiter's* bridge. In *Kashmir*, the third ship, Commander H.A. King, RN, second-in-command of the flotilla, thought it was *Jupiter* that had been torpedoed and ended up following Thew. Showered with oil from *Javelin*, King's night vision was ruined by his own gun salvos. After about five minutes, he realized he was senior officer, and immediately put his ships into line ahead, and headed south towards Brest at 28 knots to try to get ahead of the enemy, who had disappeared into the darkness. With nothing in sight by 0800, King doubled back northward hoping to meet the German destroyers, but they had won the race south and the seas were empty. In *Javelin*, wallowing and helpless with her bow and stern blown off, Mountbatten had watched with frustration as his ships disappeared into the night. Sturdy *Javelin* was saved, but 46 died and she was out of action until January 1942.[17] Mountbatten was awarded the Distinguished Service Order for again saving a stricken ship.

The German side of the action was far more straightforward and demonstrated both their skill and the nature of the "shoot and scoot" doctrine. The force comprised the destroyers *Karl Galster*, *Hans Lody* and *Richard Bietzen*, under command of *Kapitän zur See* Erich Bey, who was *Führer der Zerstörer* (*FdZ*), the *Kriegsmarine's* senior destroyer officer. A seasoned officer, Bey had led the minelaying campaign off the English coast during the first winter of the war, and had then presided over the disastrous defeat at Narvik.[18] Like Mountbatten, Bey was riding in another officer's ship, von Bechtolsheim's *Karl Galster*.

Bey's ships had just started their run back to Brest after sinking two small trawlers when they sighted "five or six destroyers." Two minutes later, at 0543, the same moment that Mountbatten altered course, Bey ordered his ships to fire torpedoes as they wheeled southwest to open from the enemy. Given the high speed and relative point-blank range of their targets – about 1,500 yards – there was only time for a snap shot; nonetheless two of *Hans Lody's* torpedoes found *Javelin*. Unwilling to grapple with a superior force and straddled by British shellfire, Bey disappeared at 35 knots behind the cover of smoke. The three destroyers suffered no casualties and sustained only minor splinter damage. The Commander, Naval Group West described the action as "a complete success" that would boost confidence.[19]

While German sailors celebrated, there was tooth-sucking in England. Five

At daybreak, *Javelin* wallowed helplessly, her bow and stern blown off by German torpedoes. The fact that she survived to fight in the June 1944 action is testament to her sturdy construction and the determination of her damage control team. (Esquimalt Naval Museum)

modern destroyers had met a lesser force under promising circumstances, but had suffered serious damage to one ship while the enemy escaped unscathed. At the Admiralty, which was responsible for identifying and disseminating operational lessons, criticism focused on Mountbatten's delay in getting into action. Citing the RN doctrinal bible, *The Fighting Instructions*, which laid out guidance for officers to follow under various tactical scenarios, the senior operations officer at the Admiralty, Captain Cecil Harcourt, thought Mountbatten had displayed poor judgment:

> It would appear that the instructions in paragraph 8 of the Fighting Instructions were not sufficiently borne in mind. Here it is stated: "In any action at night, the primary object is to develop the maximum volume of gun and torpedo fire before the enemy can do so, and all other considerations are of secondary importance. Results at night will depend on the action taken in the first minute or so…." A good deal of time seems to have been wasted in manoeuvring for position and in disposing the destroyers on the correct line of bearing, instead of going straight for the enemy and engaging with all weapons.

The ACNS(H), Captain A.J. Power, agreed: "The outstanding feature is that we apparently had the advantage of *first sighting*, but held our fire for at least three minutes."[20]

Engaging first was seen as essential but other comments on the action indicate that night fighting was in a state of flux with disagreement over matters that used to be considered fundamentals. C-in-C Western Approaches thought that Mountbatten should have used searchlights to maintain contact with the enemy, but the Admiralty disagreed, one officer noting, "A searchlight badly used or outside effective range is a known and proven danger…." This view held, and searchlights, the primary visual night fighting aid in the First World War, were not utilized in any of the actions discussed in this study. Staff officers also complained about Mountbatten's decision not to use line ahead and the confusion caused by "fiddling about on lines of bearing at night." The importance of radar was also discussed, regret being expressed at its lack of reliability and the fact it was not fitted in all destroyers. Finally, there was common agreement that the night fighting efficiency of the 5th DF was substandard, and that more specialized training was required.[21] As events three years later demonstrated, senior RN officers remained far from agreement on the essentials of night fighting, and some of the shortcomings from November 1940 continued to fester.

The radar race

The German invasion of the Soviet Union in June 1941 caused the *Kriegsmarine* to shift its thin destroyer resources to northern waters and the Baltic, and left night operations in the Narrow Seas to E-boats and other light forces. But in 1943 new Type 39 fleet torpedo boats emerged from German shipyards to be deployed into the western Channel with the task of escorting coastal convoys between ports like Brest and Cherbourg. Small destroyers, armed with four 4.1-inch guns and six 21.7-inch torpedo tubes and capable of about 28 knots, fleet torpedo boats – "Elbings" to the British – proved to be a thorn in the side of Allied naval forces. However, even though they were effective platforms, like the entire *Kriegsmarine* they lagged behind their RN counterparts in the critical area of radar.

Although Germany had pioneered the development of naval radar – the pocket battleship *Graf Spee* took a set to sea in 1936 – inter-service rivalry stifled development. The *Luftwaffe* had priority over equipment and research, forcing the navy to rely upon obsolete hand-me-downs ill suited to naval warfare. Even late in the war destroyers retained 1940-vintage FuMO25 or FuMO28 metric sets, which were limited in range and accuracy.[22] A-scans, where echoes formed vertical spikes on a horizontal range scale, were used for display and the aerial's location at the base of the foremast caused a 30° blind spot astern, a weakness known to the Allies.[23] *Kriegsmarine* action reports from the 1943-44 period reveal that although radar usually provided accurate range data, the sets had limited search capability, making it difficult to establish an accurate plot.

In contrast, the RN maintained consistent progress. When Mountbatten duelled with Bey, the understanding of radar among most operational and staff officers was about as rudimentary as *Javelin's* unreliable Type 286. That changed, and by the end of 1942, according to one historian, "The Navy had beaten the early radar into submission and attitudes had changed dramatically. Radar was not just another burden, it was essential equipment."[24] By 1943 most RN destroyers had two or three types: Gunnery (GA), Warning Combined (WC) and Warning Surface (WS).[25] For gunnery, most fleet destroyers had a variant of Type 285. Designed as a high-angle set against aircraft, Type 285 had evolved into the standard fire control set for destroyers. It provided excellent ranges and was accurate enough to detect the shell splashes of "overs" and "unders" and, to the mortification of operators, incoming rounds. The Type 285's "yagi" dipole aerials were located atop the gun director and echoes appeared on an A-scan display.[26] For Warning Combined, destroyers had Type 291, another well-tried unit that could detect surface contacts out to 9 miles. A significant weakness was that it was easily monitored by the enemy and for this reason it was seldom switched on until after action was joined. Unfortunately at that point it took up to 20 minutes to warm up.[27]

The most effective search radars were the Warning Surface sets. Unlike 285 and 291, which operated on decimetric and metric wavelengths respectively, the WS sets were centimetric, whose narrower beam produced superior range, discrimination and accuracy. By the fall of 1943, most fleet destroyers were fitted with a variant of the legendary Type 271. Performance was good – a destroyer could be detected at about 9 miles – but the aerial was usually manually rotated and after each sweep the operator had to reverse direction so the cord would not become tangled. A more serious drawback came from a problem that often occurs when new systems are strapped onto ships not specifically designed for it. The 271 aerial had to be located next to its power source, which was too heavy to be mounted on the handsome but weak tripod foremasts of British fleet destroyers. Nor could it be put atop the bridge superstructure without removing fire control equipment essential for accurate gunnery. As an alternative it was placed in the searchlight position amidships. This left it only about 15 metres above the waterline, which not only reduced range but the forward superstructure "wooded" the beam, creating a significant 54° blind spot ahead. Under some circumstances Type 285 or 291 could be used to search this gap or ships could zigzag but that would make main armament tracking difficult. No matter how ships tried to compensate, the blind spot caused serious consequences.[28]

Teething troubles are inevitable in such instances, and one must remember that radar was still a relatively new, fragile vacuum tube technology. Breakdowns were common, especially under the pounding from hard steaming or shock from main armament blast, and performance was often impaired by climatic conditions. Skilled operators and maintainers were also in short supply. Nonetheless, Allies held a significant advantage over the *Kriegsmarine* to the extent that historian Arthur Hezlet observed German sets "were to the Allied sets as a pocket torch is to a car headlight."[29] Nonetheless, an action that one British participant called "The classic balzup of the war" revealed that although RN possessed superior radar, the tactics of how best to take advantage of that, as well as how to account for the enemy's use of it, still had to be ironed out.[30]

Trials and tragedy

The arrival of the Type 39 torpedo boats in mid-1943 allowed the Germans to deploy homogeneous formations of destroyers into the western Channel, something British commanders could not do. Since 1941 the home commands had mainly relied upon Hunt-class destroyers to counter German light forces in the Narrow Seas. Near equivalents to the Type 39s, the Hunts "arose from the need for an escort vessel capable of convoy AA defence but with a good turn of speed so that it might also, when required, undertake a limited number of destroyer functions."[31] The result, designated a "Fast Escort," was a compromise and, as so often occurs when that becomes a factor in warship design, the final product was found wanting in certain areas.[32] Although the Hunts had a strong main armament of four to six 4-inch high-angle guns, they either had no or limited torpedo strength, and at 25 knots were relatively slow.[33] Although they proved effective as coastal escorts, they lacked the speed and punch to counter German destroyers or torpedo boats. This was well understood by operational authorities, and in October 1943, when Plymouth Command began to mount offensive sweeps where Type 39s would be the probable opponents, they reinforced the Hunts with any available fleet destroyers or cruisers. This resulted in formations of warships with varying capabilities led by officers not used to working together. Although the exigencies of war sometimes left little choice but to follow this path, the risks associated with using scratch forces soon became evident.

On the evening of 3 October 1943, the Hunts *Limbourne*, *Tanatside* and *Wensleydale*, buttressed by the fleet destroyers *Grenville* and *Ulster*, steamed out of Plymouth for an anti-shipping sweep along the north Brittany coast. Commander

C.B. Alers-Hankey, RN, in *Limbourne* was senior officer. At sea providing distant cover to a German coastal convoy were the Type 39s *T-22*, *T-23*, *T-25* and *T-27* of the *4. Torpedobootflotille*. At their head was *Korvettenkapitän* Franz Kohlauf, a young, dynamic officer who had gained considerable night operational experience earlier in the war as captain of the fleet torpedo boat *Seeadler*. On this night Kohlauf had the advantage of a light horizon to seaward where any British force would be expected to emerge, and darkness along the coast behind. He sighted Alers-Hankey's ships first and immediately launched a series of unsuccessful long-range torpedo attacks, which escaped notice by the British. It was not until 0100, fully 30 minutes after Kohlauf gained contact, that *Tanatside*'s Type 271 detected the German destroyers at about 10,000 yards. Alers-Hankey led his force around to southeast but to make an "undetected approach" – not realizing he had already been sighted – closed at 12 knots for five minutes, before increasing to 15, 20 and, finally, to the Hunts' maximum speed of 25 knots. When starshell burst overhead, Kohlauf withdrew at his full speed of 28 knots; the Hunts, who had to zigzag to allow their Type 271 radar to track ahead, were soon left behind and a running gun battle developed between the four torpedo boats and the two fleet destroyers. Starshell was fired to good effect by both sides. *Grenville*'s CO, Lieutenant Commander Roger Hill, RN, coolly noted "Enemy illumination was excellent…. I read a signal on the bridge." Unfamiliar with one another, *Grenville* and *Ulster* launched torpedo attacks individually rather than as a unit, which would have likely proved more effective. Their gunfire also sprayed inaccurately. On the other hand Hill described "a large number of small splashes continuously round the ship, and occasional water on the bridge [from near misses]."[34] German shells soon hit home and *Grenville* abandoned the chase when a large fire broke out on her stern. *Ulster* continued the pursuit alone until hits also forced her to break off.

Formal after-action analysis – "lessons learned" in today's parlance – is seldom immediate. In this instance, Plymouth Command had little time to absorb any lessons, and the Admiralty none at all, before they mounted the next operation. There undoubtedly would have been discussion between the commanding officers involved and staff at Plymouth, but since the next operation, dubbed Operation TUNNEL, proved remarkably similar to the previous action, except that the German torpedoes hit home, it is apparent that little remedial action was taken. One step that the C-in-C Plymouth, Admiral Sir Ralph Leatham, did take was to bolster the strike force by adding the cruiser *Charybdis*. But even that proved a mixed blessing. Although *Charybdis* and her sister *Scylla* had names derived from the classics, they scarcely represented classic cruiser design. Sailors dubbed them the "Toothless Terrors" because their relatively light main armament of eight 4.5-inch guns and six 21-inch torpedo tubes was closer suited to that of a destroyer than a cruiser. Experience had demonstrated that large warships like cruisers could be a liability in the hurly-burly of a night action fought in confined waters, but their powerful armament – usually in the form of 6- or 8-inch guns – sometimes made the risk acceptable. Sadly, *Charybdis* brought all the burden of size but none of the benefit of punch.

The objective of Operation TUNNEL was the blockade-runner *Munsterland*, which the Germans were attempting to send up the Channel from Brest to Cherbourg. Admiral Leatham intended that *Charybdis* and the fleet destroyers *Grenville* and *Rocket* tackle the escort, leaving *Munsterland* to four Hunts. Presumably because he had never previously conducted a TUNNEL, nor sailed with any of the other ships, the cruiser's CO, Captain G.A.W. Voelker, RN, elected to steam in line ahead even though it was considered unwieldy for more than five ships. Cohesion was not helped by the fact that the second in command of the force, Commander W.J. Phipps, RN, had replaced Alers-Hankey in *Limbourne* just two days prior to the operation.[35] Moreover, he missed most of the operational briefing, and what he did observe did not impress. "Very hurried show run by Voelker," Phipps complained. "I hadn't the least idea of his intentions and could not get anything out of him. Very unsatisfactory conference."[36]

On the night of 22/23 October minesweepers provided close escort to *Munsterland* while Kohlauf led five torpedo boats on a distant screen to seaward. Again Kohlauf received good warning of the approach of the enemy, this time from shore-based radar confirmed by a sharp-eared hydrophone operator in *T-25;* again he had the advantage of light with a rising moon to seaward; and again he manoeuvred his force into an advantageous tactical position. Radar reports suggested that there might be two groups of enemy ships, so after sighting the larger *Charybdis* against the lightened horizon at 0146, Kohlauf, at the head of the line in *T-23*, ordered torpedoes to be fired, then wheeled into the murk towards the coast.[37]

Proceeding westward, Voelker also had ample notice that the enemy was about (see map, page 197). The Hunts *Talybont* and *Wensleydale* warned they had detected R/T transmissions that indicated an enemy force was manoeuvring close-by but he apparently did not appreciate the value of this information. At 0130 *Charybdis* got a radar contact ahead at 14,000 yards, but Voelker did not immediately relay that information, and because their Type 271 was masked ahead none of the six destroyers picked it up themselves. Finally, according to Leatham, at 0135 "Charybdis signalled that she had a radar contact on a bearing 270° at 8800 yards. This signal was followed by another ordering the force to turn together to 280°, speed 18 knots, but it was not received by any of the destroyers except *Stevenstone*. Thus at 0130 the situation in our force was that *Charybdis* knew that there was an enemy force 7 miles ahead and closing, but did not know its composition; while *Limbourne* and *Talybont* knew that there was a force of 5 ships (probably destroyers by the [R/T] procedure) in close vicinity, but did not know where."[38] Not realizing he was entering a torpedo trap, Voelker held his course until 0145 when the range was down to 4,000 yards, then opened fire with starshell. At almost that precise instant *Charybdis*'s lookouts sighted a torpedo just before it exploded against the port side amidships. As more torpedoes surged through the British line, *Charybdis* was hit again and another found *Limbourne*. Both Voelker and Phipps were disabled – the latter was blown off his bridge by the blast – and the force fell into disarray. According to Lieutenant Commander Hill in *Grenville*, "for the next fifteen minutes the ship was manoeuvring to avoid collision." *Charybdis* went down with 481 officers and men, including Voelker, while the mortally wounded *Limbourne* had to be put out of her misery by friendly fire. Kohlauf escaped without a shot being fired in his direction.

In his war diary Kohlauf concluded, "fortune indeed smiled on us." That, and dollops of skill. Erich Bey, still *Führer der Zerstörer*, was bountiful in his praise. Fleet torpedo boats had proved to be excellent warships: "With their speed, handiness, and small silhouette they appear to be suitable as sleuth hounds in our coastal routes in the Western Region, where they can strike with surprise and destructive force because of their strong torpedo armament. Subsequently they can open out from the enemy making use their superior speed and their outstanding handiness in connection with their excellent sea-keeping properties which have been proven in the Bay of Biscay." Equally impressive were their commanding officers who, "as well as their trusty and courageous ships' companies have achieved a further success by light surface forces that is unique in naval history":

> Despite changes of experienced and proven officers and ratings which had become necessary recently, the Flotilla was able to maintain its unity in its greatest but also doubtless its most difficult and dangerous moments. Luck combined with sound reasoning enabled the Commander of the *4. Torpedobootflotille*, *Korvettenkapitän* Kohlauf, to strike the British with destructive force in the span of a few decisive seconds. The Knights Cross of the Iron Cross is a tangible expression of gratitude and respect with which this outstanding officer is held and recognition for the manner in which he has proven himself repeatedly in the face of the enemy and also for the men of his flotilla.[39]

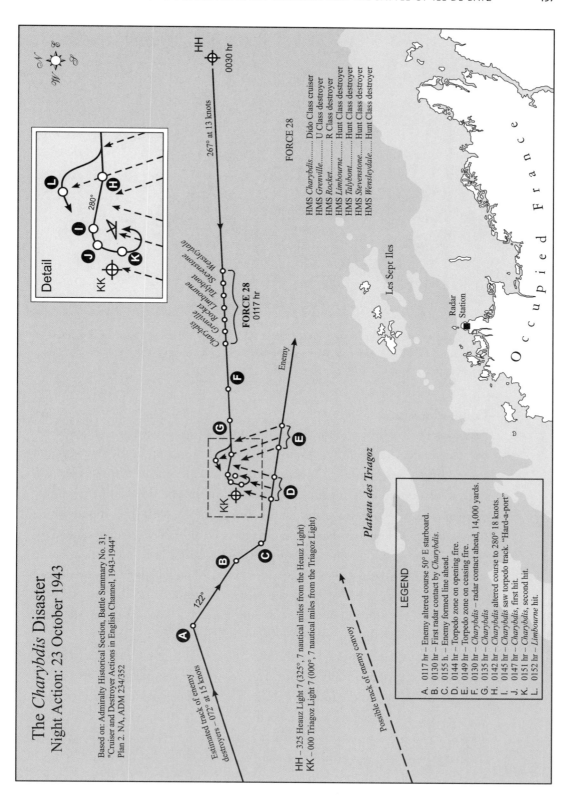

The *Charybdis* Disaster
Night Action: 23 October 1943

Based on: Admiralty Historical Section, Battle Summary No. 31,
"Cruiser and Destroyer Actions in English Channel, 1943–1944"
Plan 2. NA, ADM 234/352

Detail

FORCE 28

HMS *Charybdis*....... Dido Class cruiser
HMS *Grenville*....... U Class destroyer
HMS *Rocket*....... R Class destroyer
HMS *Limbourne*....... Hunt Class destroyer
HMS *Talybont*....... Hunt Class destroyer
HMS *Stevenstone*....... Hunt Class destroyer
HMS *Wensleydale*....... Hunt Class destroyer

Les Sept Iles

Occupied France

Radar
Station

Plateau des Triagoz

Estimated track of enemy
destroyers—072° at 15 knots

HH – 325 Heauz Light 7 (325°, 7 nautical miles from the Heauz Light)
KK – 000 Triagoz Light 7 (000°, 7 nautical miles from the Triagoz Light)

Possible track of enemy convoy

LEGEND

A. 0117 hr – Enemy altered course 50° E starboard.
B. 0130 hr – First radar contact by *Charybdis*.
C. 0155 h. – Enemy formed line ahead.
D. 0144 hr – Torpedo zone on opening fire.
E. 0149 hr – Torpedo zone on ceasing fire.
F. 0130 hr – *Charybdis* – radar contact ahead, 14,000 yards.
G. 0135 hr – *Charybdis*
H. 0142 hr – *Charybdis* altered course to 280° 18 knots.
I. 0145 hr – *Charybdis* saw torpedo track. "Hard-a-port"
J. 0147 hr – *Charybdis*, first hit.
K. 0151 hr – *Charybdis*, second hit.
L. 0152 hr – *Limbourne* hit.

HMS *Charybdis*, lost with a massive loss of life on the night of 23 October 1943. The defeat forced British naval leaders to take a cold, hard look at their night fighting capability in the Channel. (Imperial War Museum, FL-22618)

Countermeasures

The *Charybdis* calamity marked the turning point in the struggle for night supremacy in the Chops of the Channel. Analysis of the action from Plymouth and the Admiralty was scathing. Lack of training, inexperience and poor cohesion were all offered up as factors in the defeat, but in the end the Admiralty pinned the blame on Voelker. Rear-Admiral E.P. Brind, the ACNS(H), pronounced the final verdict: "Although it may be that the force was ill assorted and ill trained together, I do not consider the loss of the two ships can be attributed altogether to this cause. The loss was due in large part to a lapse of judgment by an experienced and very well thought of officer – the captain of CHARYBDIS. To continue on a steady course towards unknown vessels almost head on was dangerous, particularly in view of the unfavourable conditions of light – a rising moon."[40]

Plymouth Command and the Admiralty agreed on one thing, the situation had to improve dramatically if they were to achieve the degree of sea control necessary to safeguard the invasion on the horizon. Since taking over Plymouth in August 1943, Leatham had pressed the Admiralty for modern fleet destroyers to form a homogeneous strike force but had been rebuffed on the grounds they were more urgently required elsewhere.[41] With control of the Mediterranean largely won, the Russian convoys represented the main drain on destroyers. Essentially fleet operations with battleship and cruiser covering forces, the convoys sailed with sometimes more than a dozen fleet destroyers to counter the capital ships and destroyers the Germans had deployed in the north. Two events changed the strategic picture. In September 1943 British midget submarines damaged the battleship *Tirpitz* in her anchorage in Kaa Fjord, knocking her out of the war for six months. Then, on Boxing Day 1943, the Home Fleet sank the battle cruiser *Scharnhorst* in the Battle of North Cape. With the elimination of German capital ship strength in the north, fleet destroyer assets could be shuffled to other theatres.[42]

Discussion now turned to the best type of destroyer for the Plymouth strike force. Leatham wanted Tribal-class destroyers but staff officers at the Admiralty preferred the "J" or Jervis class for the role. The distinction was important and foreshadowed a debate between Plymouth and London over weaponry and tactics that would continue as long as the flotilla remained in being. In requesting Tribals Leatham expressed a clear preference for the gun. The Tribals had been designed in the mid-1930s from a desire for more powerful destroyers that could fulfil some of the requirements of cruisers. The standard inter-war types – the "A" to "I" – had a balanced armament of four 4.7-inch guns and eight 21-inch torpedo tubes. Tribals boasted twice the gun power with eight 4-7s in new twin turrets and half the torpedo armament in one quadruple mounting. Larger then previous designs they were informally dubbed "pocket cruisers" on account of the potent gun armament packed into a destroyer hull.[43] The Js, and their sisters the Ks and Ns, followed the Tribals on the design board and marked the return to

a more balanced design. Armed with three twin 4.7 turrets and ten 21-inch torpedo tubes in pentad mounts, the Jervises were slightly smaller than the

Admiral Sir Ralph Leatham while Deputy Governor of Malta in 1953. After the *Charybdis* disaster, Leatham pressed for a flotilla of powerful Tribal-class destroyers at Plymouth and then moulded them into a superb night fighting force. (Imperial War Museum, A-7229)

Tribals and had a lower silhouette. The Admiralty thought their torpedo strength – the most powerful of any British destroyer – made them perfect for the Channel, but Leatham won out. In early January 1944 the Admiralty transferred five Tribals, the British *Ashanti* and the Canadian *Athabaskan, Haida, Huron* and *Iroquois,* from the Home Fleet to Plymouth Command and designated them the 10th Destroyer Flotilla.[44]

Special ships to the RN, the Tribals were pure deity to the Canadian naval establishment, who envisioned them as the foundation of a stronger, more secure service. Since its birth in 1910, the RCN had experienced a hand-to-mouth existence, receiving little popular or budgetary support. After witnessing how quickly its small ship, anti-submarine navy was dismantled after the First World War, Chief of the Naval Staff, Rear Admiral Percy Nelles, determined to build a force substantial enough to withstand future political or budgetary threats. According to navy scuttlebutt, when Nelles saw a photograph of the first RN Tribal, he exclaimed, "I want those for my navy!" Nelles thought the powerful Tribals perfectly suited to the RCN's pre-war coastal defence role, which envisioned armed merchant cruisers as the major threat. More importantly, the government's stake in the big destroyers would be such that they could not easily be cast aside. The outbreak of war added weight to his arguments, and in its early wartime expansion programs the government ordered four Tribals to be built in Britain, and later added four more from Canadian yards.[45] The British thought the RCN should concentrate on less sophisticated destroyers but when confronted by that argument from one senior RN officer Nelles stubbornly held his course, emphasizing he was working to achieve two goals:

Object I: To win the war.
Object II: Before the finish of the war to have a number of Tribal destroyers in the Royal Canadian Navy, fully manned by Canadians. These he feels could not be wiped off the slate by whatever Canadian Government is then in power, as might be the case if only worn out older Canadian Destroyers existed.[46]

The RN, who wanted the four new Tribals for themselves, made other attempts to get the RCN to lower their sights but Nelles would not budge.

The RCN establishment also wanted to use the Tribals to kindle a fighting tradition for the Canadian navy. There had been no glory in the First World War, the RCN's first conflict; indeed in the only action with the enemy the captain of an anti-submarine trawler chose discretion over valour when he encountered a U-cruiser on the surface off Newfoundland.[47] Canadian sailors had brought

The sleek Tribals had the most powerful gun armament of any British destroyer of their generation, and the four 4.7-inch guns on the fo'c'sle, showed to good effect by HMCS *Iroquois*,
proved well suited to the new night fighting tactics utilized by the 10th Destroyer Flotilla.
(DND Canada, UK-012)

credit to the navy in the first years of the Second World War, but it was mainly on
convoy defence, which was not the image that naval leaders wished to cultivate.
They desired a more traditional fighting heritage, and thought the best way to
achieve that was to ensure their most powerful warships participated in offensive
operations overseas.[48]

When the first RCN Tribals neared commissioning in late 1942, Prime Minister Mackenzie King mentioned he might prefer the cream of the RCN to be
deployed for home defence on the West Coast or for convoy operations on the
North Atlantic. Naval leaders were aghast, and the Director of Plans in Ottawa,
Captain Harry DeWolf, wrote a primer on destroyer policy in an attempt to

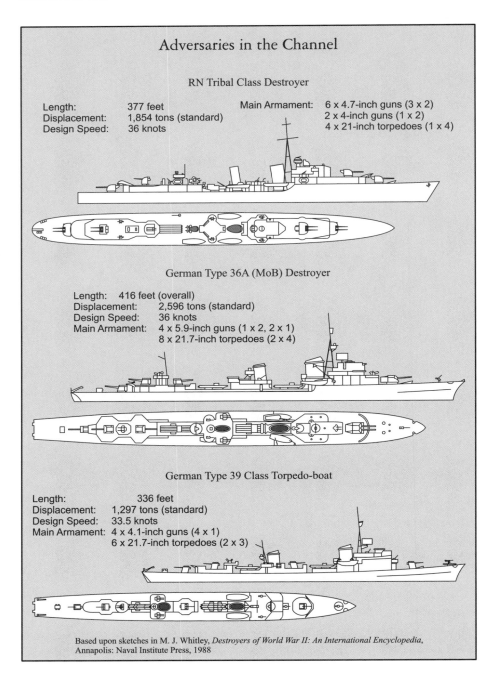

Adversaries in the Channel

RN Tribal Class Destroyer

Length:	377 feet	Main Armament:	6 x 4.7-inch guns (3 x 2)
Displacement:	1,854 tons (standard)		2 x 4-inch guns (1 x 2)
Design Speed:	36 knots		4 x 21-inch torpedoes (1 x 4)

German Type 36A (MoB) Destroyer

Length:	416 feet (overall)
Displacement:	2,596 tons (standard)
Design Speed:	36 knots
Main Armament:	4 x 5.9-inch guns (1 x 2, 2 x 1)
	8 x 21.7-inch torpedoes (2 x 4)

German Type 39 Class Torpedo-boat

Length:	336 feet
Displacement:	1,297 tons (standard)
Design Speed:	33.5 knots
Main Armament:	4 x 4.1-inch guns (4 x 1)
	6 x 21.7-inch torpedoes (2 x 3)

Based upon sketches in M. J. Whitley, *Destroyers of World War II: An International Encyclopedia*, Annapolis: Naval Institute Press, 1988.

check such heresy. "The Tribal is essentially a fighting Destroyer," he advised. "It is the largest and most heavily armed of all Fleet destroyers. It is especially powerful in surface and anti-aircraft gunnery. There are few such ships to meet the heavy demands in the fighting theatres of war, and every unit must be employed to best advantage. It would be most uneconomical to use a Tribal in North At-

Haida manoeuvring with another Tribal and the cruiser HMS *Black Prince* in February 1944. Her Type 271 surface warning radar juts up from the after canopy between the searchlight and the mainmast, and one can see how the forward superstructure blocked its beam when searching ahead. (DND Canada, R-1044)

lantic convoy escort when its guns are so urgently required elsewhere." Moreover, the Pacific had stabilized and "Under present conditions, to tie up Tribals on our West Coast would be even more wasteful than employing them in North Atlantic escort work." The proper course was to place the ships under British operational control: "where the need is greatest can best be decided by the Admiralty and it is strongly recommended that the Canadian Tribals be placed at their disposal without restriction. Only in this way can they contribute to the general cause."[49]

DeWolf's forthrightness won out – coincidently much to his ultimate advantage – and the Tribals were deployed to "the fighting theatres of war." By September 1943 all four served with the Home Fleet, the RN's main strike force, and after a hectic autumn on the Russian convoys they were transferred to Plymouth Command in early 1944.[50] *Iroquois* returned to Canada in February for refit, leaving *Athabaskan*, *Haida*, *Huron* and their RN sisters *Tartar* and *Ashanti* as the core of the new 10th Destroyer Flotilla. The flotilla became operational on 9 February 1944, and *Tartar*'s captain, Commander St. John Tyrwhitt, RN, became D10.

Moulding an effective force

With ships finally in hand, Leatham took steps to mould them into an effective strike force, which included ensuring they had the appropriate equipment, proper training and the opportunity to build up experience. The first step was to upgrade their radar. A myth long existed in Canadian naval circles that as the RCN's elite ships the Tribals received the best kit and crews. As with many

myths this was far from reality. As just one example, *Haida* and *Huron* were not
completed with modern search radar. Throughout the autumn of 1943 they were
forced to rely solely upon Type 291, the obsolescent metric set that could be easily
monitored by the enemy. This made life particularly difficult for Commander
Harry DeWolf, who in August 1943 had moved from the corridors of power to
the bridge of *Haida*. On three Russian convoys, including the one attacked by
the *Scharnhorst* during the Battle of North Cape, he led the destroyer division
responsible for protecting the rear section of the convoy in the perpetual Arctic
night but could not utilize 291 for fear of revealing the convoy's position and so
was essentially blind. Soon after arriving at Plymouth *Haida* and *Huron* were
finally fitted with modern Type 271Q but it was mounted in the searchlight position amidships so had the same 54° blind spot ahead that had plagued destroyers
in the October 1943 actions.

The Admiralty had recognized the problems caused by the *ad hoc* fitting of
radar in destroyers and in mid-1943 began to replace tripod masts with stronger lattice structures that could support centimetric radar aerials enabling clear
sweeping ahead. The conversion involved a major refit but *Tartar, Ashanti* and
Athabaskan were among the first destroyers to feature the new masts.[51] They also
received new Type 272 or 276 surface search radar. Besides improved performance – destroyers could be detected out to about 12 miles – the sets had power
rotation, which allowed consistent scanning, and echoes were displayed on a
Plan Position Indicator (PPI) instead of an A-scan, which enabled operators to
continuously monitor the positions of all contacts.

Other equipment contributed to the flotilla's mastery of the night. Navigation, notoriously difficult in the Channel, was simplified by the radio navigation aid QH-3. A variation of the RAF's GEE system, QH-3 took cross bearings
from a chain of shore-based radio transmitters. According to *Ashanti*'s navigator,
Lieutenant John Watkins, RNVR, QH allowed him "to pinpoint their position
virtually at the touch of a button."[52] Each destroyer also received the monitor-
ing device QD, popularly known as HEADACHE, which allowed specially embarked German-speaking personnel to eavesdrop on enemy radio communications. HEADACHE created an obvious advantage but it had to be treated with care.
DeWolf recalled one occasion when intercepts thought to be from destroyers
about to launch a torpedo attack turned out to be chatter among minesweepers
forming up to enter harbour dozens of miles away.[53] The ships were also fitted
with the latest IFF (Identification Friend or Foe) gear to help ascertain the identity of radar contacts. Finally, for improved gunnery, the Tribals were fitted with
target indication gear to improve radar-controlled blind fire and received larger

allocations of starshell for illumination, 4.7-inch tracer to improve spotting and flashless cordite.[54]

Group training was also critical to attaining success. Within days of their arrival at Plymouth, Leatham, perhaps to make a point, sent *Haida*, *Athabaskan*, *Iroquois* and *Ashanti* on a sweep with three Hunts. Predictably, problems arose with command and control, and in his report DeWolf "strongly recommended that the Plymouth forces exercise night encounters."[55] Well aware of that weakness, Leatham laid on an extensive sea training program consisting of high-speed formation steaming in which ships manoeuvred at close quarters, and night encounters, where they took turns launching attacks on one another. Both evolutions helped the various departments in a ship become accustomed to the demands of night operations and boosted confidence. It is difficult to enumerate the exact number of exercises the flotilla carried out since the logbooks of most destroyers were destroyed after the war; however, the cruiser HMS *Bellona* often led the training and her deck log shows that she conducted eight night exercises with the flotilla. This is quite a high number given normal operational, maintenance and training requirements, and the destroyers likely did more by themselves. Certainly, the captain of the cruiser *Black Prince*, Captain Dennis Lees,

HMS *Tartar* soon after joining the 10th Destroyer Flotilla. Her Type 276 surface warning radar is located above the crosstree on the new lattice foremast, which gives it unimpeded search ahead. Type 291 is at the head of the foremast, while the HF/DF birdcage sits atop the mainmast aft. (Imperial War Museum, FL-19719)

RN, who also exercised with the flotilla, recalled "days, weeks and months of dull, solid, slogging working up practices."[56]

Commanding officers also had access to a number of publications to coach them in night-fighting doctrine and tactics. Besides the doctrinal bibles *The Fighting Instructions* and *Destroyer Fighting Instructions*, *Admiralty Fleet Orders* contained a wealth of technical information while official handbooks like the *Gunnery Review* and the *Guardbook for Fighting Experience* distilled lessons from previous actions. Interestingly, the intense night fighting between the United States and Imperial Japanese navies in the South Pacific seems to have had little impact. Files on the actions filtered through the Admiralty and some were written up in *Fighting Experience* but there appears to have been little real effort to learn from the USN experience, and in the various analyses on the Plymouth night actions there is only one reference to the fighting in the Pacific. This seems strange considering they were grappling with similar issues in terms of tactical doctrine and the use of radar, but the USN likewise paid little attention to the RN experience; in this area at least the two navies literally remained worlds apart.[57]

Training can only achieve so much and the flotilla carried out several offensive sweeps off the French coast "to see the elephant." The enemy was not met but the operations provided valuable experience and a focus for training. They also revealed the same weaknesses that had plagued earlier operations. During a TUNNEL on 25 February 1944 confusion arose when inexperienced radar operators mistook the approach of friendly aircraft for an attack by E-boats. Then, while off the French coast the flotilla opened fire on what were thought to be German destroyers, but turned out to be a group of small islands. They left haystacks burning in their wake.[58] More serious problems arose on the night of 1/2 March when Tyrwhitt twice ordered his ships to turn away from what HEADACHE indicated were probably torpedo attacks from destroyers. The force returned to Plymouth after making only a token search for the enemy when a more determined attempt might have turned up five torpedo boats commanded by Franz Kohlauf that had stalked the Tribals and indeed launched torpedoes.[59] As a result of theses failures, Leatham replaced D10.

Officially, Commander Tyrwhitt was sent on a well-deserved rest after three stressful years of destroyer command in the Mediterranean. However, the tone of Leatham's report and the timing of the change – just a month after *Tartar* had emerged from a long refit that would have given Tyrwhitt his rest – suggest that the flotilla's performance was the deciding factor. Certainly, the change sent a strong message.[60] In his report Leatham wrote, "depending on the circumstances, which only a commanding officer can gauge, evasive action may well have to be taken, but this does not mean that the main objective is to be missed as a result."[61] The "main objective," of course, was to engage the enemy. The captains of the flotilla took his criticism to heart, and it contributed to the aggressiveness that coloured the flotilla's subsequent operations. "My feeling at the time," DeWolf recalled, "was that we had failed to go after some ships, some echoes, so from then on in the back of my mind was if we ever get an echo, we're going after it. So if we found one we did; that was just natural fear of criticism."[62]

The determination evident in DeWolf's words was one of the traits that made him an outstanding naval officer. A graduate of the Royal Naval College of Can-

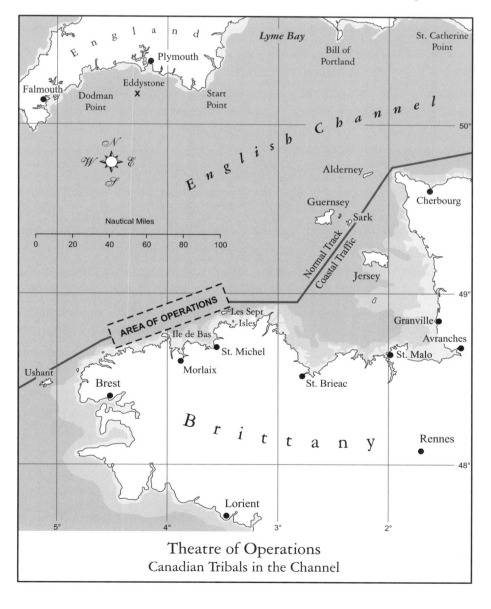

Theatre of Operations
Canadian Tribals in the Channel

ada, DeWolf's career had been typical, involving big ship training with the RN and plenty of sea time in Canadian destroyers where he earned a reputation as a brilliant, confident ship handler. Shortly after the outbreak of war DeWolf was appointed captain of the destroyer *St. Laurent,* where he distinguished himself in convoy operations out of Halifax and then in European waters during the dark days of 1940. In particular, DeWolf ordered the first shots in anger in RCN history when *St. Laurent* engaged German artillery while covering the evacuation of British troops from France, and later rescued some 800 survivors from the torpedoed liner *Arandora Star.* To his regret, DeWolf's sea time was cut short to serve ashore in headquarters in Halifax and, as we have seen, as Director of Plans in Ottawa. In August 1943 he was appointed CO of *Haida.* Quiet and reserved, DeWolf was nonetheless a brilliant leader. Captain Pat Nixon, RCN, who had been one of his watch keepers in *St. Laurent,* remembered "that he was not given to making inspiring speeches or pep talks to his men; in fact he did not need to because his professional skill was always so apparent that it earned him their automatic respect."[63] That skill and determination, combined with the fortune of opportunity that blesses many great men, continually placed DeWolf at the forefront of destroyer night actions during the spring of 1944.

"The enemy doubtless knew…"

In the aftermath of his drubbing of the *Charybdis* force, Franz Kohlauf expressed concern about the security surrounding his flotilla's activities. Having encountered enemy forces on successive missions in October 1943, Kohlauf complained, "The enemy doubtless knew about the operation in detail. Secrecy in Brest was very poor … and establishments not involved knew about its commencement. In order to achieve surprise, which is of great importance for our Fleet Torpedo Boats which will always be the inferior force in an engagement, the departure for such an operation must be organized differently the next time."[64]

Kohlauf was correct in his hunch that the Allies knew of his activities in advance, but the cause was not loose tongues in Brest. Since 1941, cryptographers at Bletchley Park had been able to decipher the *Kriegsmarine's* "Home Waters" Enigma, which controlled surface ship movements. By 1943, German destroyers in the western Channel operated primarily in the defence of convoys, and as F.H. Hinsley observed, Ultra "provided a reliable picture of the enemy's routines – showing which swept channels the convoys normally used, where they spent the night, the times at which they made and left harbour, and when and where they met their escorts – and thus a means of detecting those divergences from routine which indicated that especially important convoys were to be expected."[65] From this information analysts could easily unravel escort movements, which gave the Allies a critical advantage. It did not always mean that interception was guaranteed, nor that an operation would be successful – the October 1943 actions were proof of that – but it did mean Admiral Leatham, who was indoctrinated into Ultra, could position his ships to take best advantage of this invaluable intelligence.

Such a situation occurred in the last week of April 1944. Signals intelligence confirmed by aerial reconnaissance indicated that the Germans planned to send a convoy westward from St Malo to Brest. Leatham ordered a series of TUNNELS, the first of which was carried out on the night of 24/25 April. The cruiser *Black Prince* with three Tribals obtained radar contact with the convoy, but operators could only watch their radar displays in dismay as the enemy, warned of their presence by coastal radar, slipped into one of the many harbours along the coast.[66] Despite receiving similar warning the next night, the Germans would not escape.

On the night of 25/26 April, Force 26 – *Black Prince* (SO, Captain Dennis Lees), *Ashanti*, *Athabaskan*, *Haida* and *Huron* – were ordered to head due south from Plymouth until 10 miles off the Brittany coast, then to sweep eastward to confront the westbound convoy.[67] At 0130, before they began their eastward leg, coastal radar warned Kohlauf, at sea in *T-29* with *T-24* and *T-27*, of their position; however, he assumed they would continue south to engage the convoy to the west of his destroyers so he held his course. Coastal radar failed to inform Kohlauf of Force 26's eastward turn until 0201, by which time they were just 11 miles ahead on a collision course and had the torpedo boats on radar (see map, page 211). Presumably to draw the Allies away from the convoy, rather than closing the range to fire torpedoes as he had done so successfully against *Charybdis*, Kohlauf cranked on full speed and reversed course. *Vizeadmiral* Leo Kreisch, the new *FdZ*, criticized this decision, noting, "a headlong attack launched without changing the westward course might have brought a hard fought for success. In situations that appear hopeless, the boldest decision leads in most cases to success."[68]

Kohlauf's "scoot" initiated a chase that tested new tactics adopted by the 10th Destroyer Flotilla. Instead of line ahead, Force 26 spread out in open formation with *Black Prince* in the centre and a sub-division of two Tribals deployed 3,000 yards, 40 degrees off each bow. Based on the harsh lessons of the *Charybdis* action, and practical because the cruiser and half the destroyers had PPIs, the formation limited the cruiser's exposure to torpedoes while enabling her to illuminate targets for the destroyers. This gave the Tribals more freedom of movement and allowed them to use all forward 4.7s against the enemy.

At 0220 *Black Prince* began illuminating at 13,600 yards, and the Tribals opened fire seven minutes later. Although the range was long and visibility was impaired by German smoke, Type 285 provided accurate ranges, and because the torpedo boats did not have flashless cordite their return fire provided an easy aiming point. *T-27*, the rearmost German destroyer, was immediately straddled. *Korvettenkapitän* Werner Gotzmann could not return fire effectively because Force 26 lay beyond starshell range. At around 0236 *T-27* sustained hits on the stern that ignited fires and reduced speed to 12 knots. Kohlauf ordered her to withdraw southwards towards the coast. This move was detected but ignored by the starboard subdivision of Tribals – *Haida* and *Athabaskan* – allowing Gotzmann to fire his full outfit of torpedoes as Force 26 swept by to the north. All six missed but they forced *Black Prince* to break off to the north, and the cruiser played no further role in the battle. *T-27* eventually escaped into Morlaix with 11 dead and eight wounded.[69]

With Lees out of the action, command devolved on DeWolf. The Tribals gradually overhauled the torpedo boats, and at 0320 salvos from *Haida* and *Athabaskan* hit *T-29*, severing steam lines and reducing speed. Ten minutes later a hit on the bridge mortally wounded Kohlauf, and then the ship was smothered by shells when it emerged from smoke about 4,000 yards off *Haida*'s starboard bow. By 0335 *T-29* lay stopped with flames engulfing her upper works. DeWolf moved closer still before prompting his gunnery officer "There you are. Give him the works."[70]

Like all German warships, *T-29* demonstrated the capacity to withstand tremendous punishment. For over 45 minutes *Haida* and *Athabaskan*, joined by *Ashanti* and *Huron* after they abandoned their search for *T-24*, pounded her at close range. With Leatham's admonishments fresh in his memory, DeWolf did not want to leave any chance of survival. *Athabaskan*'s CO, Lieutenant Commander John Stubbs, described a scene of some confusion: "Although by this time burning fiercely, the Elbing maintained a constant fire of close-range weapons as we were circling her. *Huron* and *Ashanti* joined and there was a certain amount of dangerous cross fire although this was unavoidable. Fighting lights had to be switched on to avoid collision."[71] On *T-29* the situation was grim. War correspondent *Leutnant* Gasde, who was the only surviving officer, recalled, "A large portion of the ship's company gathered on the after portion of the bridge after the fire had also spread in the after portion of the ship. This concentration was highly dangerous in the heavy gunfire that was continuing. The reason that very many men were quite reluctant to abandon ship was often the fact that the destroyer closing on the starboard side was firing heavily at the ship's side and

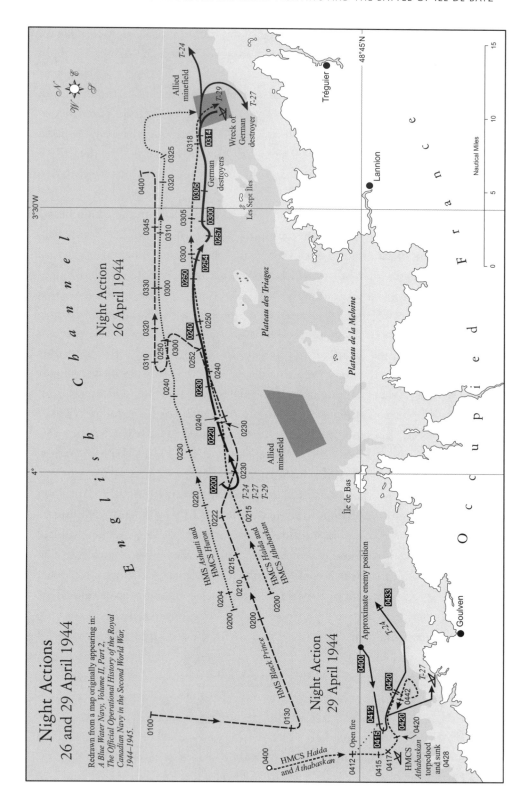

Night Actions
26 and 29 April 1944

Redrawn from a map originally appearing in:
*A Blue Water Navy: Volume II, Part 2,
The Official Operational History of the Royal
Canadian Navy in the Second World War,
1944–1945.*

Night Action
26 April 1944

Night Action
29 April 1944

water line with light weapons (4 cm tracers) and that another destroyer lay off 200-300 metres on the starboard beam and also commenced firing with light weapons."[72]

When this withering barrage failed to sink *T-29*, the Tribals tried to finish her off with torpedoes. Incredibly, as a result of failures in drill or miscalculation in deflection, all 16 torpedoes missed the immobile target. Finally, at 0422, after more of "the works," DeWolf signalled "Enemy has sunk." As the Tribals reformed for home, *Ashanti* collided with *Huron*, but both ships were able to make 25 knots.

Neither side had much opportunity to absorb lessons from the action before they met again. The damage from the collision between *Ashanti* and *Huron* kept both ships out of action until mid-May. With *Tartar* in refit, the operational load fell on *Haida* and *Athabaskan*, whose sailors had been scheduled for a much-needed rest. *Athabaskan*'s Leading Writer, Stuart Kettles, recalled in his diary that by the end of April "fatigue was no joke, for about in a week's time, we were able to get a total of about seven hours sleep, and at about this stage of the game, it was beginning to tell on the dispositions of the various lads."[73] After screening an invasion exercise the night following the action, *Haida* and *Athabaskan* finally got a night off. Matelots sought diversions in bombed-out Plymouth, while officers played poker in *Athabaskan*'s wardroom. Lieutenant Ray Phillips, *Haida*'s plotting officer, won ten quid but his winnings stayed in "Athabee." As the two ships slipped out of Plymouth next evening to cover a minelaying operation, Phillips probably wished he had his winnings in pocket.

The *4. Torpedobootflotille* also suffered lingering effects from the 26 April battle. Not least they mourned the death of their leader. In *T-27* Gotzmann mustered his sailors "to proudly mourn our flotilla commander. *Korvettenkapitän* Kohlauf with his superior ability and his considered, steady leadership was an officer who we followed with the utmost confidence in all situations." Gotzmann took over and led *T-27* and *T-24* out of St Malo for Brest and its superior repair yards. Due to the damage suffered two nights previously, Gotzmann instructed *T-24*'s CO, *Kapitänleutnant* Wilhelm Meentzen, "to head for the coast and avoid combat" if they encountered Allied warships.[74]

Ultra warned Plymouth of this activity but nature also lent a hand. Occasionally, super-refraction caused by abnormal climatic conditions enabled shore-based radar at Plymouth to track shipping moving along the Brittany coast 100 miles distant. Unfortunately for *T-27* and *T-29* this was one of those nights, and Plymouth tracked them intermittently as they headed westward. At 0307, Admiral Leatham ordered *Haida* and *Athabaskan*, patrolling in mid-Channel, to proceed southward at full speed. Like air controllers vectoring fighters onto a distant bogey, Plymouth guided the Tribals to a perfect interception, and at 0359 *Athabaskan* gained radar contact with the torpedo boats at 14 miles.

The Canadians were heading 170°, almost due south, while the enemy was steering 260° or almost due west (see map, page 211). As an Admiralty analysis concluded, "the ideal position for torpedo attack was developing automatically." DeWolf, however, rejected that option. "My first object," he explained in his action report, "was to prevent the enemy getting past to westward." With the 26 April action in mind the last thing he wanted was another long chase, and at initial contact the torpedo boats were just 20 miles from Ushant, uncomfortably close to the strong defence network at Brest. Moreover, to mount an effective torpedo attack DeWolf would have to close to about 5,000 yards, and by then he had to assume he would have been detected by German coastal radar if not by the torpedo boats themselves, opening the door for their escape if he turned on the parallel course to launch torpedoes. Weighing all this on *Haida's* bridge, DeWolf decided to close the enemy and engage with gunfire – "Straight on at 'em, I presume, sir?" Pugsley had prompted Mountbatten in November 1940.

For reasons the Germans never ascertained – they suspected, wrongly, Allied jamming – coastal radar provided no warning of the approach of the Tribals, and at 0412 sailors on *T-27* and *T-24* were taken completely by surprise when starshell burst overhead. Warned that friendly minesweepers might be in the area, Gotzmann immediately fired recognition flares. When a lookout erroneously warned of torpedo tracks, he first turned towards the enemy then reversed course towards the coast, both *T-27* and *T-24* firing torpedoes. The manoeuvre was marred by poor drill, and of 12 torpedoes only three from *T-24* were launched from the correct side. When DeWolf observed the turn away, he "altered towards to avoid torpedoes but limited the turn to thirty degrees to keep A arcs open." This allowed illumination of the enemy with the 4-inch guns in X turret, enabling the 4.7s of A and B mounts to fire HE, but DeWolf probably also wanted to overcome Type 271's blind spot ahead – with Type 285 being used for gunnery and Type 291 not switched on for fear of revealing their presence, he would have wanted Type 271 available to plot both enemy ships simultaneously. At 0417, before the turn was complete, a torpedo exploded against *Athabaskan's* hull and a large fire erupted aft. *Haida* laid a smokescreen to cover her sister and continued after the fleeing torpedo boats. Minutes later a large explosion marked *Athabaskan's* demise.

Haida concentrated on Gotzmann's *T-27* and her accurate shooting made quick work of the destroyer. Blind fire pierced the smoke that shielded *T-27*.

Two shells exploded on the water line, and then a series of hits burst against the superstructure. Losing speed, down by the bow, flames igniting ready-use ammunition and *Haida* blocking his escape, Gotzmann put his ship aground.[75] Once he realized his adversary was hard on the rocks, DeWolf briefly searched for *T-24*, which had disappeared eastward, before returning to where *Athabaskan* had last been seen afloat.

The torpedo that hit *Athabaskan* at 0417 caused such grievous damage that despite heroic efforts it was impossible for the crew to save their ship. Stuart Kettles recalled in his diary, "a diesel oil fire breaks out at the [torpedo tubes], enveloping after canopy and stern. Flames forty to fifty feet high. Pompom ammunition explodes in all directions." The explosion wrecked the aft 70-ton pump, letting the fire spread largely unchecked while damage control crews tried to wrestle another pump astern. Meanwhile, Lieutenant Commander Stubbs ordered his crew to stand by abandon ship stations, but at 0427, her deck crowded with sailors, *Athabaskan* blew up in a thunderous explosion.[76] When she arrived at the scene *Haida* found the sea spotted with stunned, shivering survivors, and in a courageous act so close to an enemy coast with daylight approaching, DeWolf stopped for 18 long minutes rescuing 42 sailors before reluctantly heading for Plymouth. German warships later picked up another 85, while six more made it home after an adventurous passage in *Haida*'s motor cutter. One hundred and twenty-eight died.[77]

Rethinking tactics

In four nights the 10th Destroyer Flotilla had sunk two destroyers against the loss of one of their own. Plymouth Command was pleased to finally achieve success, but staff officers at the Admiralty, who had a professional responsibility to be coldly objective, were less impressed. With the 25/26 April action they criticized the escape of two enemy destroyers, and deplored the fact that 16 torpedoes fired from close range failed to hit the motionless *T-29*. They were far more scathing on the second engagement, described as "An unsatisfactory action in which, quite unnecessarily, we swapped a TRIBAL for an ELBING." Criticism of DeWolf's tactical decision-making was especially harsh. Captain St John Cronyn, a senior tactical arbiter at the Admiralty, could barely contain himself: "HAIDA's para. 23 states 'No torpedoes fired'. This bald statement cannot be permitted to hide the actual situation. The two forces were closing at a 90° track angle on an approximately steady bearing. The most casual appreciation must have shown that the ideal position for torpedo attack was being rapidly approached, and conversely that if we were not ourselves going to fire we must be aware of the German attack. Our force, however, was so wrapped up in the picture of a gun action that it seems never to have contemplated the use of the torpedo." As a final shot, an exasperated Cronyn noted, "It is deplorable to realize that it was the German who sized up the situation."[78]

Cronyn had fired the opening salvo of a debate over tactics that lasted throughout the summer. Traditionalists at the Admiralty believed that the torpedo was the most effective weapon in surface action, and that tactical manoeuvring should always have the goal of using that weapon effectively. DTSD admitted in August that he had "long been perturbed at this state of affairs" and circulated a report summarizing torpedo failures in the eight night actions fought by Plymouth Command between 10 July 1943 and 30 June 1944. His superiors, ACNS(H), Rear-Admiral E.D.P. McCarthy, and DOD(H) Captain C.T.M. Pizey, agreed, and in September 1944 the Admiralty issued a formal complaint to Leatham: "Considering our superiority in all but one of these [eight] actions, the enemy losses are disappointing. The main reason appears to be a general lack of appreciation of the use of the torpedo. It would appear that on several suitable occasions torpedo fire was withheld for no apparent reason, and on some occasions when torpedoes were fired results were disappointing."[79]

These officers did not fully appreciate the degree to which radar had changed the nature of night surface actions. Due to ship- and land-based radar, it had to be assumed that the enemy would have warning of the approach of any force. Moreover, the German "shoot and scoot" tactic, coupled with the high speed of their destroyers, made it difficult to engage them with torpedoes let alone get in the first licks so important to success at night. This was a case where technological development had outpaced the formulation of doctrine, and sailors at the sharp end had to work out tactics while on the job. Innovation was required, and Commander Basil Jones, who succeeded Tyrwhitt as D10 on 15 March 1944, had both the imagination to arrive at a solution and the will to see to its implementation.

Jones was a seasoned destroyer officer, and, probably not coincidentally, a gunnery specialist.[80] He had commanded four destroyers over his career, and had experienced what could be termed a good war. Most germane to his future activities at Plymouth, in early 1943 Jones had led a number of night destroyer sweeps out of Malta, and in April 1943 had engaged two Italian destroyers, sinking both, for which he was he was awarded the DSO. Admiral Leatham was then Deputy Governor of Malta and probably came into contact with Jones; at the very least he would have known of his reputation. That made him a logical, and likely handpicked, successor to Tyrwhitt.

Although Jones did not actually lead the 10th DF in action until June 1944, he had implemented new tactics. In this he undoubtedly had the support of both Admiral Leatham and the Captain (D) Plymouth, Acting Captain Reginald Morice, RN, who had presided over destroyer operations at Plymouth since 1940. In defence of the accusation from the Admiralty that his flotilla had overlooked the torpedo, Jones wrote a succinct rebuttal that explained the tactical thinking employed by his flotilla throughout the summer of 1944. "The problem presented to the 10th DF," he wrote, "was, in the main, the destruction of a faster enemy who turned and fled while firing a greater number of torpedoes than we carried ourselves. The enemy's approved policy of turning to run gave him a position or opportunity of torpedo advantage." Radar-controlled torpedo firing had been tried but had proved ineffective in such situations. Jones therefore decided to rely upon the strength of his flotilla's radar-controlled gunnery. "The four-gun forward armament of the 10th DF gave us a gun advantage over the enemy when advancing against his retreat. The fact that the enemy ships were faster than us made any delay in turning towards him generally unacceptable. The positively Elizabethan method of projecting torpedoes at right angles to own ship's fore and aft line is a cause of delay. Our policy was not to turn away nor intend to waste any time at all in closing the enemy." Thus, guns were the best way to engage first effectively. "This attitude," he emphasized, "was not due to lack of appreciation of the use of the torpedo, which is undeniably a more effective weapon than the gun *if it can be applied*."[81]

To gain maximum gunnery advantage, the 10th DF had "to press on into the enemy during his turn away." Although line ahead made for easier station-keeping, Jones thought it unsuitable for the head-on encounters prevalent in the Channel because ships were prevented from entering action together and the destroyers at the head of the formation screened the radar of those behind. Instead:

> It was desirable that all destroyers should have their forecastle guns bearing, their Radar unimpeded ahead, and ships capable of individual action to comb enemy torpedoes. Only a reasonably broad and shaken-out line of bearing formation could satisfy these conditions. It was realised that cruising at night for lengthy periods in such a formation was a strain as regards station keeping, although the P.P.I. removed much of the strain [for those who had it]. Accordingly Line Ahead for comfort, and Line of Bearing for action, was the order of the day.[82]

Although Jones did not participate in either engagement, the open formation utilized in the 25/26 April action and DeWolf's choice of guns over torpedoes three nights later demonstrate the extent to which his thinking had taken hold. At the Battle of Île de Batz, Jones, with the confidence that his tactics were sound, would grow bolder still.

Reinforcing the flotillas

The loss of *Athabaskan* reduced the flotilla to four Tribals but additional strength arrived throughout May in the form of two British and two Polish destroyers. HMS *Eskimo* was the only Tribal but like the other RN newcomer, the J-class HMS *Javelin*, had been recently refitted and had modern Type 276 search radar on a lattice mast with PPI. Both ships had four 4.7-inch on their fo'c'sles. The Polish destroyer ORP *Piorun* was an ex-RN N-class, nearly identical to *Javelin*, but with same radar suite – and the same blind spot ahead – as *Haida* and *Huron*. The final newcomer, the Polish *Blyskawica*, was unique. Built to a special design in the early 1930s, *Blyskawica* had escaped to England after the fall of Poland. Slightly larger than her flotilla mates, the 38-knot ship was reputed to be the fastest Allied destroyer in the theatre, and her armament included eight 4-inch guns – four in twin turrets on her fo'c'sle – and the same radar as *Haida*, *Huron* and *Piorun*.

The new commanding officers were all experienced seamen. *Blyskawica's* Commander Konrad Namiesniowski and *Piorun's* Lieutenant-Commander Tadeusz Gorazdowski had escaped to Britain at the outset of war and had commanded Polish destroyers in the Battle of the Atlantic and on the Arctic convoys.[83] *Eskimo's* Commander Errol Sinclair and *Javelin's* Peter North-Lewis, were both in their third wartime commands and had seen plenty of action in the Mediterranean and English Channel respectively; on the night of 22/23 October 1943, Lewis had been in the rear of the line in *Stevenstone* when *Charybdis* was torpedoed. But as we have seen, command experience did not necessarily translate into success, particularly in a situation where innovative tactics were being implemented. *Blyskawica*, the last of the new destroyers, only arrived at Plymouth two weeks before the invasion, which left no opportunity for the comprehensive training regimen that the original ships had profited from. To compensate for that, Jones decided to concentrate his four battle-hardened veterans – *Tartar*, *Ashanti*, *Haida* and *Huron* – in the 19th Division and put the newcomers in the 20th under Namiesniowski.[84]

This heaped much of the onus for success squarely upon the shoulders of the 19th Division's commanding officers. We know of Jones and DeWolf; *Huron's* Lieutenant Commander Herbert Rayner, RCN, was another solid professional. Reserved and spiritual, Rayner had followed DeWolf as CO of *St. Laurent* earlier in the war and had been awarded the DSC for "courage and enterprise in action

Lieutenant-Commander Herbert Rayner, Rear-Admiral H.E. Reid, Commander Harry DeWolf and Captain Harold Grant in Plymouth in June 1944. All would be future Chiefs of the Naval Staff. Rayner, DeWolf and Grant, who commanded the cruiser HMS *Enterprise*, had a hand in the destruction of six German destroyers. Reid witnessed the 9 June action from *Haida*'s bridge, and advised DeWolf to break off the chase of *Z-32*. (Library and Archives Canada, PA-191705)

against enemy submarines." Lieutenant Commander John Barnes, RN, universally known as "Jimmy," drove *Ashanti*. The Tribal was Barnes's first command but, according to her first lieutenant Terry Lewin, after some initial trepidation among her wardroom, he proved to be "a highly experienced destroyer officer and expert ship handler."[85] Those were prized qualities shared by Jones and DeWolf.

The problems associated with absorbing new ships into the 10th DF seem almost trivial in comparison to the challenges facing their opposition. In recent months the German destroyer force in the western Channel had been decimated by what historian M.J. Whitley described as "a steady attrition." In December 1943 ten destroyers and fleet torpedo boats engaged the British cruisers *Glasgow* and *Enterprise* – the latter commanded by Captain Harold Grant, RCN – in a long-running daylight battle in the Bay of Biscay, losing *Z-27*, *T-25* and *T-26*.[86] Then in April *T-29* and *T-27* had been lost, and *Z-37* was sidelined for major repairs after a collision. Other ships were hung up by maintenance delays, leaving only four destroyers of the *8.Zerstörerflotille* available on the western flank of the Channel.

The flotilla was a mixed bag. *Z-32* and *Z-24* were modern Type 36A destroyers, popularly known as Narviks. Displacing approximately 3,000 tons, capable of 38 knots and armed with five 5.9-inch guns and eight 21.7-inch torpedo tubes, Narviks were larger, faster and packed a heavier punch than any British destroyer. The smaller, slower *ZH-1* was the captured Dutch destroyer *Gerard Callenburgh*, armed with five 4.7-inch guns and eight 21.7-inch torpedo tubes, and was about the equivalent of an RN "A-I" class. To bolster its strength the *8.Zerstörerflotille* would be reinforced by the 10 DF's old adversary, *T-24*, which was smaller and slower than her mates.[87] All four destroyers possessed superb torpedo control gear, which made the torpedo their most potent weapon. Radar was still of 1940 vintage with limited search capability but the Narviks had modifications to their masts to overcome the 30° blind spot astern that afflicted *ZH-1* and *T-24*. The greatest collective weakness of the ships was their wide range of capabilities. If they were to operate effectively as a unit, they would have to conform to *T-24*'s performance, but that would deprive the Narviks of one of their greatest assets – speed.

Training was another deficiency. Since 1941, there had been a steady turn-over in destroyer personnel as experienced officers and ratings were transferred to the U-boat arm. Officer complements had especially been savaged, leaving few seasoned junior officers. This only increased the need for sea training but by 1944 that had become increasingly difficult to carry out since fuel was scarce and Allied air power made ships vulnerable to attack whenever they left harbour.[88] As a result, the four destroyers not only had green crews but they had never been to sea together. Senior leadership could only do so much to overcome these weaknesses. Since April 1944 the flotilla had been led by *Kapitän zur See* Baron Theodor von Bechtolsheim, the veteran destroyer officer who had commanded *Karl Galster* in numerous night operations during the first three years of the war. He had then served as chief of staff to the *Führer der Zerstörer* and was thus abreast of the latest trends in destroyer warfare. Of the four COs, all but *ZH-1*'s *Korvettenkapitän* Barkow had commanded their ships in action; *Kapitänleutnant* Meentzen in *T-24* had a unique edge in that he had twice survived battles with the 10th DF, and his ship had sunk *Athabaskan*.

Although it is difficult to pin down the precise state of morale in the flotilla, it

The Type 36A destroyer *Z-38* in RN livery after the war. As with her sisters *Z-32* and *Z-24*, high speed and a powerful armament of five 5.9-inch guns and eight torpedoes made the "Narviks" formidable adversaries. (DND Canada, PMR 92-707)

was undoubtedly fragile. No matter what the condition of their ships and sailors, von Bechtolsheim and his COs knew their opponents held a distinct numerical advantage. They were accustomed to that, but more disconcerting was the realization that the Allies had gone a long way in penetrating the cloak of darkness German destroyermen depended upon for success. *Kriegsmarine* leaders had suspected for some time that the Allies had blind fire capability and confirmation finally came with the sinking of *Athabaskan*. The Germans were well aware that the invasion was in the offing, and the capture of 85 sailors from a modern destroyer operating out of an invasion port represented a potential intelligence bonanza. *T-24* had picked up about 50 *Athabaskan* survivors and their demeanour impressed Meentzen. "The majority of the rescued make a very good impression. They are disciplined and reserved. They make very few statements. All that we can learn is that they come from a Canadian destroyer of the 'Tribal class' that had been sunk by a torpedo hit."[89] More ruthless interrogators, uninhibited by respect for fellow men of the sea, prised out valuable intelligence, particularly about Allied cryptographic procedures. They also elicited details of *Athabaskan's* gunnery radar, including effective ranges, the fact that it could discern shell splashes "to within 5-10 m," and that it provided accurate range and bearing data. "These facts," the staff reported, "clearly confirm the enemy's substantial superiority in the field of radar and in particular in blind fire." With evident bitterness, the report concluded, "It is most unfortunate that British gunnery, which learned how to shoot from the old Imperial Navy and copied its methods several decades ago, has now achieved such a significant lead because of technical superiority."[90]

Theodor von Bechtolsheim recognized the bleakness of the situation con-

A skilled, professional sailor with plenty of successful night fighting experience: *Kapitän zur See* Baron Theodor von Bechtolsheim while in command of the destroyer *Karl Galster* early in the war. (Courtesy Götz von Bechtolsheim)

fronting his destroyers. Despite having been in harm's way many times earlier in the war, it was only shortly after taking command of the *8.Zerstörerflotille* that he felt compelled to compose a farewell letter to his wife. He knew the odds were stacked against him and did not expect to survive.[91]

D-Day

When D-Day arrived von Bechtolsheim's sense of foreboding would have been even more deeply rooted had he been aware that within a few hours of the first assault troops crashing ashore in Normandy, NEPTUNE naval commanders knew of his intentions on an almost real-time basis. Attacks by Allied air forces and French partisans in the weeks leading up to the invasion had disrupted German communications networks in France, forcing them to use radio instead of se-cure landlines. As an example of how vulnerable this left them to code breakers, Group West's invasion alarm, broadcast at 0402B on D-Day, was intercepted, de-crypted at Bletchley Park, forwarded to the Operation Intelligence Centre at the Admiralty and then transmitted to naval commanders at 0435B, or just 33 min-utes after it was transmitted by *Admiral* Theodor Krancke.[92] Later that morning when Krancke ordered *Z-32*, *Z-24* and *ZH-1* north from the Gironde to join *T-24* at Brest, and then later broadcast the route and timetable they were to follow, the signals were decrypted and forwarded to Plymouth within hours.[93]

Despite this advantage, uncertainty plagued Allied decision-making. Initially, Leatham intended the 10th DF to attack the German destroyers in the Bay of Biscay before they reached Brest, but when he learned of a massive U-boat de-ployment into the Channel he changed his mind, deciding instead to rely upon a Coastal Command anti-shipping strike force. Leatham's apparent dilly-dallying angered the First Sea Lord, Admiral of the Fleet Sir Andrew Cunningham, im-patiently scrutinizing operations from the Admiralty war room. "C-in-C Ply-mouth, appears to be afflicted with infirmity of purpose," he groused to his diary. "A most mistaken sense of values for which we may pay dearly. I will give him snuff in the morning although it's little to do with me."[94]

Despite Cunningham's feistiness, it is hard to fault Leatham's thinking. On 6 June dozens of U-boats began to leave their Biscay bases for the western Chan-nel, and a destroyer force would be vulnerable to attack from that quarter. There was also a chance that they would be mistakenly engaged by the scores of Allied anti-submarine patrol aircraft blanketing the area; far better if the pilots flying the Liberators, Wellingtons, Sunderlands and Halifaxes knew with certainty that every contact could be considered hostile. But most importantly, the Coastal Command strike force that Leatham could call upon was a potent maritime weapon in itself. Flying Bristol Beaufighters armed with 25-lb armour-piercing rocket projectiles (RPs) and 20mm cannon, the strike wing comprising No. 404 RCAF and No. 144 RAF had ravaged German coastal shipping off Norway throughout late 1943 and early 1944. Experience had shown that RPs fired into the sea just short of a ship maintained enough momentum to pierce the hull before exploding below the waterline, causing potentially devastating damage to thin-plated vessels, including smaller warships like destroyers. In April 1944, the strike wing moved south to Davidstow Moor in Cornwall to help defend against enemy surface vessels attempting to attack the western flank of the invasion.[95] Calling upon this force instead of risking his destroyer strike force in the face of numerous U-boats was a sensible use of maritime assets by Leatham.

At 1820 on D-Day, 14 Beaufighters from No. 404 and 17 from 144 took off from Davidstow Moor. Wave hopping at about 150 feet, distracted by minesweepers and a U-boat, they found their objective off St Nazaire at 2030. The fighter bombers attacked out of the low sun, and exploding flak, gun flashes, rocket trails and wheeling aircraft crowded the horizon. In the face of heavy anti-aircraft fire the Beaufighters inflicted only minor damage, and one was shot down. Early the next morning, a smaller follow-up strike of six 404 and five 144 Beaufighters shot up the destroyers just before they put into Brest at the cost of one aircraft.[96] Overall, the results of the two attacks were not as decisive as Leatham had hoped. The two Narviks suffered punctured fuel bunkers and other hull and superstructure damage, with *Z-24* the worse for wear. *ZH-1* sustained only light damage. Repair parties crawled over the destroyers as soon as they put into Brest, welding patches and fitting additional light anti-aircraft guns scrounged from other ships. After a delay of about 36 hours they were ready to sail, their morale undoubtedly a little more shaky.[97]

Defending the invasion corridor

Since D-Day the 10th Destroyer Flotilla had carried out defensive patrols in mid-Channel, one division on station while the other replenished in Plymouth.[98] Invasion excitement initially ran high, and sailors huddled around radios listening to updates on the BBC. After three days of uneventful patrolling, boredom started to set in and they began to wonder if they would make any real contribution to NEPTUNE. They need not have worried. Intelligence analysts kept a close eye on Brest, and on 8 June Ultra decrypts informed them that the *8.Zerstörerflotille* were heading out that night to attack invasion shipping between Portsmouth and Normandy, intending to put into the shelter of Cherbourg at dawn. But more than their intent, Ultra also revealed their exact course and speed, which enabled

Leatham to concentrate the entire 10th DF – dubbed Force 26 – across the route he knew the enemy would take.[99]

At 1637 Leatham ordered Force 26 to reach position KK (49° 00' N, 3° 56' W) by 2145 and then to sweep westward at 20 knots to LL (48° 52' N, 4° 23' W), to patrol the 20-mile line between the two until 0400 9 June, at which time they were to return to Plymouth. Later that evening, probably at 2236, Leatham, informed by another decrypt, ordered Jones to shift his patrol line 5 miles northward and to extend his front by positioning the 20th Division 2 miles north of the 19th. Jones had no trouble achieving this disposition and at 0114 on 9 June the 19th Division was zigzagging on a base course of 255° at 20 knots in staggered line ahead. The 20th Division, led by Commander Namiesniowski in *Blyskawica*, bore 000° at 2 miles distance, which put it slightly abaft the 19th Division's starboard beam. The sky was overcast with intermittent rain, a light breeze blew out of the southwest and the sea was calm. Visibility was variable between 1 and 3 miles, with a rising moon.

At 0114, *Tartar*'s Type 276 detected a contact directly ahead bearing 241° at 10 miles (see map, page 226). At 0120, after allowing the plot to develop, Jones increased to 27 knots and two minutes later ordered Force 26 to execute a White turn to course 290° to spread his ships across the enemy's bearing, and then at 0125 ordered Blue 240°. In the 19th Division, each Tribal turned towards the enemy under the port quarter of the ship ahead so they ended up in a loose line of bearing, bow on to possible torpedo fire. Jones, Barnes, DeWolf and Rayner manoeuvred their destroyers with practised ease but the success was not repeated in the 20th Division. Although *Blyskawica*'s signal log recorded Jones's signals for the White and Blue turns, Namiesniowski later wrote after each entry, "**I have no rec**ollection of having had this signal reported to me," and complained of congestion on internal communications circuits. No matter what the cause, the friction that so often accompanied night action had begun to settle. Consequently, instead of eight destroyers heading towards the enemy in line-of-bearing, only the 19th Division was in the intended formation, while the 20th maintained line-ahead behind *Blyskawica*. This was not critical as the 20th Division still occupied an effective position but the problems foreshadowed a far more serious breakdown to come.

Having passed an uneasy passage from Brest, von Bechtolsheim was not surprised to encounter Allied warships. From 2200 radar detectors indicated as many as five aircraft were shadowing his ships, but it appears that only one, a Liberator from No. 547 RAF was actually in contact. Shortly before midnight von Bechtolsheim forecast dim prospects. "I cannot imagine," he recorded in his war diary, "that the British are going to allow us to complete our passage to Cherbourg unscathed. On the other hand, we can make life unpleasant for them in the Bay of the Seine. I expect to encounter enemy destroyers no later than reaching the Channel Islands." He did not have that long to wait. At around 0100, just after receiving reports of enemy radar transmissions, *Z-32* gained fleeting radar contact with surface contacts. Then, at 0125, Jones's Blue turn exposed the 19th Division to moonlight and von Bechtolsheim reported, "Shadows to port," then "Enemy in sight."[100]

At contact he immediately turned south to place his ships against the dark horizon and in the hope that the rocky coast might disrupt enemy radar. However, before *Z-32* completed the turn the quartermaster warned him that a minefield lay down that track so he hauled around to seaward, ordering his force to fire torpedoes as they wheeled. *Z-32*, *Z-24* and *ZH-1* were able to pick out clear targets and each launched four torpedoes at the charging 19th Division – the 20th Division was still undetected to the north. HEADACHE in the Allied ships now paid dividends. The German torpedo firing order – "Toni Dora" – was known from previous actions and HEADACHE operators warned their COs what was on the way. That forewarning, along with the flexibility of the line of bearing formation, allowed the 19th division to avoid the torpedoes.[101]

Jones sought a close-range "pell-mell battle," and that is precisely what unfolded after the 19th Division, followed closely by the 20th, opened fire. The German destroyers were in the midst of their turn to port crossing the bows of the charging 19th Division at about 3,500 yards in the order *Z-32*, *ZH-1*, *Z-24*, *T-24*. As the northernmost ship in the 19th, *Tartar* initially engaged *Z-32*, hit-

ORP *Blyskawica* anchored with her flotilla mates in Plymouth Sound. Although the Polish ship was reputed to be the fastest Allied destroyer in the European theatre, her speed was not a factor in the June action. (Library and Archives Canada, PA-180512)

ting her four times, but when she sped off northwards Jones left her for the 20th Division and joined *Ashanti* against *ZH-1* and then *Z-24*. *Haida* first engaged *Z-24*, then DeWolf, much to the chagrin of his gunnery officer, Lieutenant Murray Heslam, who had to cease fire to lay on a new target, joined *Huron* against *T-24*.[102] From this point the battle devolved into confusion and the best way to untangle the events of the next few hours is to follow each German destroyer after the initial clash.

Jones had let *Z-32* escape northward, confident the 20th Division would settle her fate. Instead the enemy benefited from the division's lack of night fighting experience. The encounter began well enough. Steaming in the order *Blyskawica*, *Eskimo*, *Javelin*, *Piorun*, at 0130 the division sighted *Z-32* directly ahead at about 5,000 yards, still illuminated by the 19th Division. If the 20th had been spread out in line-of-bearing, *Z-32* would have been in dire straits indeed; as it was the usual problems of line ahead materialized. The two sides exchanged fire and *Z-32* reported hits forward while *Blyskawika* and *Javelin* were straddled. Before the division could press home their advantage, *Blyskawica's* HEADACHE operator reported the "Toni Dora" torpedo firing order but rather than continuing towards *Z-32*, Namiesniowski laid smoke and hauled around to starboard. The rest of the division, thinking they were wheeling for a torpedo attack, followed with *Eskimo* and *Javelin* firing torpedoes as they turned. All missed, although their wakes were sighted in *Z-32*.

At this point whatever cohesion existed in Force 26 evaporated. Instead of pursuing *Z-32*, Namiesniowski continued northeast away from the action for 17 minutes. He gave no explanation as to why he withdrew, although it appears he was concerned about engaging Jones's division by mistake. Certainly the other ships were in the dark. *Javelin's* Lieutenant-Commander North-Lewis presumed that Jones had ordered Namiesniowski to clear the area so the 19th Division could mop up. In *Piorun*, which lost contact with her division mates during the turn to the northeast, Lieutenant-Commander Gorazdowski assumed that Namiesniowski would return to close the enemy, and displayed good initiative by turning back to the sound of the guns at 0142, or eight minutes before *Blyskawica* finally headed back.[103]

If nothing else the incident accentuates the value of experience and training. It had been a long-standing RN doctrine that it was imperative to avoid torpedoes by turning towards the enemy so as not to lose contact.[104] Although Namiesniowski was trained in a different navy, he would have been briefed on this tactic by Jones but for some reason did not follow through. Experience in night action would have demonstrated the necessity of this manoeuvre and training would

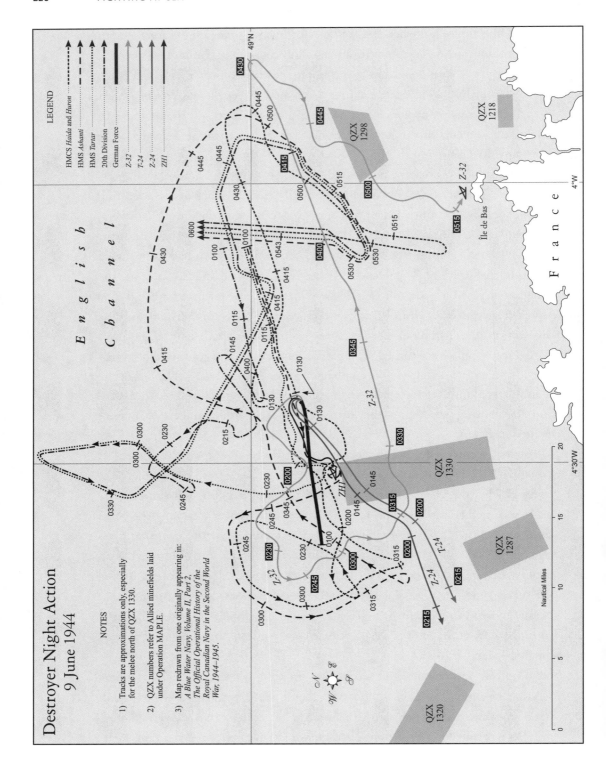

Destroyer Night Action
9 June 1944

NOTES

1) Tracks are approximations only, especially
   for the melee north of QZX 1330.

2) QZX numbers refer to Allied minefields laid
   under Operation MAPLE.

3) Map redrawn from one originally appearing in:
   *A Blue Water Navy, Volume II, Part 2,
   The Official Operational History of the
   Royal Canadian Navy in the Second World
   War, 1944–1945.*

LEGEND

HMCS *Haida* and *Huron* ┅┅┅
HMS *Ashanti* ┄┄┄
HMS *Tartar* ┈┈┈
20th Division ▬▬
German Force ———
Z-32
T-24
Z-24
ZH1

English Channel

France

49°N

4°W

Île de Bas

4°30′W

QZX 1218

QZX 1298

QZX 1330

QZX 1287

QZX 1320

Nautical Miles
0   5   10   15   20

have made it easier to accomplish. But since *Blyskawica* had only been with the flotilla for two weeks there was neither time nor opportunity for either. Leatham, Morice and Jones knew this, which casts doubt on the soundness of their decision to have Namiesniowski lead the 20th Division.

Pandemonium

Having escaped the clutches of the 20th Division, von Bechtolsheim found himself immersed in a close-quarters brawl that more resembled the age of sail than modern night combat. Heading west in an attempt to reform his force, at 0138 *Z-32* sighted *Tartar* at close range on her port quarter and quickly scored three hits on her bridge superstructure.[105] Hugh Meicklem, *Tartar's* navigator's yeoman, described the devastation wrought by the 5.9-inch shells:

> When the actual crashes came upon us the A.I.C. [Action Information Centre] was plunged in darkness and a brilliant flash pronounced the end of the Type 293 [sic] Radar set. Shrapnel rattled around in all directions and soon the small compartment filled with choking smoke. Pandemonium reigned for a few minutes on the bridge immediately above us, and from the wheelhouse adjoining our action station came the voice of the Coxswain shouting loudly, "Someone's been hit."… Curling smoke swirled everywhere, and the stench of blood was sickening.[106]

Commander Jones, slightly wounded and perhaps regretting the wisdom of seeking a pell-mell melee, reported:

> Hit 1. – Started fire before the foremost funnel.
> Hit 2. – Was an H.E. shell which burst above the Bridge causing casualties, killing the P.C.O. [Principal Control Officer] and both Torpedo Control Communication ratings, and wounding a number of others.
> Hit 3. – Went through the foremast wrecking all Radar and W/T aerial gear.

In the midst of this damage and confusion, individual acts of bravery, steeped by years of training and experience, helped lessen the impact of the damage and enabled *Tartar* to regain her fighting posture. Splinters rattling around the forward boiler room fractured steam and water lines and perforated the steam drum. Despite the choking smoke and scalding steam, Stoker Petty Officer Tom Ogden remained at his post to ensure that steam pressure and electrical power were not disrupted. Above him on deck, fire had spread to both sides of the fo'c'sle break, threatening to prevent the Forward Damage Control Party from fighting the flames engulfing the bridge superstructure. According to a later commendation,

Although of poor quality, this photo nonetheless reveals Commander Basil Jones' fatigue after the June action. (Author's collection)

Chief Stoker Bertram Payne "was the first of his party at the fire. Rallying the party, at least half of which had never been in action before, he started them fighting the fire, and himself went into the Chief Stoker's Store above and forward of the galley, which was also on fire and full of smoke, and succeeded in bringing out some spare fire extinguishers although almost overcome by smoke." Later, when *Tartar* was on her way home, Payne "was untiring in his efforts" to distribute the fresh water supply, which Ogden had helped preserve, amongst parched, exhausted sailors.[107]

Amidst the chaos *Tartar*'s sirens jammed in the on position. Jones recalled this "made a noise such that orders and communications had to be delivered at the maximum lung power."[108] Yelling would not enable Jones to maintain contact with Plymouth or the rest of his force, and *Z-32*'s shell-fire had damaged his communications. *Tartar*'s sparkers worked hard to restore them but the initiative of a Canadian sailor ensured their labour bore fruit. Slipping out of *Huron*'s W/T office to catch a peek of the action, Chief Telegraphist Len Stone saw flames engulfing *Tartar*'s superstructure and guessed that her communications had been affected. On his own initiative for the rest of the night Stone had his team rebroadcast *Tartar*'s weak transmissions, keeping Plymouth in the picture and allowing Jones a modicum of control over Force 26. For their professionalism under the most demanding conditions, Ogden, Payne and Stone each received the Distinguished Service Medal.[109]

"As big an explosion as I have ever seen"

After *Z-32*'s tussle with *Tartar*, fortune again smiled upon the Narvik. *Tartar* had hit her three times and von Bechtolsheim disengaged to assess damage. As he attempted to escape, *Ashanti*, which had been attracted by gun flashes and a HEADACHE report that an enemy destroyer was heading towards the burning *Tartar*, brought *Z-32* under fire. However, before any decisive damage was incurred, the thick pall of smoke from *Tartar*'s fires cloaked *Z-32*. Before *Ashanti* could sort out the confusion, *ZH-1* emerged from the smoke, "wallowing and helpless."[110]

At the outset of the action *ZH-1* had followed *Z-32* around to port but *Korvettenkapitän* Barkow lost contact with *Z-24* and *T-24*. According to *ZH-1*'s reconstructed war diary, this had "a tragic effect" when Barkow sighted ships he thought hostile but chose not to engage for fear they were actually his flotilla mates. Under no such handicap, *Ashanti* and *Tartar* mauled *ZH-1*. Among a deluge of hits, shells penetrated the machinery spaces, cutting propulsion and electrical power, and bringing the destroyer to a stop. When the cloud of smoke and steam that shrouded *ZH-1* lifted, she lay naked before her assailants. Despite her own damage, *Tartar* raked *ZH-1* from point-blank range while *Ashanti* put a torpedo into her, blowing off most of the bow.[111] Realizing *ZH-1* was doomed, Barkow, in likely his last order before dying, directed his engineer to place scuttling charges in critical parts of the destroyer. At 0230, *Oberleutnant zur See* Hansen, the senior surviving officer, recalled, "Three cheers on board *ZH-1*. The ship's company leaps over board, most of the floats had been launched. I move clear with the cutter. *ZH-1* is fired on by an enemy destroyer to port, and a short time later by another Tribal from astern. *ZH-1* receives numerous hits. Scraps of the upper works are blown overboard, there are hits on the floats in the water." At 0240 *ZH-1* blew up in a massive blast that reverberated around the western Channel; in *Eskimo* Lieutenant-Commander Sinclair described "as big an explosion as I have seen." Hansen and 28 others reached Île de Batz safely in the cutter, while an RN escort group picked up another 141 survivors the next day.[112]

The chase west

To the southwest, *Haida* and *Huron* pursued *Z-24* and *T-24*. At the outset of the action, *Haida* had opened fire on *Z-24*, the third ship in the German formation, from about 4,000 yards, "A" mount firing rapid salvoes while "B" provided illumination. According to DeWolf's after-action report, the target, "just then turning away, very quickly started to make smoke and zig zag at fine inclinations. Some ten or fifteen salvoes were fired at this target and several possible hits were scored before another target was observed to the left."[113] *Z-24*, which was also briefly engaged by *Tartar* and *Ashanti*, suffered damage and casualties from hits to the bridge, engine room and forward gun mount. Claiming he had seen *Z-32* turning away to westward, *Korvettenkapitän* Birnbacher headed that direction at full speed.[114]

The situation was equally confused in *T-24*. She had been totally surprised when starshell burst overhead and 4.7-inch tracer streaked past. *ZH-1* was being pummelled ahead, and the torpedo boat narrowly avoided collision with the burning destroyer. When a Narvik was seen withdrawing westward, *Kapitänleut-*

nant Meentzen followed, assuming it was von Bechtolsheim. The torpedo boat was engaged by *Huron* before *Haida* joined in, and although surrounded by near misses, *T-24*'s luck against the Canadians held and nothing hit home.[115]

The battle now took a familiar turn for *Haida* and *Huron*. In the 26 April action they had become embroiled in a long chase; now they pursued *Z-24* and *T-24* as they fled southwest, *Huron* faithfully conforming to *Haida*'s movements. Conditions were unfavourable and DeWolf later reported the enemy destroyers

> were engaged with the wind dead ahead and rain squalls were frequent. Cloud base was never more than 1000 feet and often as low as 500 feet. Consequently illumination was poor and starshell were generally half burned before they effected any illumination whatsoever. The enemy made excellent use of smoke throughout and continuously took avoiding action thus making spotting at times well nigh impossible.[116]

The performance of search and gunnery radars was also hampered by the poor climatic conditions. Nonetheless, the two Tribals pounded southwest at 32 knots and might have overhauled the enemy – *Z-24* had slowed to 27 knots so *T-24* could keep pace – had fate not intervened.

Throughout the spring of 1944, under Operation MAPLE, the Allies had laid a series of offensive minefields along German coastal shipping routes to restrict the movements of U-boats and surface vessels against the invasion.[117] On this night they hindered rather than helped. At 0150 plots kept in both Canadian warships indicated that the Germans were entering minefield QZX-1330, which contained some 150 mines.[118] *Haida* and *Huron* followed orders to skirt the area while *Z-24* and *T-24* steamed through unaware and unharmed. By the time the Tribals resumed direct pursuit they had fallen 9 miles behind, and they lost radar contact shortly thereafter. At 0214, his position "with regard to own forces and remainder of the enemy was obscure," DeWolf reluctantly abandoned the chase to seek out *Tartar*.[119]

Z-24 and *T-24* played no further role in the action. A hit on the Narvik's bridge had killed nine men and caused considerable damage, knocking out much of her communications gear. Birnbacher later asserted that he held on for Brest because it was the last course ordered by von Bechtolsheim, and he had received no amplifying reports due to battle damage. *T-24* did receive updates but Meentzen neglected to pass them on to Birnbacher even though the two were exchanging visual signals. After the action a frustrated von Bechtolsheim concluded that had Meentzen relayed the information, Birnbacher "would have grasped the situation and the subdivision would have made an attempt to close me."[120] This is the fallout one expects from forces not used to operating together; as it was *Z-24* and *T-24* maintained their course for Brest unbeknownst to von Bechtolsheim, who thought they remained in the vicinity and thus continued to count on them in his tactical deliberations.

DeWolf versus von Bechtolsheim

By this time the situation was thoroughly confused with both commanders unsure of the position of their own forces let alone the enemy. At 0237, Jones, using *Huron* as a communications link, attempted to gain a semblance of control, signalling his ships to concentrate on *Tartar*. To the west, von Bechtolsheim made similar efforts and headed "on a southern course in order not to get too far away from the battle area."[121]

Meanwhile, *Haida* and *Huron*, already on their way back to Jones, proceeded cautiously northeast unaware of *Tartar*'s precise location. Visibility was obscured by rain squalls while climatic conditions and the shock from gun blast and high-speed running made radar imprecise and unreliable. At 0223 both ships obtained a firm contract bearing 032° at 6 miles. Because their plots indicated that *Tartar* should bear 040°, both DeWolf and Rayner thought this was their leader. IFF provided no confirmation but they could not be certain if *Tartar*'s set had been damaged. According to DeWolf:

> Made identification by light and ordered Plot to carry out radar search for other ships which might be concentrating. Ship in sight replied to our signal by light, but his signals were unintelligible. Main armament was brought to the ready and the challenge made, but the reply was again unintelligible. I still considered it might be TARTAR with damaged signalling gear and [wounded] personnel. The ship made smoke and turned away to the west and south but was not plotted by Radar and range was opened to 9000 yards before this move was appreciated.[122]

Z-32, the ship encountered by the Canadians, was equally in the dark. Von Bechtolsheim, looking for *Z-24* and *T-24*, recalled, "Individual shadows are sighted. Exchanges of recognition signals by blinker gun, and even by night identification signal, do not lead to any identification. The fact that, despite German recognition signal interrogation, these shadows do not fire, however, causes me to make the decision not to use my weapons."[123] *Z-32* initially accelerated away to the northwest but gradually swung around to an easterly course. DeWolf followed but still had doubts about its identity. Finally, at 0254, starshell revealed the distinctive silhouette of a Narvik and both Tribals opened fire.

As DeWolf and von Bechtolsheim considered each other across the Channel murk, they stood as professional peers with remarkably similar careers. They joined the navy within five years of each other – DeWolf in 1918, von Bechtolsheim in 1923. More importantly, throughout the interwar period they learned their profession in small services that had something to prove. They received their first destroyer command within months of each other, and used that opportunity to establish reputations as outstanding fighting commanders. That had been followed by important staff positions ashore. Both were quiet men with superb professional qualities that earned them the unreserved respect of their sailors. But no matter what their shared qualities, modern naval combat is coldly impersonal. Second World War sailors seldom saw the people they fought, and in night actions in particular the enemy remained virtually invisible, their position marked only by the glare of star shell, stabs of light from hits or gun flashes, or by the stark blip on a radar screen. DeWolf and von Bechtolsheim did not know one another's identities. The German was unaware his opponent was a Canadian, or even that he was a destroyer man, as he had misidentified his adversaries as Glasgow-class cruisers. Still, the action evolved into a duel between two skilled, confident commanders, each determined to achieve his objective.

DeWolf's aim was straightforward, the destruction of the enemy. That outcome was not critical for von Bechtolsheim, who wanted to keep his force intact as a threat to the invasion. In his war diary von Bechtolsheim explained his rationale when he had encountered *Haida* and *Huron*:

> As I am operating alone with *Z-32* I can achieve nothing against the overwhelmingly superior enemy, particularly as I must first load my spare torpedoes and there is no longer sufficient gun ammunition for a prolonged engagement. Proceeding eastwards must remain the aim! As I can no longer reach Cherbourg before first light, I decide to make for St. Malo. I hope to get the other destroyers to re-join during the passage and before reaching the Channel Islands. I therefore order the following: "Break through to the East. Goal St. Malo, join me."[124]

Although HEADACHE operators had monitored numerous enemy transmissions throughout the night, the intercepts revealed nothing beyond the fact that the Germans were confused about each other's whereabouts. Nonetheless, DeWolf had a far better grasp of the situation than von Bechtolsheim. In response to his query about a massive explosion that had lit up the horizon at 0252, *Ashanti* informed him it was an enemy destroyer blowing up. "Nice work," DeWolf replied. He had chased two other destroyers towards Brest and realized from past experience they were unlikely to rejoin the battle. That left just the destroyer bathed in his starshell and DeWolf knew he could concentrate on her exclusively.

Shooting conditions remained poor as the Canadians began their pursuit, but that only made their technological advantage more telling. Dense smoke laid by *Z-32* made spotting difficult but tracer was followed for line and Type 285 provided precise ranging – von Bechtolsheim assumed the accurate fire was due to flares dropped from aircraft. The Tribals scored several hits but before they had any effect, minefield QZX-1330 again intervened on the enemy's behalf.

Z-32 unwittingly entered the minefield at 0311, and *Haida* and *Huron* were forced to alter around it to the northeast. By the time they recovered on their previous course a half hour later *Z-32* was 10 miles to the southeast. Minutes later, they lost radar contact. The sense of frustration must have been deep and one can only imagine the strong oaths uttered amongst Canadian sailors. At this point DeWolf, under instructions to reform on *Tartar*, could have broken off the pursuit without any fear of criticism; indeed a senior officer who was present recommended he do exactly that. Rear-Admiral H.E. Reid, the RCN's senior representative in Washington, who had fortuitously picked that night to ride in *Haida* as an observer, urged DeWolf to abandon the chase. It was testament to DeWolf's confidence and determination that he ignored Reid's advice. Although he thought the enemy might escape into the port of Morlaix, he doggedly continued the chase with *Huron* matching his every move.[125]

Like DeWolf, von Bechtolsheim remained optimistic. Although *Z-32* had taken "several heavy and light hits," the damage was "not serious enough to compel avoiding a fresh engagement." He predicted this confrontation would occur close to the Channel Islands, but believed he would have a good chance of success as *Z-24* and *T-24*, thought to be only 12 miles astern, should have joined by then. This expectation was dashed at 0420 when Birnbacher reported they were actually 25 miles to westward, had sustained heavy damage, and wanted to put into Brest. Von Bechtolsheim's reaction to this setback reflects his professionalism: "With a heavy heart I must therefore decide to abandon the mission ordered. I cannot achieve a breakthrough to the East with *Z-32* alone in these circumstances. Whether I will be able to break through to the West remains to be seen. Moreover, I suspect the presence of warships, as there is a shadower to the Northwest." Lest these be interpreted as self-serving words crafted to preserve his reputation, at 0445 he radioed those precise intentions to Group West. Like his other transmissions that night it was quickly broken by Bletchley Park.[126]

If *Z-32* had possessed effective search radar von Bechtolsheim would not have had to divine what lay to the west, but his suspicions proved correct.[127] Since

Bristol Beaufighters of No. 404 Squadron RCAF with 25-lb armour-piercing rocket projectiles slung under the wing. Beaufighters from the Canadian unit and No. 144 RAF harassed von Bechtolsheim on his passage to Brest, reinforcing the notion that it would court disaster to be caught in the Channel during daylight. (DND Canada, PL-41049)

0412, *Haida* and *Huron*'s Type 271Q indicated they were slowly overhauling *Z-32* but at 0432 the range began to drop rapidly. DeWolf initially suspected that the enemy was trying to reach the safety of the coast but it soon became apparent that he was heading west, and so the Canadians altered to the south to cut him off. Meanwhile, Commander Jones had concentrated the rest of Force 26 about 6 miles to the north, where he was in position to block any attempt to escape eastward. *Z-32* was trapped.

At 0444 *Haida* and *Huron* opened fire from 7,000 yards. Von Bechtolsheim, thinking he was under attack from two cruisers, turned south, returned fire and launched his remaining torpedoes. The underwater projectiles missed their mark and although several 5.9-inch shells burst close to the Tribals, they caused no damage. On the fo'c'sles of the Canadian destroyers, gun crews worked hard slamming 50-pound shells and 35-pound cartridges into the breeches of the four 4.7-inch guns. DeWolf initially ordered rapid salvoes but, dismayed by the apparent results, changed to more accurate salvoes at 0452. Even with that reduction, five or six salvoes of semi-armour piercing shells rocketed towards the enemy every minute.

Ships' companies fight as a team, and although gunners could inflict damage directly on the enemy, as DeWolf's and Rayner's post-action commendations emphasize, many in *Haida* and *Huron* were responsible for their ship's performance. In *Haida*, Lieutenants John Crispo Annesley and Ray Phillips displayed "outstanding coolness and skill" in coordinating the navigational and tactical plots in close proximity to Allied minefields and the French coast. Lieutenant Charles Mawer, double-hatted as Radar Officer and Gunnery Control Officer, was "untiring in his efforts to improve the efficiency of the ship's radar organization." Gunner (T) Lloyd Jones kept electrical equipment functioning throughout the action. Leading Seaman Robert White assured accurate gunnery ranging by keeping the Type 285 tuned and calibrated. In the engine room – "down below" – Chief Stoker George Lang's care of the main engines enabled them to exceed the revolutions achieved on *Haida*'s original full-power trials.

Huron has always been overshadowed by her more famous sister but her sailors performed just as effectively. The "coolness and ability" demonstrated by GCO Lieutenant Alan Watson and Starshell Control Officer Lieutenant James Oppe was vital to the ship's gunnery performance. Type 271 operator Able Seaman Ben Honsinger "was of invaluable assistance in gaining and maintaining touch with the enemy." When "A"-mount jammed from a loading error, Chief Ordnance Officer James Haywood, "working with great coolness and efficiency under fire, succeeded in clearing the jam." Rayner reserved special praise for *Huron*'s coxswain, the ship's senior rating and a key member of the leadership team: "Although an elderly rating, fifty years of age, Chief Petty Officer [Charles] Burch has always carried out his duties at the wheel faultlessly and with no signs of fatigue. He has exerted a good and steadying influence on the lower deck and displayed a hearty enthusiasm for action at all times."

Despite the confidence in their sailors, DeWolf and Rayner were unsure if they were hitting *Z-32* as they chased her towards the coast but von Bechtolsheim attested to the accuracy of their shooting. He had altered southwards "to get out of the excellent straddle coverage of the enemy gun batteries. The ship is constantly caught by hits. The way things are going, my running won't last long." Except for one brief interruption *Haida* and *Huron* maintained their withering fire. *Blyskawica* briefly joined in from the north but appears to have scored no hits. Sometime around 0500, *Z-32*'s port engine quit and three hits put "Anton," the forward turret, out of action. Hoping that the "tremendous quantities" of shells fired by the enemy would cause them to run out of ammunition, von Bechtolsheim attempted to escape in the shadow of the coast but at 0513, in the midst of more hits, the starboard engine lost power. Realizing the end was at hand, von Bechtolsheim ordered the ship, now engulfed by flames, run aground. *Haida* and *Huron* fired a few more salvoes but checked fire when they realized *Z-32* lay hard on the rocky shore of Île de Batz.[128] Theodor von Bechtolsheim nearly drowned when the current sucked him out to sea as he swam ashore, but he was picked up exhausted by a rescue vessel. Hours later, Beaufighters from Nos. 404 and 144 thundered in to complete the job they had started three days previously, pulverizing the wreck with rockets and cannon fire to prevent any attempt at salvage.

Aftermath

The defeat on 9 June 1944 dashed any German hopes of interceding against the western flank of the invasion with major surface forces. Not only had they lost *Z-32* and *ZH-1*, the damage to *Z-24* took weeks to repair and there was little that *T-24* could do alone. The Germans attributed the defeat to their poor state of training, the withdrawal of *Z-24* and *T-24*, and overwhelming odds. Von Bechtolsheim's performance received praise. The *FdZ*, *Vizeadmiral* Leo Kreisch, lauded him as a "daring, experienced and resolute commander, with excellent tactical skills, exemplary offensive spirit and a clear perspective of the battle." Despite the failed mission, Kreisch thought von Bechtolsheim "brought honour to the destroyer arm" and awarded him the Knights Cross of the Iron Cross.[129]

The Allies, of course, celebrated. Jones in particular came in for accolades. Leatham reported to the Admiralty the "successful outcome was due primarily, in my view, to the correct and immediate action of Commander Jones ... to force close action, while at the same time avoiding the enemy's torpedo fire." DeWolf, Rayner and Barnes also received praise, as did many of their sailors. But there was also unease. Admiral Sir Bertram Ramsay, the naval commander for NEPTUNE, grumbled to his diary, "I wanted all to be sank." But for the 20th Division's "inexcusable" turn away, Leatham and Admiralty staff thought the 10th DF would have routed the enemy. This seems unlikely since *Z-24* and *T-24* still would have escaped if *Tartar*, *Ashanti* and the 20th Division had joined the chase and they showed no inclination to return to the battle under any circumstances. All that likely would have changed is that the fate of *Z-32* would have been settled much sooner. Beyond Namiesniowski's performance, comment largely focused on technical matters. There were complaints about the minefields that had twice allowed the enemy to outdistance *Haida* and *Huron*. The senior mine warfare officer at the Admiralty could only explain that unless "the position of mines and of the ships is known with considerable accuracy," which was considered virtually impossible when mines were laid in enemy-controlled waters, "this hampering effect is bound to occur."[130]

From a strictly Canadian viewpoint, *Haida* and *Huron*'s role in the battle demonstrated the skill and offensive spirit of Canadian sailors. Newspapers at home gave full accounts of the victory, garnering the publicity the navy coveted

Z-32 hard aground on the rocks of Île de Batz. Note that her main armament is directed down the bearing from which *Haida* and *Huron* pounded her with their final salvoes. (DND Canada, CN-6870)

Hemorrhaging oil, *Z-32* is ravaged by Beaufighters on the morning of 9 June 1944. (DND Canada, A-78)

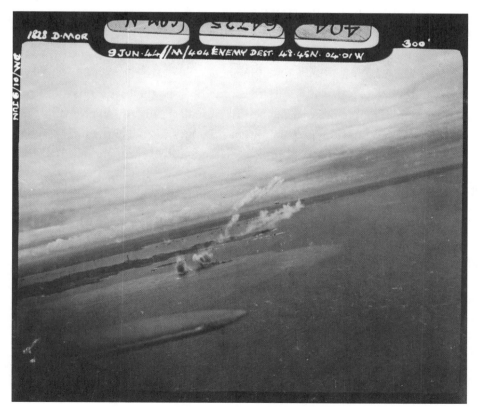

when it pushed to have the Tribals deployed to the "fighting theatres of war." *Haida* and DeWolf received the most acclaim, and to this day *Haida*, preserved in Hamilton, Ontario, remains Canada's most famous warship and "Hard Over Harry" DeWolf, who went on to an illustrious career culminating as Chief of Naval Staff, her most renowned fighting sailor.

As for the 10th Destroyer Flotilla, having achieved its primary mission in the Battle of Île de Batz, for the remainder of the invasion summer it applied its hard-won night expertise against German shipping moving close inshore. It was a different type of warfare that brought new challenges, but again sailors absorbed hard lessons, and in three actions in August ships of the 10th DF virtually annihilated lightly defended convoys caught scurrying along the Biscay coast. Only one destroyer was encountered on these operations, but, again, *T-24* escaped unscathed after a brief skirmish with HMCS *Iroquois*. The torpedo boat's luck finally ran out on 24 August when Nos. 404 and 144's Beaufighters sank her and *Z-24* in the Gironde. When the Allied armies finally chased German forces out of northwest France, the 10th DF lost its *raison d'être*, and Basil Jones took the British ships out to the Far East, while the Canadian ships returned home for refit or re-joined the Home Fleet.[131]

The Battle of Île de Batz provided ample evidence of why sailors are so uncomfortable fighting at night. Even with an edge in training, experience, skill and technology, as well as knowing precisely when and where the enemy would appear, a lot could still go wrong, and the margin between success and failure was slight and fragile. As in naval warfare across the ages, the battle and the events leading up to it emphasize that the best naval leaders can do is to give sailors the best tools of war, and the preparation and training to enable them to succeed. Night fighting in the Chops of the Channel undoubtedly brought new challenges but demonstrated, nonetheless, that hoary old lessons of naval warfare, such as attacking effectively first and keeping forces intact so they can gain experience collectively, still survive as war winners.

"Dining-in Night," HMCS *Huron*, summer 1944. Despite being overshadowed in the torrent of publicity accorded *Haida*, *Huron* played an important role in the actions on 25/26 April and 8/9 June; a fact perhaps being celebrated on this night by *Huron*'s officers. Horatio Nelson, whose portrait is on the bulkhead behind LCdr Rayner, would have approved of the comradery. (Author's collection)

PRELUDE

A New Kind
of U-Boat War

THIS LAST CHAPTER, by Malcolm Llewellyn-Jones, examines how U-boats tried to adapt to the growing power of Allied counter-measures. By the summer of 1944 older U-boats were still the only ones available to attack Allied shipping, but their vulnerability made operations extremely hazardous. The only measure that seemed to offer a chance of survival was to fit them with a device enabling them to run their diesel engines while submerged, showing only a tube above the water. A lifesaver, literally, the schnorkel made U-boats much harder to detect as they could now avoid spending time on the surface. As with most boons, however, the U-boats paid a price when they used the schnorkel – they now travelled only at the slow speeds these old designs were capable of submerged. In addition, life below the surface in these boats proved rather uncomfortable, but survival trumped discomfort.

Without the speed offered by surface travel, U-boats had to change their operating areas. Unable to travel fast enough to intercept targets on the open ocean, they now searched in areas where they could hope to ambush them. The only places this could happen predictably were much closer to shore than U-boats had operated for years. Initial hesitation to move to patrol areas near the United Kingdom was overcome when operations after the Normandy invasions demonstrated the significantly enhanced survivability of schnorkel-equipped U-boats in the English Channel. A crash program to fit all U-boats with schnorkels ensued, as well as the deployment of increasing numbers of U-boats into coastal waters.

The surprising survivability of schnorkel-fitted U-boats arose from several inter-related factors. The novelty of the new German tactics was certainly one factor. U-boats had found operating in coastal waters so dangerous for so many years that Allied antisubmarine forces had grown rusty in the skills associated with operating there. Rustiness proved important because hunting U-boats in coastal waters was significantly different from the open ocean. Shallow water teemed with fish, the hulks of old wrecks, and rocks, all of which provided echoes that could sound far too much like a U-boat. The tidal streams found in many areas could even result in a Doppler-like effect for these immobile targets, lending even greater verisimilitude. Precision navigation to plot these non-sub sites became highly important, along with skilful plotting of movements of all ships in the group – skills of much less relevance to open-ocean ASW. Aircraft found the sudden disappearance of their prey from the surface hardest to adjust to, as radically new technologies were needed for them to find submerged targets reliably. Whereas aircraft had been effective in both scarecrow and killer roles in the mid-war years, the introduction of schnorkel sharply reduced their ability to find and kill U-boats. Nonetheless, aircraft remained important as a threat to keep U-boats submerged.

Fortunately for the Allies, shallow-water ASW skills could be re-learned. Even more important, many of the command and control procedures developed by the Royal Navy could be adapted to shallow waters. Experience proved its worth here as well. The years spent bringing together disparate pieces of information and intelligence allowed the more experienced teams to produce informed estimates of likely U-boat destinations or operating areas, and these shrewd assessments based on sometimes sketchy information proved vital to Allied success.

Extensive Allied resources were available to hunt the often-small numbers of U-boats operating in coastal areas. The disparity in numbers between the hunted and hunters might seem to make Allied success inevitable, but, as always, numbers alone were seldom enough. What the odds help to highlight is the desperate bravery of the U-boat crews as they skulked along the coasts of their enemy. There were few illusions of their chances by this stage of the war, and the reason that *BdU* continued to send boats out was less for the small amount of damage they continued to inflict, but more to divert Allied resources, such as four-engined aircraft, away from attacking Germany directly. It is doubtful that U-boat crews realized at the time how steep the odds they faced were, but there can be little doubt that they understood that their prospects were poor. Nonetheless these men, including the doomed crew of *U-247*, fought with tenacity much greater than their tainted cause deserved.

Their opponents, the crews of the many ships and aircraft ceaselessly patrolling the waters around England, were experienced but also often tired. The apparently endless number of false contacts, each requiring a rigorous assessment and prosecution, wore away at their reserves of energy. The large numbers involved in the Allied side makes keeping track of individuals a daunting challenge for an historian, but it is important to recognize that individuals still did make a real difference. In the end it was a small group of determined men that sealed the doom of *U-247*, and their story is well told in the pages that follow.

On Britain's Doorstep

THE HUNT FOR *U-247*[1]

Malcolm Llewellyn-Jones

ON 8 JUNE 1944, the Tracking Room in the Operational Intelligence Centre (OIC) learned that a U-boat *en route* to western France had instead been ordered to an area that straddled the entrance to the North Minch.[2] The change of plan for the U-boat was a reaction to the invasion in Normandy, and by the following day the OIC had, from a further decrypt by Bletchley Park, learned that the U-boat was the schnorkel-fitted Type VIIC *U-247*. Her mission was to tie down British forces and *BdU* thought that, owing to the invasion, the defences between Cape Wrath and the Butt of Lewis were likely weak.[3] The OIC passed this information to Vice Admiral H. Harwood, Admiral Commanding Orkney and Shetland (ACOS), at Lyness, but for the next month nothing more was heard about the U-boat. Nevertheless, by a process of dead reckoning the OIC Situation Report for 12 June reported that *U-247* had reached its operational area.[4]

Apart from regular Coastal Command patrols from 18 Group flown over the area, no specific naval reaction seems to have taken place until the beginning of July. The main antisubmarine (A/S) forces, other than those involved directly in the Channel area, had been on A/S sweeps controlled by CinCWA since D-Day, where on 26 June, HMS *Bulldog*, part of a scratch group, had sunk *U-719* when following up a HF/DF bearing from HMCS *St. Thomas*.[5] *U-719* had been operating in the approaches to the North Channel between Ireland and Scotland and, like *U-247*, had been told to distract British forces (see map, page 242). The *ad hoc* escort group was then moved northwards, where, under the command of Captain H.T.T. Bayliss, RN, in the escort carrier (CVE) HMS *Vindex*, they were lent to ACOS for a search of the area north of the Minch.[6] The Swordfish and Hurricane aircraft of the CVE's composite 825 Squadron had been hampered by indifferent weather and contaminated aviation fuel.[7] Moreover, with no definite

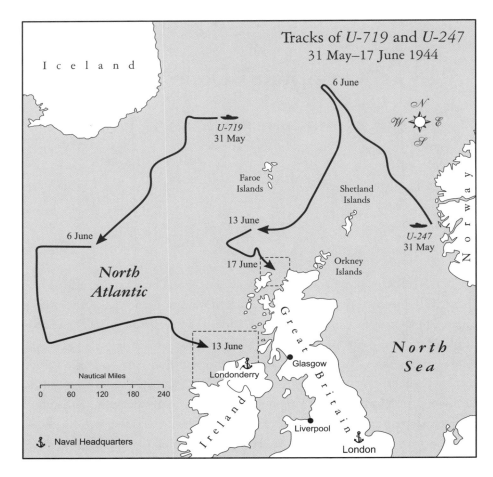

Tracks of *U-719* and *U-247*
31 May–17 June 1944

intelligence of the whereabouts of the elusive U-boat, the search had so far been unproductive.

This was a different sort of antisubmarine warfare. Towards the end of 1943, the ubiquity of Allied maritime aircraft had forced U-boats to seek safety by operating primarily submerged. Until that point in the Battle of the Atlantic, U-boat tactics had relied on surface travel, where they could use their diesel propulsion to achieve high tactical mobility. But now forced into submerged operations, where they had to rely on their limited battery power for movement, they were tactically immobile. Before there was a significant threat from aircraft in mid-Atlantic, the U-boats could achieve, if unmolested, perhaps 250 miles per day. Forced to operate largely submerged, this distance was cut, the British estimated, to some 70 miles per day (allowing several hours on the surface to recharge the batteries). Adoption of the *Schnorchel* (literally "nose"), anglicized to schnorkel or snort, allowed the U-boats to avoid even this small surface exposure, but further cut their mobility, allowing them to make little more than 50 miles per day.[8]

These schnorkel-fitted U-boats, operating primarily underwater, achieved a
high degree of immunity from detection by A/S units, especially from aircraft.
Professor E.J. Williams and L. Solomon, of the Admiralty's Department of Naval Operational Research (DNOR) calculated that the effective path swept by
a searching aircraft was about one-third of a mile by day, falling to a twentieth
of a mile at night. (By comparison the swept width for an aircraft searching for
a surfaced U-boat was about 5 miles by day and 3-7 miles by night.) In August
Captain R. Winn in the Admiralty's OIC came to the judgment that the "evolution of the Snort U-boat will be found to have affected profoundly the balance of
power between hunter and hunted...."[9] The German advantages gained by this
new technology were, however, mitigated by the general lowering of expertise in
the U-boat arm, and inexperience in operation of the schnorkel.[10] Antisubmarine
aircraft could, at least, keep the U-boats down, and this shifted the onus for locating and destroying them to surface escorts. However, the Allied antisubmarine
vessels, searching with their asdic (sonar), were achieving a swept width of barely
300 yards on each side of the ship.[11]

The relative immunity from air search meant that the U-boats could once
more operate in British coastal waters, which were closer to their operational bases, an important factor now that their transit speeds were so low. However, from
the Allied point of view, searching for the schnorkel-fitted U-boats in shallow,
coastal wasters was made difficult by poor asdic conditions and the inexperience
of operators in discriminating between U-boat echoes and the abundant acoustic
returns from wrecks, rocks, fish and thermal layers. Furthermore, U-boats now
operated singly, for, as a direct result of schnorkel operations, they lacked the
communications and tactical mobility necessary to operate in packs. The shift
to inshore operations thus resembled a guerrilla campaign, in which the U-boats
tried to get onto the shipping routes and lie in ambush for passing shipping. The
lack of co-ordinated pack operations meant, in turn, that the volume of U-boat
radio traffic was much reduced. In the days of the great convoy battles in the
spring of 1943, U-boats, on average, sent about six signals per month (not including those sent during convoy actions), while by September 1944 this average had
fallen to just one signal per month.[12] Allied intelligence was thus largely deprived
of intelligence direct from sea. Even so, on the rare occasions when U-boats did
transmit, the comprehensive Allied High-Frequency Direction-Finding (HF/
DF) chain was able to intercept and provide an approximate fix of the U-boat's
position.[13] After D-Day, the action by French forces in the interior, Allied bombing and their advance deep into France, led to a rapid disintegration of the German naval command structure. So difficult had the situation become that on 27

August, as *U-247* sailed, the Captain U-boats West, the local German command, was dissolved and operational control was removed to the U-boat Command Headquarters (*Befehlshaber der Unterseeboote*, or *BdU*).[14] As a result, communications via landlines was practically nonexistent, and this forced the enemy to rely on radio traffic for U-boat control. This, in turn, provided an opportunity for the British to gather a great deal of operationally relevant intelligence from intercepted signals.[15]

The resultant intelligence picture remained frequently obscure as to the exact movements of the U-boats.[16] When precise intelligence was gained, however, the British, by mid-1944, proved far less reticent in using the information to directly deploy A/S units. Instructions to these aircraft and surface vessels were sometimes camouflaged by reference to High-Frequency Direction-Finding (HF/DF) radio intercepts, or, more often, sent with no amplifying information.[17] There were, of course, a few officers who had been indoctrinated for staff appointments who then returned to sea, and were able to "read between the lines". The Coastal Command Manual of Anti-U-Boat Warfare also gave a broad hint, when it stated:

> Although it is not possible for reasons of security to reveal our knowledge of U-boat movements, aircrew can take it as certain that patrols are ordered in areas in which there are known to be U-boats operating or on passage.[18]

Special Intelligence (SI) material, based on the Ultra code-breaking at Bletchley Park, was passed to Flag Officers' headquarters ashore (and a few, selected Senior Officers afloat), and some were even provided with personnel who acted as the direct link to the OIC. Such liaison had existed for several years for CinCWA and the collocated 15 Group, Coastal Command. As D-Day approached this system was extended to cover, for example, C-in-C, Plymouth and 19 Group.[19] Such close liaison did not, however, exist in Harwood's headquarters in Lyness.

For Harwood, the uncertainty over the location of the U-boat in his area was first dispelled at 2120 hours on 5 July, when an urgent telephone message was received by the Duty Staff Officer at ACOS Headquarters, saying that an unknown vessel was reporting by radio that a submarine had surfaced some 10 miles west of Cape Wrath. However, from the garbled transmissions, it was impossible to tell the exact position of the occurrence, or whether an actual attack was in progress. Only later did it become apparent that a U-boat had been shelling a group of fishing trawlers, one of which, the *Noreen Mary*, was sunk (see map, page 249).[20] In the meantime, Harwood's staff telephoned the OIC and asked whether it was likely that a U-boat was still present in the Cape Wrath area. The OIC continued to think that *U-247* was in the area and that the reports of an attack by a U-boat "were most probably true."[21] Harwood thereupon immediately requested air support and Air Officer Commanding, 18 Group, was contacted via Commander-in-Chief, Rosyth, but no aircraft could be provided as all the airfields were fog-bound.[22] AOC, 18 Group, however, passed on the request to the adjacent 15 Group and they were able to provide a Leigh-light Liberator, which searched for 11 hours overnight without result.[23] Meanwhile, Harwood also despatched orders for a striking force of local A/S trawlers to raise steam and head to the area under the direction of HMCS *Huntsville*, a Squid-fitted Castle-class corvette. These ships were at sea within an hour of the attack on *Noreen Mary*.[24] C-in-C, Home Fleet, at Scapa Flow, had also been contacted and he despatched Captain D3's destroyers *Milne*, *Marne* and *Verulam* to the area.[25]

Harwood also told *Vindex*, some 80 miles to the northwest of the Minch, but Bayliss had intercepted the trawlers' distress messages and pre-empted Harwood's implied directive by immediately scrambling two deck-alert Hurricane fighters to the area and ordering a strike of A/S Swordfish biplanes to follow them.[26] These aircraft were new Swordfish Mk III's with the improved ASV MK XI radar, and the experienced observers had practised with the new radar against Captain SM3's submarines in the Clyde before sailing for this operation.[27] The operational effectiveness of the equipment initially suffered from poor maintenance and about a third of sorties were lost due to equipment failures. However, as the ship's maintenance teams grew more experienced, so the availability of the aircraft increased to some 90% by the end of the operation.[28] The radar was able to locate a U-boat's schnorkel, or periscope, at relatively long ranges, but it could also detect small fishing boats, or, if the sea was very calm, pieces of floating wreckage. At night, or in poor visibility, it would be difficult to distinguish among these various contacts. Because of its short 3-cm wavelength, submerged U-boats would be unable to detect the approaching radar.[29] However, if sighted by the U-boat, the slow Swordfish would rarely be able to close before the U-boat went deep. Contact would then be lost, for, at this stage, none of the aircraft had any sensor which could be used to detect a fully submerged U-boat. Bayliss in his report was to emphasize this shortcoming and suggested that his aircraft be equipped with sonobuoys.[30] Installation of the new ASV MK XI radar in the Swordfish brought with it other limitations. The weight of the equipment meant that a telegraphist/air gunner could not be carried, if the aircraft was to carry a realistic depth-charge load. This put added pressure on the observer because, if he was to take full advantage of the improved detection capability of the MK XI radar, he had to concentrate on its operation to the detriment of aircraft navigation. The 825 Swordfish thus had to rely, more than earlier variants, on direction from the ship.[31]

The Hurricanes reached the area of the attack on the trawler very quickly, but soon got out of radio touch with the carrier and in the darkness and poor visibility lost their way. Both fighters, running short of fuel, managed to make the coast and crash-landed in a football field near Castletown.[32] As for the Swordfish, they arrived in the area before midnight, and at 0024 hours one of these aircraft, flying over a calm sea, obtained a good contact 7 miles northeast of the Butt of Lewis, but the contact disappeared before it could be identified (see inset map, page 249).[33] Meanwhile, Captain D3 was approaching from Scapa Flow at 30 knots and reached the search area at 0200 hours. He then carried out an asdic search to the southward of the attack position and in the process sighted *Huntsville* searching for her trawlers in the patchy fog.[34] The poor visibility, with forces from various commands all arriving at different times during the first night, did not lend itself to a well co-ordinated search. Then, shortly after 0300 hours, *Bulldog*, having been detached by Bayliss, approached the area of the Swordfish contact, where she also detected a radar contact. But this turned out to be one of a group of small trawlers which had been fishing in that area for the previous five hours.[35]

On the morning of 6 July, Harwood and his staff were planning their next move. It seemed very unlikely that the U-boat would evade southwards into the Minch, or, for that matter, remain in the area of the attack on the trawlers. The most plausible escape routes were either to the northwest or northeast. The exact position of the attack on the trawlers now seemed to be some 10 miles west of Cape Wrath, but only the "good ASV contact" by *Vindex*'s Swordfish at 0024 hours that morning seemed to provide a sound datum for the subsequent search and indicated a westerly movement by the U-boat.[36] News of *Bulldog*'s contact on the trawler had not been received by ACOS Staff, which might have caused Harwood to reassess the Swordfish's report. With the information available at 1044 hours, Harwood ordered *Vindex*'s group to block the U-boat's retreat northwestward and *Huntsville*'s group her escape northeastward. Both forces were to assume a submarine speed of advance of 2 knots.[37] At the same time, C-in-C, HF had just ordered D3 to break off the search by 1300 hours, unless he was in contact, as his ships were needed for other operational tasks.[38] Then, at about midday, Harwood once more consulted the OIC, who continued to favour a westerly retreat for the U-boat. Harwood accordingly signalled *Vindex* at 1224 hours on 6 July that

> When time factor overstretches area to be searched with forces at your disposal concentrate particularly on enemy retiring course between 270° and 300° from Butt of Lewis.[39]

"Hunting on a catchy scent"

But barely an hour later, a series of events occurred which were to radically alter Harwood's appreciation. They began when the staff at C-in-C, Rosyth and AOC 18 Group Area Combined Headquarters (ACHQ) telephoned his staff with the news that a Catalina had reported sighting a periscope 30 miles to the north of the Orkneys.[40] The aircraft, Catalina "H" of 330 Squadron, had been on patrol for about an hour when at 1300 hours the captain, Lieutenant Christiansen, Royal Norwegian Navy, saw a disturbance in the water, which he identified as the periscope of a U-boat on a heading of 150° and travelling at 2-3 knots (see map, page 249). As the aircraft turned to attack, the periscope disappeared leaving no trace and Christiansen aborted his attack.[41] This report seemed to strengthen the possibility that the U-boat was escaping to the northeast, and, given its reported heading, was perhaps attempting to pass between the Orkneys and Shetlands. Harwood at once telephoned the OIC to ask for their view of the authenticity of H/330's report. Their reply was that,

> while it was almost certain that there was only one submarine in the area, whose course as previously stated was most probably westerly from the last reported position, this could not be guaranteed.[42]

Harwood thus faced a quandary. Christiansen's sighting suggested that the U-boat had made off to the northeast, while the OIC's confidence that she had chosen a westerly or northwesterly route was supported by the Swordfish report off the Butt of Lewis. At this point, Harwood's uncertainty was further complicated, for *Bulldog* had arrived at Scapa Flow for fuel and reported his intelligence about the trawlers in the vicinity of the Swordfish datum. This cast doubt on the reliability of the Swordfish's report, which in turn added weight to the possibility that the U-boat had retreated northeastward. The problem was that in order to have reached the position reported by H/330 the U-boat would have had to make good a little under 8½ knots, or about 6¾ knots if the Swordfish's contact was discounted. Such speeds could only have been made if the U-boat was trying to escape on, or partly on, the surface, a possibility ACOS entertained due to the widespread fog that had shrouded the area and had limited 18 Group operations.[43] Harwood decided to hedge his bets. *Vindex* was to continue searching to the northeast, while *Huntsville* and the trawlers were to concentrate on the area around H/330's contact.[44]

Over the operational area, Canso "C" 162 RCAF Squadron had been carrying out a Creeping Line Ahead (CLA) search northwards from a line between the Butt of Lewis and Cape Wrath since the early hours of the morning of 6 July.[45] At

1450 hours, about 10½ hours into his sortie, Flight Lieutenant J.C. Wade, RCAF, the aircraft's pilot, reported a contact to *Vindex* which was only 17 miles to the west of the ship (see inset map, page 249). Several Swordfish closed the area, and Bayliss detached HMS *Inman* from the screen to investigate. But this proved abortive, for Lieutenant Commander P.S. Evans in *Inman* soon sighted trawlers in the vicinity, and C/162 classified his contact as a floating mine.[46] Meanwhile, Liberator "B" from 120 Squadron, 15 Group, was on a navigation exercise some 60 miles west of the Hebrides when, at 1453 hours, about an hour-and-a-half into the sortie, the Canadian pilot, Pilot Officer M. Weiner, sighted an oil slick. A single sonobuoy was dropped and a "warbling note and occasional metallic note heard." Weiner, therefore, followed up with a full sonobuoy pattern that detected "engine throbbing…." However, having reported his contact, Weiner was instructed to resume his patrol after only 30 minutes of investigation.[47] News of B/120's contact did not reach Harwood until just after midnight on 6/7 July, who immediately passed it to *Vindex*. If this was *U-247*, she would have had to achieve some 6½ knots over the ground, but the weather to the west of the Hebrides had been better than that to the northeast. Probably for this reason, Harwood was not convinced this was the U-boat that had attacked *Noreen Mary* and for which Bayliss had been searching for the last few days.[48]

Then at 0315 hours on the morning of 7 July, ACOS received another telephone report from the ACHQ stating that Catalina "Q" of 210 Squadron had just signalled a U-boat sighting 50 miles northwest of the Shetlands (see map, page 249).[49] Q/210 was one of 10 aircraft following up H/330's contact, when about 2 hours into the sortie her pilot, Flying Officer Campbell, saw a wake 4 miles ahead of the aircraft, which seemed to be coming straight towards him on a course of 270°. As the aircraft closed, the contact disappeared at 2 miles, too far away for an attack to be made.[50] Harwood was soon told that the "enemy submarine previously reported was on the surface and was observed to submerge."[51] Inquiries with AOC 18 Group established that the aircraft's crew were very experienced and that their report "could be relied upon to a high degree."[52] These reports seemed to confirm the idea that the U-boat "had, in fact, proceeded north until darkness permitted her to surface in order to charge."[53] As a consequence, by 0424 hours, ACOS sent all the available surface forces, apart from *Vindex*'s group, hurrying off to scour the area.[54] Around 1030 hours on the 7th, Harwood, having consulted the OIC, concluded "that there were two U-boats in the area, and accordingly he told all forces engaged to assume this fact."[55] Both U-boats were now thought to making ground to the west. Thus, the A/S trawler force with *Bulldog*, *Huntsville* and initially D3's destroyers were to carry out a sweep to the west of Q/210's contact. *Vindex* was to assume that the westerly U-boat, apparently based on B/120's contact, could have reached a position some 130–140 miles west of the Butt of Lewis by 1800 hours that day, 7 July.[56] With the two searching forces over 200 miles apart, the OIC issued an assessment of the "western" U-boat's progress, though curiously not to ACOS. They were clearly far more sceptical of B/120's sonobuoy contact of the previous day, which, they thought, "seems too far on" for the U-boat that had attacked *Noreen Mary*.[57] Harwood, however, remained committed to the idea of two U-boats and again signalled his appreciation during the evening.[58]

Harwood, with only partial and irregular access to SI, was more focused on tactical sightings by Coastal Command aircraft, whereas the OIC, with a more comprehensive view of the overall U-boat dispositions (vague though this was),

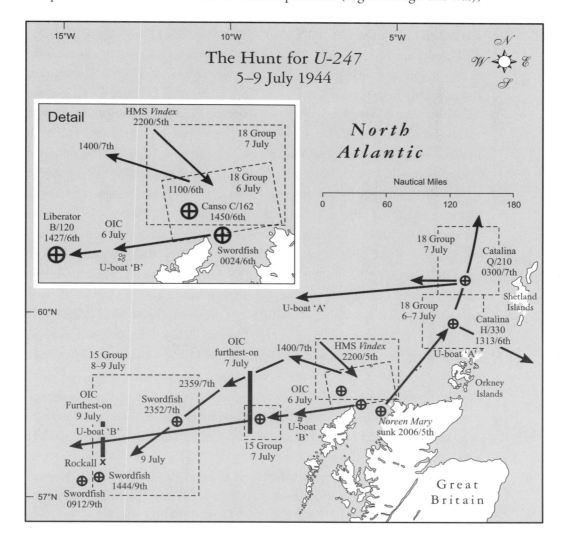

placed more weight on their assessment that only one U-boat had been operating to the north of the Hebrides.[59] As midnight 7/8 July approached, Harwood realized that the continuing low visibility in the northerly areas meant that the northern U-boat, by making a dash on the surface, was probably now well past *Bulldog's* search. The A/S ships were therefore told to abandon the northward search and instead to sweep southward towards the position of *Noreen Mary's* sinking, and then to disperse to harbour. Meanwhile, Bayliss was told to "dispose your force to best advantage against probable approach of both U-boats towards Rockall Bank."[60] Almost as this instruction was being transmitted, one of *Vindex's* Swordfish obtained an ASV contact 40 miles southwest of the ship and 70 miles northeast of Rockall (see map, page 249). The contact, which disappeared as the aircraft closed, was on the 2-knot furthest-on position from B/120's sonobuoys contact. Bayliss immediately ordered further Swordfish and four escorts to cover the area, though there were no further contacts.[61]

By mid-morning of 8 July, Harwood had abandoned the search to the northeastward and refocused the search to the westward. From *Vindex's* last contact, it was already in the area controlled by Admiral Sir Max Horton, C-in-C, Western Approaches (CinCWA,). Harwood therefore signalled Horton and suggested that he take control of the hunt. Horton readily agreed. He would take over from midday and, at the same time, direction of the air searches would be transferred to 15 Group from 18 Group.[62] The assumption of control by CinCWA brought with it several advantages. By the autumn of 1944, Horton's headquarters had been collocated with Air Officer Commanding, 15 Group in the ACHQ, in Liverpool for 3½ years. This had fostered intimate co-ordination between the surface and air A/S operations. The ACHQ also had an especially close relationship with the OIC. Constant use of their material, both in short-term tactical application and as a source of deep background data, was second nature to CinCWA' staff, at least for the small number of staff officers in the headquarters with clearance into Ultra. Nowhere was this more obvious than in Horton's anticipation at the beginning of 1944 of an inshore campaign by the U-boats after they had suffered a series of crushing defeats in open ocean convoy operations during 1943.[63]

Thus at 1225 hours on 8 July, just 25 minutes after assuming control of the operation, Horton signalled his appreciation and intentions to *Vindex* and the other ships of Force 34. Horton, probably because of his close consultation with the OIC, abandoned all idea of a second U-boat. No SI had been available on *U-247* (or any other U-boat in the Minches area) since 12 June.[64] The OIC, however, calculated that *U-247* could be about 50 miles northeast of Rockall, based on the possible contact by HMS *Vindex* at 0152 hours, on 8 July.[65] Horton, following their lead, assessed that the single U-boat was making ground to the west, and that it would pass to the north of Rockall before turning southward. He adjusted the datum for the search to the position of the Swordfish contact obtained at 0157 hours (that is, about 150 miles west of the Butt of Lewis, and some 80 miles northeast of Rockall). This assumed an overall speed made good over the ground of about 3½ knots since the attack on *Noreen Mary*. In *Vindex*, Bayliss responded by ordering his A/S escorts to carryout an expanding spiral, or "Vignot," search based on the Swordfish datum, and to establish a roughly north-south patrol line stretching from the east to the northeast of Rockall, across the U-boat's antici-pated path. Flying from the carrier was intermittent due to the poor visibility. Horton added to his intentions that a continuous patrol of four Sunderlands or Liberators from 15 Group would co-operate with the surface forces. All the aircraft would be centimetric-ASV equipped and, for the night sorties by Libera-tors, fitted with Leigh-lights.[66] This resulted in the establishing of a search in a 90 × 120-mile rectangle centred about 50 miles northeast of Rockall.[67]

15 Group, Coastal Command, had repeated their patrols of the previous day, and *Vindex* had manage to get aircraft off too. One of the Swordfish had, at 0912 hours, already obtained another ASV contact 25 miles to the southwest of Rockall (see map, page 249). Bayliss organized a hunt by her A/S escorts and two Sword-fish were kept in the air continuously. By this stage *Vindex* was running short of fuel for her aircraft, so, as the Coastal Command Sunderlands and Liberators came on task they too were directed to assist in the search. Without their help, Bayliss reported, he could not have been able to saturate the area around the ASV contact position.[68] This intense search, however, yielded no further contact and Bayliss turned his attention to refuelling his escorts. This had just begun when, at 1444 hours, another Swordfish obtained "a further good contact" close-by the earlier ASV detection. However, during the afternoon,

> the hunt was not made easier by the gale which now came upon us, accompa-nied by thick weather, in which the hunting group lost contact [with *Vindex*], and *Pevensey Castle* had considerable difficulty in meeting anyone.[69]

At 1331 hours, 9 July, the OIC signalled that *U-247* was "probably near Rockall."[70] And so matters stood as the day came to a close.

"The hounds are lifted"

Then in the early hours of 10 July, the entire picture changed. Probably because Matschulat had not seen Allied A/S forces close-by for some time, he decided at 0116 hours that morning to surface *U-247* for about 30 minutes to allow his navigator to take a fix with bearings from radio beacons and to send a two-part signal. *U-247*'s first transmission was at 0131 hours and the second five minutes later.[71] *U-247* dived eight minutes after that.[72] The signals were intercepted by the Allied DF chain, and although the bearings obtained were poor, the Admiralty DF organisation was able to construct a fix centred 90 miles northwest of the Butt of Lewis though with a radial error of 60 miles. The centre of this probability area was some 200 miles from the position currently being searched by Force 34 and Coastal Command, and showed that *U-247* had not made nearly as much ground to the west as had been supposed since the attack on *Noreen Mary*. The details of the HF/DF fix were passed to AOC 15 Group, who ordered his land-based aircraft search to be shifted to cover the probability area around the HF/DF fix. Relays of 3 VLR Liberators were to start at 0730 hours, 10 July, and continue throughout the day and be followed by 4 Leigh-light Liberator sorties overnight covering an area biased towards the southwest.[73]

At 0236 hours the OIC had also signalled the HF/DF intelligence to Force 34, and at 0245 Horton followed this with instructions to Bayliss to search the probability area.[74] "Still hunting albeit on a somewhat catchy scent," when the C-in-C's signal was received in *Vindex* at 0332 hours, "hounds were finally lifted," Bayliss wrote, "and laid on to a fresh fox."[75] As he moved to the northeast, Bayliss tried once more to refuel his escorts, but this was made difficult by the gale that now beset them. The attempts were in vain, causing fuel hoses to part and in the early hours of 11 July the procedure was abandoned and half of his escort force was detached to port for fuel. The timing could hardly have been worse, for now, as *Vindex* entered the U-boat probability area, she had a barely adequate close escort of HM Ships *Manners* and *Pevensey Castle* and HMCS *St. Thomas* (now SO Escort), which left little to act as a striking force just when it was needed, without laying the CVE open to attack.[76] This was a consequence of the RN's persistent failure to develop an effective abeam refuelling system that was more robust in poor weather (and becoming commonplace in the USN). The lack of efficient refuelling at sea persistently hampered RN operations in the Atlantic battle.[77]

Ever since the signals from *U-247* had been intercepted in the early hours of 10 July, Bletchley Park had been busy, and after about 26 hours they had succeeded in breaking them to reveal the contents of Matschulat's messages. The results were immediately passed to the OIC and by 0505 hours that morning they forwarded the information to CinCWA and HQ Coastal Command (see map, page 254).[78] The message included *U-247*'s position and also revealed that she had 17 days of provisions remaining.[79] These two facts, as will be seen, might have produced a more realistic assessment of the U-boat's movements had they been taken into account. When this information was sent to Horton, the OIC were able to add, from another intercepted signal, that on the previous evening *BdU* had ordered Matschulat to make for western France. What does not seem to have been passed on to the operational commands was the knowledge that *U-247* had been ordered "not to send another situation report until well clear of the operational area."[80] Meanwhile, Force 34, ignorant of this SI information, was approaching from the southwest overnight on 10/11 July, and 6 Leigh-light Liberators from 15 Group flew a crisscross pattern covering the most probable U-boat furthest-on positions, centred about 140 miles west of the Butt of Lewis and 120 miles northeast of Rockall. At 0648 hours, one of these sorties bore fruit when Liberator "A" of 120 Squadron, captained by Canadian Pilot Officer M. Weiner, reported that he had "attacked a U-boat at periscope depth" about 110 miles west of the Butt of Lewis (see map, page 254). This was just on the edge of the HF/DF fix probability area.[81] A/120 had been flying at 1,000 feet, just below the cloud, and was just about to go off patrol when Weiner "sighted white smoke at sea level 6 miles distant." As the Liberator closed, Weiner saw "a definite periscope feather" heading 300° at about 3 knots. As the aircraft roared in on its attacking run at a height of 150 feet "no clear view of periscope or schnorkel was obtained" and the feather disappeared 15 seconds before Weiner attacked with 6 × 250-lb. depth charges. The crew estimated that the charges had entered the water about 50 yards ahead and straddling the U-boat's track. After loitering for about 20 minutes, looking for some evidence of a successful attack, Weiner decided to drop a pattern of sonobuoys. These detected "grinding noises followed by numerous explosions."[82] By now three other Liberators from 120 Squadron patrolling the area had homed in on Weiner's attack message and began pre-arranged tactical searches around the attack position.[83]

By 0925 hours, after a false start due to *Manners* having a defective asdic, *Vindex* despatched *St. Thomas* and *Pevensey Castle* to the scene (leaving only *Manners* close to hand) as well as several Swordfish.[84] These ships arrived at the scene of the attack at 1030 hours and started a sweep through the area, just as Weiner, having remained in the area for two-and-a-half hours beyond his scheduled sortie time, was forced to leave for base.[85] The large reserve of fuel carried by Weiner's Liberator was probably due to the uncertain weather conditions, which made long diversions to an open airfield a distinct possibility. In the search area, the hunt was gradually widening, when at 1617 hours, one of *Vindex*'s Swordfish obtained a contact 22 miles southeast of A/120's attack (see map, page 254). Allowing for a U-boat speed of 2½ knots this contact, Bayliss thought "seemed very promising."[86] He immediately diverted *St. Thomas* and *Pevensey Castle* to

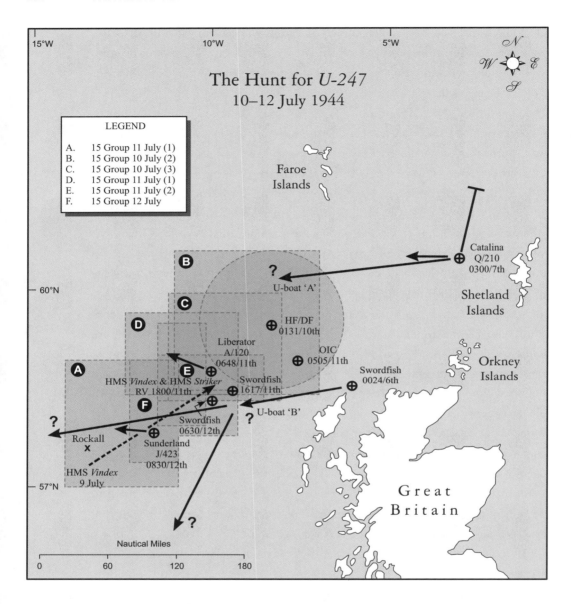

The Hunt for *U-247*
10–12 July 1944

LEGEND

A.	15 Group 11 July (1)
B.	15 Group 10 July (2)
C.	15 Group 10 July (3)
D.	15 Group 11 July (1)
E.	15 Group 11 July (2)
F.	15 Group 12 July

investigate and also maintained half a dozen Swordfish in the air in an attempt to swamp the area of the sightings.[87] At the same time, HMS *Striker*, another CVE, was on her way to relieve *Vindex*, and hearing of A/120's attack had flown off protective Swordfish patrols even as she left the Minch. Now, at 1700 hours on the 11th, she despatched a strike of two Swordfish to join the 6 Swordfish from *Vindex* searching an area out to 14 miles from the Swordfish datum.[88] *Striker* rendezvoused with *Vindex* at 1835 hours and the former's commanding officer, Captain W.P. Carne, RN, now became the Senior Officer of Force 34. It was agreed that *Vindex* would remain in company until 13 July, at which point her petrol supplies would run out. While the two carriers were operating together, *Striker* was to fly the daytime sorties and *Vindex* those at night (probably because the 825 Squadron Swordfish in *Vindex* had the superior radar, and the ship had developed techniques for landing-on aircraft at night and in poor visibility).[89] By now the focus of the search shifted to the Swordfish contact at 1617 hours. During the evening there were 10 Swordfish and 9 Coastal Command aircraft searching within 30 miles of the datum, while the two CVEs were positioned on the northern edge of the search area.[90]

By 1251 hours on 11 July, the OIC had concluded that A/120's attack was probably on *U-247*.[91] In the light of subsequent data, it now seems likely that the Liberator actually attacked a williwaw, and indeed the U-boat Assessment Committee's analysis later concluded that there was "Insufficient evidence of the presence of a U-boat."[92] But at the time, a more optimistic assessment of this attack held sway. Moreover, in planning for the subsequent air searches on 12 July, the evidence of the *Vindex*'s Swordfish contact biased the planning to a more southwesterly line of retreat for the U-boat. Then, just after midnight 11/12 July, CinCWA sent his intentions for Force 34 and 15 Group. Barely camouflaging the new information passed by the OIC, Horton signalled that, "In default of further information," the carrier aircraft were to search to the southward of the Swordfish contact at 1617 hours on 11 July, assuming that the U-boat was making good some 2-3 knots. Coastal Command aircraft would cover the possibility that the U-boat had withdrawn to the westward with six Sunderlands flying an intense crisscross search over an area centred 80-90 miles northeast of Rockall (and southwest of the Swordfish datum). In addition, the CVEs were, however, to operate so as to be able to support both their own carrier-borne aircraft and those of Coastal Command with a surface striking force.[93]

One of the Coastal Command aircraft flying on the east-west part of this search on 12 July was Sunderland "J" of 423 Squadron, RCAF, captained by Flying Officer Ulrich. J/423 had been airborne for some three hours in overcast weather, with the main cloud base at 1,300 feet and a visibility of 15 miles, but interspersed with patches of low cloud down to 100 feet, which cut the visibility to a mile. At 0826 hours, the Sunderland was a little over 70 miles east-northeast of Rockall, flying on an easterly track, at 1,200 feet, just on the lower limit of the full cloud cover and in a clear patch, when Ulrich sighted a small wake or feather on the surface of the sea bearing Green 30° at a range of 4 miles from the aircraft (see map, page 254).[94] Intent on making an immediate attack, he paused only long enough to change his position from the second to the first pilot's seat, where he could operate the weapon switches, before starting a shallow power dive towards the sea.[95] A few minutes later, the Sunderland passed a short distance to the north of the wake, and Ulrich could see about 4-5 feet of a U-boat's periscope exposed, leaving a wake 8 feet long, broadening to 12 feet wide astern of the periscope, from which he estimated the U-boat's speed at 3 knots and her course to be 275°. J/423's radio operator sent a "Flash" enemy contact report giving the U-boat's position. Meanwhile, Ulrich, with the throttles wide open, began to haul the 30-ton Sunderland round to the right aiming to approach along the U-boat's wake from a mile astern at a height of 50 feet.[96] As the Sunderland thundered in on its attacking run with just over a quarter of a mile to go, the periscope disappeared. Ulrich pressed on, aiming for the residual swirl and just four minutes after sighting the periscope, he dropped eight depth charges, set shallow and spaced 60 feet apart. The depth charges entered the water eight seconds after the periscope disappeared and while the swirl was still visible. About a minute after the attack J/423 was back over the aiming point. Two parallel oil streaks were seen near the froth patch caused by the depth charge explosions. Within a few minutes the oil had dispersed to such an extent that its edges could not be distinguished, suggesting that oil was not continuing to rise. The U-boat Assessment Committee would later conclude that in this attack the U-boat was "Probably slightly damaged."[97]

As soon as the attack had been made, J/423 had laid a datum marker and began transmitting homing signals for at least three hours. No signals were received from the surface forces, though they were heard transmitting on ship-to-ship R/T. Nevertheless, J/423 was soon joined by four other Sunderlands, some of whom had seen the naval force approaching the scene. First they circled "J"'s marker and then began a "hunt to exhaustion" for the U-boat, consisting of square searches around the area. The idea was that if a continuous patrol could be kept over the possible area in which the U-boat could be, it would be sighted when it was forced to schnorkel again – or to surface due to damage.[98] The first part of the hunt was automatically initiated. J/423 would cover Square "A", within 9 miles of the datum. The next two aircraft to arrive, P/423 and F/423, were to patrol Square "B" out to 16 miles from the datum, which, together with Square "A", covered the possible area which the U-boat could reach in the first four hours. Ideally, four more aircraft would arrive to patrol Square "C" out to 24 miles from the datum and covering eight hours of U-boat travel.[99] In this case, however, only two more Coastal Command Sunderlands arrived, which diluted the search. All these aircraft had been on task when J/423 attacked, and all reached the prudent limit of their endurance at about 1630 hours, eight hours after the attack.[100] But other events were also to compromise the integrity of the post-attack search.

The chase to the south

At this point it is worth reviewing the actual track of *U-247* after her attack on *Noreen Mary*, from what is now known, but what was only imperfectly understood at the time.[101] Immediately after sinking the fishing boat, *U-247* moved off to the northwest but, although schnorkelling, was not detected by *Vindex's* Swordfish during the early hours of 6 July. The Swordfish contact reinforced the British presumption that the U-boat would head westward *en route* to one of the Biscay ports as part of the general German effort to reinforce the area to maintain attacks on Allied shipping in the Channel. The general analysis was right, but their assumption that the U-boat would immediately begin its passage to Biscay was wide of the mark. *U-247*, with no orders to the contrary, remained in her operational area north of the Minch for a full five days before she was told to make for the west coast of France. Meanwhile, the British calculated that the U-boat was moving at some 50-60 miles per day, in a westerly and then southwesterly direction. This was a reasonable assumption, given that, as the OIC knew, the U-boat had been at sea for nearly six weeks, and by the time she completed the three-week passage to Biscay, her provisions and crew endurance would be practically exhausted.

The OIC seems to have been confident that only one U-boat was operating in the Minch. In this they were on firm ground, for although the precise position of operational U-boats was, for the most part, obscure, the OIC could be reasonably confident of the numbers at sea and their general deployment. However, in the naval and air command headquarters the "secret" plot had to be constantly reconciled with the operational picture generated from sighting reports, particularly from aircraft. The major challenge was that which beset most A/S operations: the ability to correctly classify contacts. This was made particularly difficult in the operations against *U-247* (and other U-boats at this time) for Allied units operating in adverse weather conditions alternating between rough seas and low cloud or fog.

Moreover, the enemy had only just begun to operate schnorkel-fitted U-boats. Although their advent and the consequent move to inshore operations had been anticipated, the opportunity for Allied air and surface units to become familiar with these new tactics by early July had been limited. In practical terms, too, detection of schnorkel-fitted U-boats relied on visually sighting a periscope or schnorkel head. Radar was occasionally successful, and on rare occasions sonobuoys gained contact. All these forms of detection proved much less reliable than those obtained during the main convoy battles of 1941-43 on the much larger and more obvious shape of a surfaced U-boat. This was exacerbated by the relatively few U-boats deployed on operations. The low density of U-boats in any operational area gave the impression, as Horton later remarked, of "ghost" U-boats, rarely revealing their location. Finding them became

> largely a matter of guesswork in sorting the "sheep" from the "goats" from the large number of moving and bottomed contacts, disappearing radar blips, positive snorts, sonobuoy contacts, transit A/C sightings and other wonders of the deep.[102]

There was, it is clear, in this early stage of the inshore campaign, a somewhat uncritical willingness to accept aircraft reports at face value. It is ironic that some of these reports, such as that by the Canso of 162 RCAF Squadron on 6 July, only 17 miles west of *Vindex*, which attracted several Swordfish and a surface escort was quickly abandoned when the aircraft reported that the sighting had been of floating mines. *U-247* was, in fact, close-by and witnessed much of the action and may even have attempted to close on *Vindex*.[103]

Many of the spurious contacts could have been, and later were, ruled out when full information was analyzed. But the speed and capacity of the available British communication systems proved inadequate to resolve the classification problem in near real time. Operational headquarters were therefore challenged to combine SI material only rarely based on current decrypted signals with a plethora of sighting reports from aircraft ill-practised in recognizing U-boats from "other wonders of the deep." Combining and sifting this data was made more difficult when the naval and air headquarters were physically separated, as was the case of Harwood and 18 Group. This all helps to explain why an erroneous, but at the time reasonable, assessment was made that there could be two U-boats operating to the north of the Minch. When this was added to the presumption that the two U-boats were bent on retreating, the resultant dilution of British A/S assets is understandable. As Harwood relinquished control of the hunt to Horton on 8 July the search for the second U-boat was evaporating, and Horton was able to concentrate his available forces against a single westbound U-boat.

As seen by the OIC, the U-boat transit routes between Biscay, the Channel and Norway had historically been beyond longitudes 17-25°W, that is, some 200-450 miles west of Ireland.[104] It was on the inner boundary of this swath that the OIC estimated *U-247* would pass. But, once Matschulat signalled his situation report late on 9 July, and the OIC received the decrypt early on 11 July, these assumptions do not seem to have received serious revision. The decrypt also exposed the fact that the U-boat had not made substantial ground to the west and revealed that Matschulat calculated he had only 17 days of provisions.

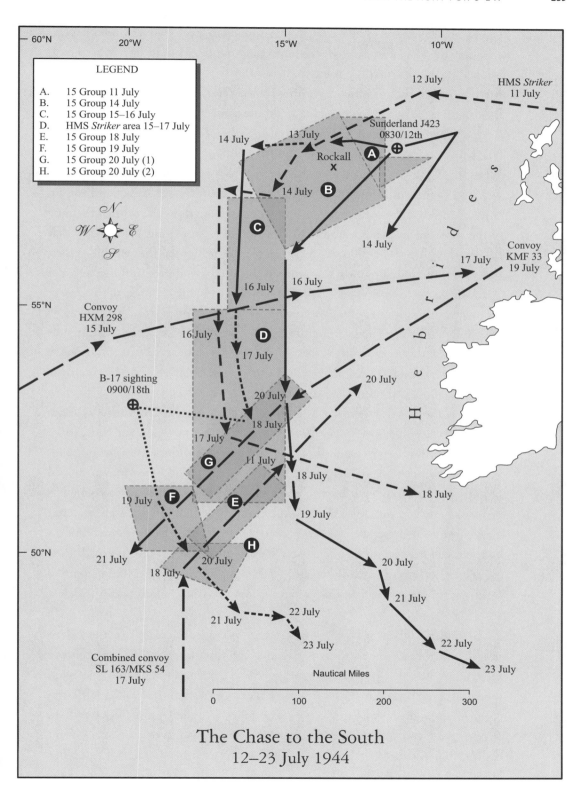

The Chase to the South
12–23 July 1944

Unless, the crew were to be rationed, this implied that the U-boat would have to abandon any idea of a passage far out into the Atlantic. This, in turn, would have implied the U-boat taking a route to the southeast of Rockall rather than passing to the north of the rock before turning south. The OIC, in particular, clung to the notion that the U-boat would take a route some 200-300 miles to the west of Ireland. This was between 60 and 100 miles further to the west than the actual U-boat's route. Horton, and consequently 15 Group, seem to have initially tried to deploy the CVE and Coastal Command forces to cover the U-boat's escape to the west and to the south from A/120's contact, and those by *Vindex*'s Swordfish, on 11 July. However, the consistent OIC assessments soon drew the centre of gravity of the subsequent search towards the west. It seems surprising at first sight that the attack by J/423 on 12 July did not succeed in realigning the searches. However, when the initial reports are reviewed the ability to distinguish between the attack by A/120 (on a non-sub) and that by J/423 (actually on *U-247*) was almost impossible. Both were on "positive" visual sightings of "periscopes" and therefore seemingly indistinguishable one from another. Ironically, given the existing presumption of a westerly-going U-boat, both attacks were on U-boats assessed as travelling on a roughly westerly heading. Only weeks later, when detailed analysis had been completed, were these attacks correctly assessed.

When the sighting report was received from J/423, *Striker* (with *Vindex* in company) was some 85-90 miles northeast of the attack position. She had a strike of two Swordfish airborne and these were despatched to the scene of the action. However, at 0915 hours, Ulrich reported that he had sighted "two torpedo-shaped objects 12-15 feet long light brown in colour...."[105] The report was interpreted in *Striker* as a whale sighting about a mile from the initial U-boat sighting, and Carne immediately recalled his Swordfish.[106] He preferred to concentrate his air and surface forces on following up ASV reports by *Vindex*'s Swordfish made during the early hours of 12 July. The latest of these, at 0630 hours, was some 10 miles southwest of the earlier Swordfish report at 1617 on 11 July, which had at the time seemed so promising.

By midday, when *Striker*'s Swordfish were on task, the search area had been moved a further 10 miles southwest (to take account of presumed U-boat movement), but was still 50 miles northeast of the continuing Coastal Command search around J/423's attack.[107] At this time 15 Group Control asked Ulrich to confirm his sighting of a periscope. The CinCWA then prompted Carne to "flood" the area of the attack, and as a result Swordfish were despatched from *Striker*, followed an hour later by *St. Thomas* and *Pevensey Castle*.[108] These ships arrived on the scene some 10½ hours after J/423's attack. The ships covered the quadrant between south and west out to 15 miles from the datum and the search was completed by 0045 hours, 13 July. The ordering of the search makes little sense, for by the time it was started the U-boat could easily have travelled some 20 miles or more, and so was likely to be outside the search area.

On 12 July, the OIC concluded that J/423's attack had probably been on *U-247*, and by the following day they were able to tell CinCWA that the U-boat had been told specifically to make for Brest and was then passing just to the north of Rockall.[109] On the next day, the OIC also announced that since the middle of 13 July, "only partial Special Intelligence [was] available." They, nevertheless, assessed that *U-247* continued "probably moving slowly westwards" and was then some 60-70 miles west of Rockall.[110] Over the following two days, 14-15 July, the OIC assessed on the barest intelligence that *U-247* had turned to the southward but that her position was extremely vague.[111] Carne, of course, was only partially aware of the details of these intelligence assessments. He had now been joined by the 3rd Escort Group, under Commander R.G. Mills, RN, in HMS *Duckworth*. However, the best deployment of his force was difficult for by this stage the "whereabouts of the U-boat was obscure...."[112] Nevertheless, Carne used part of EG3 first to scour the area to the south of Rockall, and then to sweep to the north and west of that rock.[113] Meanwhile a sporadic flying programme was maintained by *Striker*, though limited by poor visibility and light winds in which the Swordfish, which could not be catapulted, were unable to operate. Simultaneously, 15 Group was searching an area extending 50 miles either side of a line that was gradually being extended in a south-westerly direction from Rockall. The sorties were flown by Liberators of 120 and 59 Squadrons, and, during the day by Sunderlands of 422 and 423 RCAF Squadrons, with up to six aircraft simultaneously on patrol.[114] None of the searches was successful.

On 15 July, three days after J/423's attack, CinCWA signalled that "Information of whereabouts of U-boat is very vague."[115] Horton's assessment echoed the increasingly uncertain estimates being produced by the OIC, with whom he (or his staff) was in regular touch by telephone.[116] Horton decided that Coastal Command was, for the time being, to continue searching a rectangle centred about 120 miles southwest of Rockall and covering the estimated track of *U-247* for 15-16 July.[117] To the south of this area, *Striker* was to cover an area extending over 200 miles to the southward.[118] Between these two areas lay the planned route of convoy HXM 298 (see map, page 259). The 9-knot, 117-ship convoy was escorted by the Canadian Escort Group C4 under Commander E.W. Finch-Noyles, RCN, in HMCS *Wentworth* and three merchant aircraft carrier ships.[119] Finch-Noyles had received warning earlier in their passage of U-boats which might affect his convoy, one of which was from the position of A/120's attack. Not only was this not actually against *U-247*, but the fix was mis-timed by 24 hours.[120] This had the effect of making the U-boat's position on 16 July seem to be somewhere to the south of the convoy's route, and therefore less of a threat. In any case, none of Coastal Command aircraft or units from Force 34 were diverted to support the convoy, even though it was passing through the (albeit rough) estimated positions of *U-247* on 16 July.[121]

As *U-247* continued her southward passage and into *Striker's* search area, the OIC admitted that the U-boat was "entirely unfixed" and Coastal Command seems to have been largely withdrawn from the hunt.[122] However, by 18 July, the OIC assessed her position very tentatively as posing a threat to the north-bound combined trade and operational convoy SL 163/MKS 54. As a result, Coastal Command air escorts, by two Sunderlands of 422 Squadron, were laid on to cover the area of supposed interception, even though the escort by Escort Group B3 was already supported by the CVE HMS *Campania*.[123] Then, at 0930 hours on 18 July, a transiting B-17 sighted a U-boat some 330 miles west of Ireland and well to the west of *Striker's* patrol area (see map, page 259).

The information seems to have taken some time to filter through to the OIC, but when it did on 19 July, they assessed that this could be *U-247* actually travelling even further west than they had supposed.[124] However, the previous evening, *Striker*, along with her escorts, had already been withdrawn from the operation and sometime around midnight on 17/18 July Force 34 swept across the path of *U-247*, but too far away for either side to be aware of the other's presence.[125] Meanwhile, 15 Group despatched 5 Sunderlands to sweep over an area on the new predicted route for the elusive U-boat.[126] The search, without any result, was extended into the following day with four Sunderlands. But such was the uncertainty of the U-boat's position that cover was also provided for the fast outbound convoy KMF 33 by a string of Sunderlands also provided by 422 and 423 RCAF Squadrons.[127] By 20 July, the Coastal Command search for *U-247* was finally abandoned, when *U-247* was actually 130 miles southwest of Ireland and some 200 miles east of Coastal Command's search.

9th Escort Group enters the picture

U-247 was just one of several U-boats which held the attention of C-in-C, Plymouth, and AOC 19 Group in their ACHQ at Mount Wise. In particular they were concerned over the U-boats estimated to be making for the Brittany port of Brest. These included *U-247* as well as others returning from operations in the Channel. In an attempt to intercept these boats, the Allies began to tighten their blockade of the port by stationing escort groups close off the entrance with Coastal Command patrols over the approaches. Apart from large numbers of Canadian airmen in RAF squadrons, some 40% of the offensive surface escort force was provided by the RCN. Moreover, these operations between July and September 1944 (as well as those which were to follow) clearly showed the full integration both in the air and at sea of Canadian units and individual personnel into the Allied antisubmarine campaign in British coastal waters.

For example, at 2216 hours on 19 July, Liberator "C" of 224 Squadron lifted off from St Eval. Half the crew were Canadian, and the aircraft's mission took her to a search area off the northern coast of Brittany.[128] After four hours airborne, C/224, was 30 miles west of Brest when at 0232 hours on the 20th the radar operator obtained a contact 2 miles ahead of the aircraft (see map, page 264). As the aircraft closed, the navigator at the front gun position saw two wakes, "one thin ahead – and large and broader wake behind," passing under the starboard wing of the aircraft.[129] About 45 minutes later the Liberator returned to the position of the initial contact and dropped a single sonobuoy, on which propeller noises were heard. After the noises faded, another sonobuoy was dropped 2 miles down the supposed southeasterly track of the contact but this obtained no contact, so additional buoys were dropped, one of which produced very weak propeller noises. Then, at 0420 hours, the crew got another radar contact at a range of 3½ miles. The aircraft dived to close at low level and when they were three-quarters of a mile away the Leigh-light was switched on. This was rather too close to the target, and the pilot was only able to glimpse a dark object with a wake passing under the aircraft's port wing as the Liberator sped past.[130] Consequently, the aircraft was too close to attack, though another sonobuoy was dropped. Although C/224 circled the position for another 3½ hours, no further contact was obtained and the aircraft departed for base.[131]

The OIC had by early afternoon concluded that C/224 had probably made contact with *U-953* and signalled C-in-C, Plymouth accordingly. The U-boat was loitering off Brest after a successful mission in the Channel, when C/224's radar detected the U-boat's masts as the she requested entry instructions by radio. By late afternoon the OIC was able to confirm that the U-boat's time of arrival in Brest would be in the early hours of the morning of 22 July.[132] Ashore in the ACHQ, Commander M.J. Evans, RN, had been in charge of A/S operations since before the invasion of Normandy.[133] For this post, Evans had again been indoctrinated into Ultra.[134] Before the OIC assessment had been received, Evans had already pieced together the rather poor DF bearing and the sighting report from C/224 with the general SI tracking of *U-953*'s progress down Channel. At

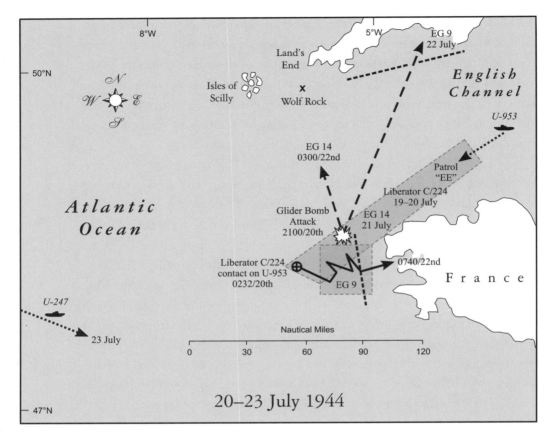

8°W

Land's
End

5°W

EG 9
22 July

50°N

Isles of
Scilly

✕
Wolf Rock

*English
Channel*

U-953

EG 14
0300/22nd

Patrol
"EE"

Liberator C/224
19–20 July

*Atlantic
Ocean*

Glider Bomb
Attack
2100/20th

EG 14
21 July

Liberator C/224
contact on U-953
0232/20th

0740/22nd

France

EG 9

U-247

23 July

Nautical Miles

0   30   60   90   120

47°N

20–23 July 1944

1009 hours he directed the 9th Escort Group, under Commander A.F.C. Layard, RN, to proceed to a 15-miles square area off the entrance to Brest with the hope of intercepting the *U-953*.[135] Although Layard had had access to "the highly secret U-boat plot" during an earlier appointment in the Admiralty, none of the very secret special intelligence was passed to him while in command of EG9.[136] The instructions from C-in-C, Plymouth, sent to Layard were less detailed, telling that:

> From sightings and DFs they think that a U/B is approaching Brest from the W, arriving either am 21st or more likely 22nd. We have been given are area in the approaches to Brest to patrol, and EG14 are to join us once they have finished topping up with oil at Plymouth.[137]

Events, however, were not to turn out as planned. The first signs of trouble appeared as EG9 approached their patrol area, when enemy aircraft were spotted. Layard reported them, and about an hour later some Mosquito fighters arrived to provide cover but left after the ships had been in the area for about an hour.[138] For nearly two hours the patrol was quiet, until about 2050 hours. Then Layard was called to the bridge of *Matane*. As soon as he arrived he saw a big explosion astern of *Meon*. At first, this explosion was thought to be a Gnat, but Layard soon realized that it had been caused by a glider-bomb launched from one of the German aircraft still visible. He also noticed that a second missile was on its way, and, as he wrote later,

> By this time we were at action stations and seeing the bomb I rang down full ahead but almost before it could take effect this winged monster was right on top of us travelling at an incredible speed in about a 70° dive. There was a shattering explosion and a frightful escape of steam and I thought we'd been hit right on Y-gun and I was expecting to step off into the water any minute but we still remained afloat....[139]

The glider-bomb had grazed the side of the ship and exploded in the water blowing in the side of the ship, wrecking the engine room, which immediately flooded and brought the ship to a stop. More bombs were launched, but none hit, though "one landed uncomfortably close to us on our port beam."[140] The heavy damage to *Matane* caused EG9 to retire to Plymouth. Meanwhile, EG14 under the SO Commander R.A. Currie, RN, sailed at 1915 hours for a high-speed passage to relieve EG9 off Brest. Currie arrived in the area at 0130 hours on 21 July, and that afternoon they too were attacked with glider-bombs, but were now provided with constant air cover which disrupted the German attacks. EG14's ships were also better armed with 4.7-inch guns and therefore able to put up a more effective barrage, which, Currie remarked, "appeared to disconcert the aircraft."[141]

During these events, *U-953* was loitering off Brest just to seaward of the patrol area allotted first to EG9 and then to EG14, and although she schnorkelled for long periods she remained undetected by the escorts, probably because the small radar echo produced by the schnorkel head was lost amongst the clutter generated by the westerly wind and sea state of 3-4.[142] Then, just as *U-953* edged closer to Brest shortly after midnight 21/22 July, C-in-C, Plymouth, ordered Currie to break off his patrol and head to the north, probably chasing after

Commander A.F.C. Layard, RN, on the bridge of HMCS *Swansea* in November 1944. Layard's command of RCN Escort Group 9 reflected both the scarcity of experienced professional sailors in the RCN and the ability of this seasoned professional sailor to work well with less experienced Canadians. (Courtesy of Raymond Layard)

another U-boat which was thought to be heading for the coastal convoy routes running between the Lizard and Start Point.[143] Shortly after Curry departed to the northward, the group intercepted an HF signal which suggested that a U-boat had arrived off Brest.[144] In fact, *U-953* entered Brest in the early hours of 22 July and by 0740 hours was safely ensconced in a concrete-covered U-boat pen.[145] Her escape was a double misfortune for the British. Not only was the chance of destroying the U-boat missed, but also of killing her experienced captain, Karl-Heinz Marbach, who was soon transferred to command of the new Type XXI high-speed U-boat, *U-3014*.[146] The combination of novel technology and veteran captains posed a threat to Allied shipping that the Admiralty took seriously. Even as the hunts for *U-953* and *U-247* unfolded, the Royal Navy was in the process of converting HM Submarine *Seraph* into a high-speed target designed to develop tactics for use by the escort groups to counter the Type XXI U-boat.[147]

On the morning of 22 July, Layard's crippled *Matane* was towed into Plymouth. Three of EG9's vessels, HMC Ships *Stormont, Swansea* and *Meon,* sailed in the early afternoon to patrol the convoy route between Lizard and Start Point.[148] Meanwhile, *U-247* was approaching from the west, but the OIC had had neither firm evidence of her position, nor an estimate of her time of arrival in Brest.[149] That changed in the early hours of 26 July. In an intercepted signal from Matschulat, the OIC learned of *U-247*'s position and her intention to enter Brest within 48 hours. In subsequent signal traffic, the OIC also learned that her escort would be available from 0400 hours on 28 July. This information was passed on 26 July to C-in-C, Plymouth, at 1359 hours and amplified three-and-a-half hours later.[150] Evans, in the ACHQ, reacted by organizing a 19 Group patrol to be set up covering *U-247*'s position and then, later, by sending the 2nd Escort Group under Commander N.W. Duck, RNR, racing southwards to an east-west patrol line astride *U-247*'s approach route to Brest (see map, page 267).[151] The group was to be on station by 1100 hours on 27 July, but although EG2 reached the patrol about this time, *U-247* had already passed to the north of the position and entered Brest unhindered at 0644 hours on the morning of 28 July.[152]

Almost three hours later, Layard, having taken command of HMCS *Swansea*, led EG9 out of Plymouth for an area in mid-Channel, before shifting to a search of the convoy route between Mounts Bay and the Lizard.[153] That day Layard also noted from the BBC news that:

> The Americans seem to be attacking in strength in Normandy and have broken well through the German lines in places. The Russians also continue to advance and they are now within 3 miles of Warsaw.[154]

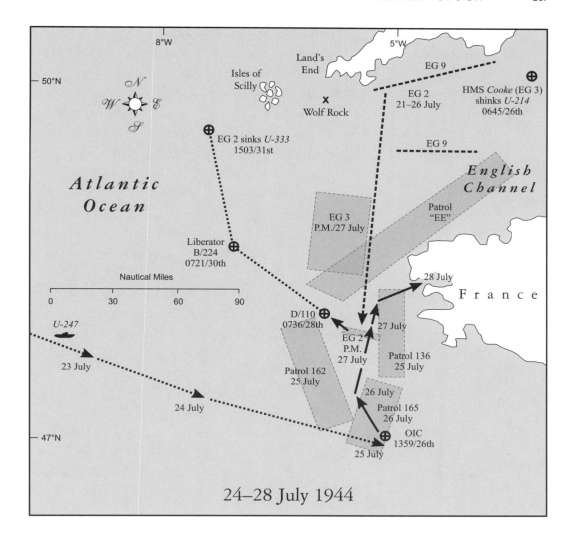

24–28 July 1944

Throughout his diary, Layard makes it plain that he was kept abreast of events as the war progressed. By contrast, Matschulat was largely insulated from the out-side world while on operations. When he sailed from Norway at the end of May, the invasion in the west was expected but had not happened. The news from the Eastern Front was probably worrying, though the Russians were still about 200 miles from Warsaw.[155] By the time *U-247* entered the U-boat pens in Brest the situation had deteriorated. The Western Allies had not only established a firm foothold in Normandy, but just a few days before the U-boat entered Brest, the Americans had also broken through the western part of the German front and were rapidly exploiting their success towards Avranches.[156]

Ever since the German occupation of the Biscay ports in 1940, U-boats had been able to operate more efficiently against Allied convoy routes. Now during the summer and autumn of 1944, as the Allies strove to build up their ground forces in Normandy, these same bases were ideally placed close to the flank of the resupply shipping routes.[157] Almost half of the initial counter-invasion U-boats had sailed from Brest.[158] By the third week of August that port still contained almost one-third of the operational, or nearly operational, U-boats in the Biscay ports, though thereafter the number of effective U-boats rapidly decreased.[159] Already some U-boats had been moved from the northern bases to Bordeaux and La Pallice, followed by a general evacuation of all the Biscay ports with most boats that were operational being sent to Norway, with a few (including *U-247*) directed to carry out patrols in coastal waters before proceeding northwards.[160] With the onrush of the American land forces into the Brest Peninsula early in August, the British considered it necessary "to move support groups as near as possible to the U-boat's bases...." These groups, together with bombing, mine-laying and strikes by aircraft, were to harass the U-boats continually "and do all that was possible to retard their departure."[161] Layard's EG9, among other groups, was engaged on this task, with their search areas (and those of supporting Coastal Command A/S aircraft) being constantly changed in an attempt to intercept U-boats entering or leaving the Biscay ports.[162]

This close blockade of the U-boat bases was to be supported by Vice Admiral F.H.G. Dalrymple-Hamilton, Admiral Commanding 10th Cruiser Squadron (CS10).[163] Dalrymple-Hamilton, as CS10, was formally part of the Home Fleet but he was indoctrinated into Ultra and specifically chosen for these operations "to ensure that maximum use was made of the available Sigint."[164] He led Force 26, consisting of the cruiser *Bellona*, two destroyers and, initially, HMS *Striker* providing fighter cover. CS10 was soon joined by the cruisers *Mauritius* and *Diadem*, each with destroyer escorts forming Forces 27 and 28.[165] These cruiser forces were primarily designed to prevent the enemy using his few remaining destroyers or evacuating troops from the Biscay ports, as American forces closed in on them from the landward side. The interaction of all these measures was noted by Layard, who mentions an engagement on 22 August some way to the north of his patrol area in which:

> Force 26 carried out some sort of anti-shipping operation ... off Lorient and sank some minesweepers. What with him and surface patrols, nothing much can move on the French Atlantic coast without being detected, and any evacuation of troops from the beleaguered ports would be quite impossible.[166]

Of course, the loss of minesweepers magnified the effectiveness of the mines laid by Bomber Command, which could not then be swept. Two days later,

Layard was on the bridge at 2020 hours when he saw a strike of 18 Beaufighters streaking across the sea at very low level heading for the Gironde.[167] These aircraft were from 236 Squadron, RAF, and 404 Squadron, RCAF, led into the attack by the New Zealander Squadron Leader E.W. Tacon. They struck the German destroyer *Z-24* and the large torpedo boat *T-24* lying in the estuary of Le Verdon. *T-24* had already been damaged earlier in the month in an encounter with Force 27, and now the Beaufighters hit both ships above and below the waterline with 25-lb. rocket projectiles and 20mm cannon fire. The attacks were pressed home to short-range in the face of intense flak from the ships and shore defences, which damaged 15 of the aircraft. The attack lasted only a few minutes, and as the Beaufighters climbed away, they left *T-24* sinking and *Z-24* so seriously damaged that she capsized and sank a few hours later. These had been the last remaining serviceable destroyers in the Biscay ports. Three of the aircraft had to put down on a French airfield, which they hoped was not still occupied by the enemy, and where Tacon also landed to pick up the crews (though one crew, badly wounded, had to be left behind). Another aircraft from 236 Squadron was also heavily damaged and successfully ditched in the sea, watched by an aircraft from 404 Squadron RCAF. This crew were picked up by *Port Colborne* and *Meon* from EG9, who were, while carrying out the rescue, fired upon by Force 28.[168] Layard was angered by the incident, but his ships had been out of position and had initially failed to send the correct identification signal when challenged. The Admiralty thought they had placed too much reliance on their IFF transponders, which were not seen by Force 28.[169]

By the end of August, the Allies were clearly winning the "battle of the build-up" of both stores and men into the Normandy battleground, which emphasized the importance of German attempts to interdict the resupply routes across the Channel. The U-boats were still striving to carry their attack into the Seine Bay area, but the OIC had calculated that this had led to some 40% of the U-boats initially sent against invasion shipping being sunk.[170] As a result, the enemy had become utterly reliant on the use of schnorkel-fitted U-boats to penetrate the Allied defences but, by the nature of these operations, had little idea of the progress being made by the U-boats or the success, if any, being achieved.[171] During August, however, the U-boats' direct attack on Allied shipping in the Seine Bay began to wane.[172] Signs of a change in *BdU's* strategy were already apparent in early August with the appearance of a U-boat off the north Cornish coast, where it sank a merchant ship and a Canadian corvette.[173] By the end of the month, the OIC's suspicion that the enemy were about to embark on a widespread campaign in British inshore waters was confirmed by indications of U-boat activity off the

Butt of Lewis, further patrols in the Bristol Channel and off Land's End, where the OIC assessed that *U-247* was to fill the billet.[174] However, before she could be made war ready, Matschulat had to repair the damage to his boat caused by the RCAF Sunderland J/423's attack off the Hebrides.

Although Matschulat could work in the apparent safety of the concrete pens at Brest, the surrounding infrastructure was under constant attack by the heavy bombers of the Eighth Air Force and Bomber Command. The Director of Naval Intelligence estimated that sustained air attacks on the U-boat bases, while not likely to cause direct damage to U-boats in the pens, contributed to substantial delays in U-boat turn-round times and, at least "must have caused some disorganisation."[175] Matschulat found that his crew's shore-side accommodation was one of targets allocated to the bombing raids, but he might have hoped for relative safety while working in the concrete U-boat shelters.[176] In this, his optimism was dramatically shattered by a series of air raids by the 617 "Dambusters" Squadron during the month in which they dropped twenty-six 12,000-lb. Tallboy bombs in precision daylight attacks on 5, 12 and 13 August, hitting the U-boat pens with nine bombs, four of which made holes in the 5.6-metre-thick concrete roof. Though these hits caused no damage to the U-boats, including *U-247*, in the shelter, it must have been apparent to the Germans that even the safety of the pens was now compromised.[177]

Simultaneously, other RAF bombers laid mines off the harbour.[178] Throughout the month of August, as *U-247* lay in the bunker, RAF and USAAF aircraft attacked Brest on 15 occasions. Then, on 25 August, HMS *Warspite*, escorted by EG14 and other destroyers, including HMCS *Assiniboine*, arrived off Brest to bombard the defences with 15-inch shells.[179] For the first time, a specialist naval shore spotting organization was deployed to ensure the accuracy of the bombardment, and 50 rounds were fired at ranges of some 15 miles.[180] This was accompanied by heavy attacks by tactical aircraft of USAAF Ninth Air Force and followed overnight by attack by No. 6 (Canadian) Group, Bomber Command.[181] These attacks, as well as continued pressure from Allied naval forces close off the Biscay coast, all helped to induce the German High Command to announce on 27 August that:

> The area off the Biscay ports is no longer to be a field of operation owing to lack of counter-measures to the extensive mine and air danger and search groups....[182]

An accidental hunt [183]

Shortly after midday on 25 August the OIC signalled to C-in-C, Plymouth, that it was assumed that *U-247* had sailed from Brest and was then some 20 miles west of the port, ultimately heading for the Channel.[184] But by the evening they had corrected their warning because from a further decrypt it became clear that defects had once more delayed *U-247*'s sailing until 2200 hours that night. The decrypt also revealed that the U-boat was to make ground to the west for about 60 miles, before turning northwards for an area around Wolf Rock, though it was not yet clear whether this was "itself the operational area or merely an approximate point on route to it."[185] This decrypt also revealed that two other U-boats, "KN" (probably *U-758*) and "KT" (probably *U-262*), which were already at sea and thought to be heading for patrols in the Channel, had now been ordered to make for the north Cornish coast.[186] Through this area ran the resupply convoy routes from the South Wales and West Country ports *en route* to the Normandy beachhead. Although the immediate battle of the build-up had been won, these convoy routes still carried heavy traffic to support the breakout now in progress on the American sector of the Normandy front. The convoy routes, therefore, received considerable attention from C-in-C, Plymouth's staff, particularly as one U-boat, *U-667*, had already operated in the Bristol Channel in early August and had sunk a merchant ship as well as HMCS *Regina*.[187]

Continuing delays, including air attacks on the ships intended to escort her out of the harbour, bedevilled *U-247* and kept her in harbour until the late afternoon of 26 August, when she finally sailed.[188] C-in-C, Plymouth, was told of this news by the OIC in the early afternoon and he also learned that only one other serviceable U-boat remained in Brest.[189] By the middle of the following day the OIC predicted that *U-247* was some 30 miles west of Brest.[190] Winn also informed C-in-Cs that *BdU* had already "expressed the intention that there should be no more U-boat operations in the Seine area of the English Channel…."[191] Rear Admiral J.H. Edelsten, ACNS(UT), estimated at least 57% of the U-boats taking part in the Channel operations to date had been lost – a relatively poor return for the few ships sunk – and ACNS(UT) concluded that "We now pass to the Battle of the Inshore Routes."[192] As the OIC refined its assessments, Winn concluded that the area for U-boat operations in the next few weeks

> probably includes St. Georges Channel and Bristol Channel and North Channel. It is doubted whether the Irish Sea is included. It would not be safe to assume that the south coast of Cornwall is excluded.[193]

As C-in-C, Plymouth, read the OIC's appreciations and contemplated his response, he watched the plot in his headquarters which showed *U-247* and at least
two other U-boats converging on the coastal convoy routes round the Land's
End area, as well as three others passing out of the Channel which would also
need attention (see map, page 273).[194] During this period too, there were several
U-boats that had recently left the Biscay ports that were *en route* for Norway. To
counter these movements, C-in-C, Plymouth, had about four escort groups and
20-25 Coastal Command aircraft patrolling areas, designated according to the
available special intelligence, in the inner Bay of Biscay and western Channel.[195]

Thus, on 28 August 19 Group laid on Patrol 104, which consisted of an elon-
gated rectangle lying along *U-247*'s track (though slanted towards an approach
from the position of T/105's contact). The patrol was to be flown by three aircraft at a time.[196] Simultaneously, C-in-C, Plymouth, shifted Commander L.A.B.
Majendie's 15th Escort Group from the northern Bay to an oblong area centred
40 miles south of Land's End and extending 40 miles east-west and 20 miles
north-south and lying directly across *U-247*'s track. Moreover, Majendie's Group
were ordered to approach the area along *U-247*'s predicted track.[197] EG15 arrived
in the new area in the evening of 28 August.[198] They had hardly got settled when
they were diverted at midday on the 29th to investigate a "possible U-boat" just
west of their area reported by Liberator "L" of 224 Squadron.[199]

This aircraft was captained by Canadian Flying Officer K.O. Moore, who
had the distinction of sinking *U-441* and *U-373* in under half-an-hour on the
night of 7/8 June 1944. Now L/224 was flying at 5,000 feet homeward bound
from a search mission in the central Bay area targeted against U-boat "KT",
when, at 1003 hours on 29 August, the aggressive Moore saw a swirl in the
water some 6 miles from the aircraft (see map, page 273).[200] Moore thought
this looked suspicious and might have been caused by a U-boat. Unfortunately,
Moore only had enough fuel to mark the position before heading for base.[201]
Ashore in the ACHQ it was realized that this sighting could have been the U-
boats "AR" or "ED" coming down-Channel, or *U-247* if she had made faster
progress than predicted towards Land's End. EG15 was therefore sent post-haste
to investigate.[202] Majendie searched the area around L/224's marker until the
early evening but found nothing, whereupon C-in-C, Plymouth, ordered him
to resume his earlier search area, though shifted 10 miles northwards to allow
for *U-247*'s progress along her predicted track.[203] Unfortunately, they did not
get any contact.

On 29 August, as EG15 was being moved over *U-247*'s route, 19 Group adjusted the air search (now Patrol 109) to cover *U-247*'s approach, moving it slightly

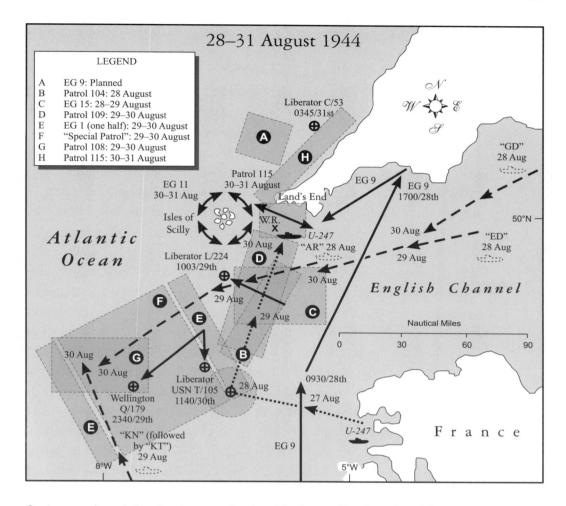

28–31 August 1944

LEGEND

A   EG 9: Planned
B   Patrol 104: 28 August
C   EG 15: 28–29 August
D   Patrol 109: 29–30 August
E   EG 1 (one half): 29–30 August
F   "Special Patrol": 29–30 August
G   Patrol 108: 29–30 August
H   Patrol 115: 30–31 August

further north and aligning it more closely with the predicted track.[204] That same day the 1st Escort Group, under Commander C. Gwinner, RN, in HMS *Affleck*, was split into two divisions and shifted to two areas some 70 and 130 miles west of Brest.[205] Sandwiched between the two halves of EG1 was a 50-mile square which enclosed the "crossroads" between likely routes for U-boats transiting from the Biscay ports or from the Channel to Norway.[206] The Liberators of the USN Air Wing (operating under 19 Group control) were to carry out a "special" search of the square using sonobuoys, while Gwinner's ships acted as pouncers for any contact obtained by the Americans. That day, however, the stiff southwesterly wind substantially reduced the effectiveness of the sonobuoys and no contacts were made.[207] Overnight Leigh-light Wellingtons took over the search and one of these aircraft, "Q" of 179 Squadron, had been on patrol for two hours when at 2340 hours its captain, Flying Officer O'Dwyer, sighted "a bluish coloured light which flickered irregularly…." After eight abortive runs over this light, O'Dwyer attacked with six depth charges.[208] Gwinner immediately ordered his ships to concentrate in the area of Q/179's contact. Communications were established with the aircraft and O'Dwyer reported that "a periscope had been sighted on a westerly course."[209] Considering the weather conditions, Gwinner "felt some dubiety about this report." However, as he noted, "the sighting was in a high probability area and therefore, a most meticulous and thorough search was carried out."[210] However, by 0900 the next morning, Gwinner was convinced that that they "were chasing a 'Will-o'-the-Wisp'" when he received a signal from C-in-C, Plymouth, saying that Q/179's sighting was "now considered doubtful."[211] It seems likely that O'Dwyer had been attacking one of the sonobuoys remaining afloat in the area from the earlier "special" search by the USN Liberators.

That morning of 30 August, the USN Liberators arrived on task while Gwinner's frigates were still clearing the area, and no sooner had they begun laying their sonobuoys than Liberator "T" of Patrol Bombing Squadron 105, captained by Lieutenant E.E. Edwards, USNR, obtained a detection and by 1105 hours, Edwards was sufficiently confident in the sonobuoy contact to make an attack at 1140 with their "special armament," a euphemism for the Mk. 24 mine, or homing torpedo (see map, page 273). The attack, Edwards reported had "satisfactory results" and was followed by a further pattern of sonobuoys.[212] Another USN Liberator, M/110, arrived and began to monitor the additional sonobuoys laid by T/105, but neither aircraft regained firm contact.[213] Forty-five minutes later, Gwinner, who was racing towards the scene, was 2 miles distant when he slowed to 7 knots as an anti-Gnat measure. The first three ships of EG1 swept through the attack position but detected nothing, until about an hour later *Bentley* obtained an asdic contact about 2¼ miles to the southwest. *Bentley* attacked with Hedgehog, and when Gwinner joined, *Affleck* too obtained a "shaky" contact. Although he classified the contact as non-sub, Gwinner attacked it with eight depth charges.[214] Further attacks followed before EG1 broke off to search out to 14 miles from the datum.

Just before dark as the ships returned to the datum, Gwinner obtained another contact, and he managed to get an echo-sounder trace of the target. This rose to 36 feet above the bottom, had no shape, and the echo "though solid, was lifeless and very 'non-sub.'"[215] Gwinner thought this contact was a bottom feature, rather than a U-boat, but "decided to make sure that whatever it was remained lifeless, and, calling in *Bentley*, carried out six further attacks.'[216] Each of these attacks was made using the echo-sounder and the charges were set to burst on the bottom, but at no time did any direct evidence of a U-boat come to the surface. At 2320, after 12 attacks in all, Gwinner abandoned the hunt. *Affleck* then set course for

Plymouth, arriving at 0800 on 31 August to rectify an asdic defect. While along-side, Gwinner visited the RN/RAF ACHQ, where he heard that T/105's contact was considered "a certain submarine."[217] He thereupon reviewed his own records. Although plotted eight hours apart and on separate plotting sheets, both contacts had been fixed accurately using QH and proved to be in the same spot. Gwinner then tried to convince himself it was "just possible that our contacts were upon a submarine destroyed by the aircraft's attack…."[218] He was not entirely alone in the conflation of weak evidence. The previous day, 30 August, OIC assessed that *U-247* could have been the target of Q/179's and T/105's attacks (if, that is, the latter proved to be on a U-boat). Alternatively, the attacks might have been on one of two boats then estimated to be leaving the Channel.[219]

Rarely was the intelligence picture less obscure than that which was available at Plymouth ACHQ at the end of August 1944. Crucially for the Allied cause, no matter how good the intelligence, units at sea still had to convert contacts into kills. For most aircraft, their main problem was that they still relied on depth-charge attacks. These were only effective against U-boats on or near the surface, where there was a visual aiming mark. Once the enemy went deep, only those few aircraft fitted with sonobuoys were able to continue to track them. Moreover, the technology was in its infancy and subject to many interfering underwater sounds that were difficult to distinguish from submarine noises. Attacking deep targets from the air was only possible using the Mk. 24 homing torpedo, with which some aircraft were equipped. For the surface escorts, the problem of clas-sifying asdic contacts was made more difficult by the habit of U-boats lying on the bottom as an evasion technique. As Commander J.D. Prentice, RCN, Senior Officer of the 11th Escort Group (EG11), discovered during operations in the English Channel, echoes from wrecks were often better than those obtained from a U-boat, which were often woolly. There was no doppler and no wake echo, both of which helped in classifying contacts in deep water. And in a strong tidal stream, trying to plot the movement of the target could be deceptive, and it was easy to shift target inadvertently from the initial target to an adjacent non-submarine ("non-sub") contact.[220]

Radar plotting and radio aids (such as QH) to accurately fix the ship's posi-tion helped greatly in maintaining contact on these tenuous asdic contacts. Accu-rate knowledge of a contact's position ought to have allowed escorts to compare the location with that of known wrecks. These, however, were very inaccurately plotted and, as a result, escort groups had to laboriously attack every suspect contact to try to bring up evidence of a U-boat. The problem was, as Commo-dore G.W.G. Simpson, Commodore (D), Western Approaches, noted, "the disintegration of the hull cannot reasonably be expected however many charges are dropped on it."[221] About four out of five attacks brought up only oil or nothing at all – not enough to differentiate between a U-boat and a wreck.[222] Moreover, attacking an indifferent asdic contact posed problems. A series of trials were carried out to find out how best to determine in which direction the U-boat was lying. Several escort groups developed attack methods based on the use of the echo-sounder to guide the attacking ship directly overhead of the U-boat to drop a small number of depth charges alongside the target.[223] Escorts often persevered with these attacks against individual contacts continuously for up to 48 hours. "It is most strongly emphasized," Captain Howard-Johnston, Director of the Anti-U-boat Division, remarked, "that *persistence* in the search or hunt is of the greatest importance until clear evidence of destruction is obtained."[224] Anti-U-boat warfare had, by mid-1944, entered into a new form, in which deliberate and persistent attacks largely replaced the dynamic engagements of the great convoy battles of 1942-43. The inshore operations in 1944 have rightly been likened to the "dawn of modern antisubmarine warfare."[225]

Contact

As Commander Evans pondered the OIC's assessment of 30 August, he may have concluded that, if the attacks had been on *U-247*, then, from their position the U-boat would arrive off Land's End several days later than initially anticipated. Alternatively, it was possible, if the attacks had been successful, that the threat to the coastal resupply convoy routes had been removed, at least for the time being. By the 30th the OIC was "almost certain" that *U-247* was not a minelayer and calculated that only one U-boat remained in Brest.[226] As a result of his deliberations, Coastal Command's Patrol 109, neatly covering *U-247*'s predicted approach track, was continued throughout the day by pairs of aircraft from 19 Group, initially amidst low cloud and rain.[227] That evening, as the weather brightened, Majendie's EG15 left his patrol area across *U-247*'s approach for a lay-over in Belfast, and the 11th Escort Group (just arrived from Londonderry), was established on a patrol along the convoy route between the Lizard and Seven Stones.[228] EG11 consisted of only two destroyers, both in desperate need of refit but driven by the experienced and determined, although very tired Commander J.D. Prentice, RCN.[229] Meanwhile, the equally exhausted Layard on board *Swansea* was alongside in Plymouth preparing to sail the following morning for another patrol. He confided to his diary, "I'm finding it more and more difficult to make the necessary effort these days...."[230] During 1944, Layard spent 211 days at sea and only 29 days on leave at home.[231] When not actually hunting a U-boat, Layard and the other escort group commanders concentrated on the training and material state of their group, whether at sea or in harbour. The mental fatigue resulting from the unremitting strain of maintaining the highest level of operational efficiency was endemic amongst escort group commanders. In some, like Layard, there was physical fatigue, while in others, like Prentice, medical symptoms were evident.[232] It was, for both Canadian and British officers, their sense of duty and pugnacious spirit which carried them through. By the summer of 1944, the progress of the war seemed to be going steadily in the Allies' favour, and the immediate news from the Continent gave cause for optimism. As Layard noted in his diary:

> The battle in France is going well. The Americans S[outh] of Paris are push-ing on West, and the British and Canadians have secured bridgeheads over the Lower Seine.[233]

At 0930 hours on the morning of the 31 August, EG9 cleared harbour and formed up to pass south of Eddystone before turning to the west into a fresh breeze. Initially Layard's orders were to patrol an area around the Scillies, which would extend the patrol recently established by EG11 to the west.[234] With EG15 now heading northwards to lay-over, the two Canadian groups, EG11 and EG9, were to provide cover for the convoy routes passing round Land's End against *U-247*, as well as the slightly more distant threat from *U-758* and *U-262* approach-ing the north Cornish coast from the southwest. These arrangements were in place just in time, for as EG9 approached the Lizard, C-in-C, Plymouth, received

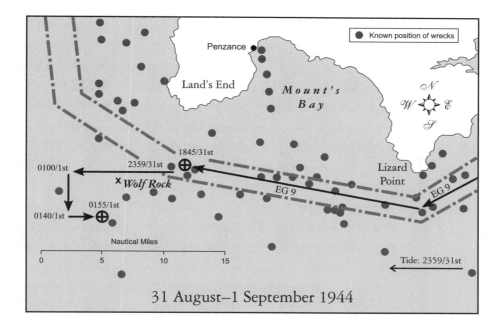

31 August–1 September 1944

the latest assessment from the OIC, which confirmed that *U-247* was "probably within 25 miles of Land's End."[235] However, C-in-C, Plymouth's neat and timely arrangements were soon disrupted by fresh intelligence. Firstly, late on 30 August two Coastal Command aircrew spotted a periscope off the north Cornish coast a little to the west of Trevose Head.[236] That evening Leigh-light Liberator "C" of 53 Squadron took off from St Eval for Patrol 115, which extended along the convoy route along that same stretch of coast (see map, page 273). At 0345 hours on the 31st, nearly seven hours into the sortie, the crew picked up a radar contact at a range of 15 miles, in position 300° Trevose Head 17 miles, and not far to seaward of the coastal convoy route to the Normandy bridgehead. The aircraft's Canadian captain, Flying Officer C. Waldrop, immediately turned towards the contact, but at a range of 12 miles it disappeared and nothing more was seen. Waldrop reported to base but could only loiter in the area for some 45 minutes because No. 2 engine began to lose power and C/53 was forced to turn for base, landing at St Eval shortly afterwards.[237]

A U-boat had operated off the north coast of Cornwall in early August, and now the OIC was warning of renewed U-boat activity in the Bristol Channel area.[238] With C/53's sighting, C-in-C, Plymouth, needed no further prompting and immediately ordered EG9 to make for the area of the Liberator's contact, sweeping along the coastal convoy route as they went. At the same time, Prentice was ordered to extend his patrol westwards to the Scillies, thus covering the convoy routes threatened by the approaching *U-247*.[239] From these deployments, it seems that C-in-C, Plymouth, had concluded from C/53's sighting that *U-247* might already have passed *through* the area around Wolf Rock and was now off the north Cornish coast. It is not entirely clear why C-in-C, Plymouth, ordered EG9 to prosecute the sighting by C/53, rather than EG11, which seems to have been nearer to the sighting, nor why, as a result, EG9 was not switched to the coastal patrol along the convoy routes from the Lizard to Seven Stones. Perhaps the command system was unable to respond with that level of flexibility given the relatively limited communications capacity and the consequent lack of up-to-the-minute positional information from the ships at sea.

Then, in the middle of the afternoon of 31 August, C-in-C, Plymouth, received the OIC's signal which assessed that *U-247* was actually to patrol the area around Wolf Rock.[240] With EG11 heading for the Scillies and EG9 planned only to pass through the area *en route* for Hartland Point, C-in-C, Plymouth, was in danger of being wrong-footed. Fortunately, EG9 had not yet passed Falmouth. Once more a considerable time would elapse while this information was assimilated and sanitized before any reordering of C-in-C, Plymouth's forces could take

Lieutenant-Commander W.R. Stacey, RCNR, the commanding officer of HMCS *Saint John*, proved *U-247*'s nemesis. The highly professional and tenacious Stacey, a merchant mariner before the war, earned high praise from Commander Layard in the hunt for the U-boat. (Courtesy of Lieutenant-Commander Ray Stacey, CD (Retd))

place. As it turned out, however, events at sea nullified the need for new orders from shore, for at 1845 hours HMCS *Saint John*, in moderating weather, obtained an asdic contact 5 miles east of Wolf Rock, and on the convoy route (see map, page 277).[241] Her captain, Lieutenant Commander W.R. Stacey, RCNR, began to manoeuvre *Saint John* as her asdic team, led by Lieutenant J.R. Bradley, RCNVR, began to sniff at the contact. Although it was intermittent, Able Seaman L. Haagenson, RCNVR, the senior asdic rating, was sure that the asdic returns exhibited enough submarine-like qualities.[242] It was these men who had brought *Saint John* to a high pitch of operational efficiency, which guaranteed that she got the lion's share of antisubmarine contacts within EG9.[243] As Bradley's team investigated the contact, the rest of the group steamed ahead of *Saint John*, but still her asdic probed this elusive contact, until after about an hour Stacey too became confident that it was a possible U-boat and called Layard on TBS to ask for the group to come back and assist in his investigation.[244]

Investigations of this type were commonplace during operations in the shallow waters around Britain. It was a laborious process, sometimes reducing the progress of the group to a snail's pace, resulting in the expenditure of large numbers of depth charges, and normally ended in a classification of non-sub. Over the previous month, for example, EG9 had investigated several such contacts almost every day at sea. Something like a third of them were reported by *Saint John* initially.[245] Why, then, should this latest contact from *Saint John* prove anything but a non-sub in an area abounding with such false contacts?[246] Worse, for Layard, who, of course, knew nothing of the anticipated arrival of a U-boat in the Wolf Rock area, the investigation of yet another contact – which he felt would probably prove to be another non-sub – would delay his passage towards a "known" U-boat position off the north coast of Cornwall. The tired Layard's "annoyance" was borne of having to face another dilemma: whether to proceed with his orders, or go back and help Stacey. He decided on a compromise, and detached *Monnow* with her division to continue along the convoy routes, while he took *Swansea* and *Port Colborne* back to help *Saint John*.[247]

About an hour and a quarter after *Saint John*'s first tentative report *Swansea* and *Port Colborne* were closing on her contact. As the ships swept through the area, *Swansea*'s asdic team picked up *Saint John*'s contact, though Layard thought it might be fish. So he decided to drop a single depth charge, but instead of dispersing the echo as expected, Stacey, following behind in *Saint John*, reported that the attack had brought up small traces of oil. The contact, however, remained difficult to hold and could only be picked up from one direction. Even so, *Saint John*'s asdic team managed to hold it long enough for Stacey to take his ship in for a Hedgehog attack. This considerably increased the patch of oil. Unable to ignore this contact, Layard now reported it to C-in-C, Plymouth. He was still far from convinced that this was a U-boat and thought it more likely to be one of the many wrecks plotted in the area.[248] The problem was that, although *Swansea* was fitted with the latest navigation aid, QH, which could fix the ship to within fractions of a mile, the plotted positions of most wrecks were somewhat vague, which made it impossible to correlate asdic contacts positively with wrecks.[249]

The contact remained elusive, but *Saint John* was able to carry out a second attack at 2300 hours, after which contact was lost completely. "Quite definitely," Layard confided to his diary at the end of the day, "I haven't the patience for this game."[250] Still, Layard persisted for another hour, but it was already too dark to see the oil patch and at midnight he decided to carry out a lost contact search. He formed the three frigates into line abreast and began to sweep on a westerly heading which took them down-tide and therefore along the most likely path of an evading U-boat using the tide to help him make as much ground as possible away from the attack site. Allowing for a U-boat speed over the ground of about 3 knots, Layard, passing to the north of Wolf Rock, swept about 8 miles to the west, before turning the frigates to a southerly heading using an "Easy Item," or "EI", turn.[251] The ships steamed at about 7 knots, as fast as was safe with a possible U-boat in the vicinity who was likely to counter-attack with a Gnat torpedo if the group passed close. The speed also meant that the group did not have to stream Cats, which would have degraded their asdic search for what had already proved to be an elusive target. Although asdic conditions were good, this only exacerbated the prevalence of unwanted bottom echoes and reverberations out to 2,600 yards, making it difficult to distinguish a slow-moving or stationary U-boat.[252] Having made ground to the south for about 35 minutes, Layard altered to the east, again by an "EI" turn, intending to return to the original datum position after sweeping south of Wolf Rock. This was entirely possible to do accurately, given that *Swansea*, at least, was fitted with QH (or "Gee" radio navigation aid, which could fix the ship to within 300 yards) and, in addition, good radar fix positions were obtained using Wolf Rock.[253]

While EG9 was sweeping westwards past Wolf Rock, in the C-in-C's headquarters the inner circle of staff officers indoctrinated into Ultra material must have been contemplating a difficult plot. EG11's patrol had already been extended to the Scillies to counter the arrival of *U-247*, or possibly one of the several U-boats known to be approaching the north coast of Cornwall, but whose time of arrival was uncertain. The pair of sightings by Liberators off this coast seemed to suggest that at least one of them had arrived, and half of EG9 was moving towards this datum. The headquarters had also received the OIC's latest assessment stating that *U-247* was to operate around Wolf Rock, and now EG9 had an intermittent contact in the area, but the tenor of his signal suggested that the Senior Officer had little confidence in the contact. As a result, Layard was told to "Remain with contact until further instruction and report your classification giving reasons."[254] But by the time this signal arrived in *Swansea* and had been decoded, the whole situation had altered, for at just before 2 in the morning of 1 September, Able Seaman Haagenson's asdic team in *Saint John* gained an asdic contact just over 2,000 yards ahead of the ship and 3 miles south of Wolf Rock.[255] The contact showed no movement, doppler or HE and was fortuitously up-tide of *Saint John*, which made it possible for Stacey to make an echo sounder run directly from his initial contact position. As *Saint John* ran over the contact, on a heading of 105° at 9 knots, the echo sounder produced an excellent trace of an object on the seabed in 42 fathoms of water (about 250 feet), which, he reported, "had all the characteristics of a bottomed submarine, with a hard sharp peak on the centre of the hull...."[256] As soon as he saw the echo-sounder trace Stacey reported this new contact to Layard, who thought it "really might be a submarine."[257]

Meanwhile, Stacey was manoeuvring *Saint John* to pass over the U-boat on a heading of 240° to confirm its orientation, before circling round to approach again on a southeasterly heading with the intention of carrying out the EG9 version of a "tin-opener" attack. This took about an hour. By 0300 hours *Saint John* was running in for an attack and, as the echo-sounder began to register the contact, Stacey ordered the first depth charge of the "tin-opener" five-charge pattern to be dropped. As *Saint John* moved away, the delayed action depth-charges exploded. Haagenson heard "a heavy tearing explosion" and the bridge team soon saw "a heavy flow of oil" come to the surface.[258] Layard had put the non-TBS-fitted *Port Colborne* on an "Observant" square search to guard against an escaping U-boat, while *Swansea* moved closer to assist *Saint John*. *Swansea*

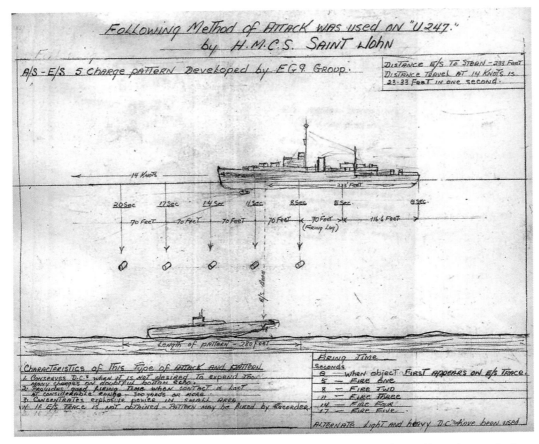

This diagram provides a detailed explanation of how HMCS *Saint John* used her echo sounder to carefully place five depth charges very close to the bottomed hull of *U-247*. That ship's attack at 1404 on 1 September 1944 produced conclusive evidence of the destruction of the U-boat. ("Following Method of Attack was Used on U-247 by HMCS *Saint John*," Diagram, n.d., National Archives (U.K.), Envelope ADU.1561/44, Enclosure Box 704, ADM 199/1462)

was, however, unable to gain asdic contact, even though "the oil was plainly visible in the moonlight."[259] Stacey's team were having better luck, and 20 minutes later Stacey moved in again for another "tin-opener" attack which

> produced a terrific tearing explosion with a further heavy flow of oil which soon extended over an area 2½ miles long, and a mile wide, but no wreckage was seen due to [the] darkness [and the, now, obscured moon].[260]

In an attempt to get more positive indication of a kill, Stacey took *Saint John*, still the only ship in intermittent contact, round again and attacked with Hedgehog at 0345, which produced yet more oil, but still no visible signs of wreckage.[261] Probably the series of attack had disturbed the sediment on the bottom, and this obscured the target to the probing asdic beam. In any case, at 0410 hours, Stacey declared that he was unable to relocate the target.[262] Though oil was still rising and provided an adequate mark of the datum position, none of the ships could gain contact. Nevertheless, EG9 continued searching for another hour, before Layard ordered all three ships to carry out an east-west patrol over the datum, during which many bottom echoes were investigated in the area, but they all proved to be non-subs. As he started this patrol, Layard reported to C-in-C, Plymouth, at 0510 hours that he had attacked this new contact and brought up large quantities of oil.[263] He may have still harboured doubts over whether this was a U-boat or a wreck, but mindful of the C-in-C's instructions to remain with the contact, he added that he was awaiting daylight.[264] In the interim Layard "got about an hour's much needed sleep" and as dawn came, he wrote, "there was the oil, rather dispersed now but definitely with a fixed source which was unaffected by tide...."[265]

Even using the oil as mark, none of the ships could get asdic contact, so at 0930 hours, Layard moved all three ships over to the first datum position 5 miles

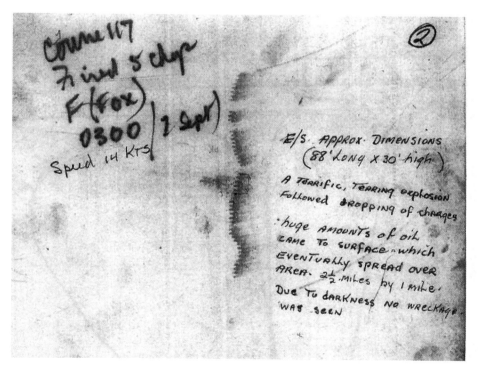

The actual echo sounder traced that was used in the first tin opening attack at 0300 on 1 September 1944 can be seen here. The amount of detail provided by the echo sounder when the attacking ship was skilfully manoeuvred immediately over top of its position can be clearly seen, as even the conning tower of *U-247* can be discerned. ("Echo Sounder Trace, [U-247]," n.d., National Archives (U.K.), Envelope ADU.1561/44, Enclosure Box 704, ADM 199/1462)

east of Wolf Rock. There some traces of oil remained, but they could get nothing but fish echoes on the asdic. The continued presence of oil at the site of the first attacks might have suggested that the two contacts were different targets. However, the initial westerly tidal stream was counterbalanced by a westerly wind which would have kept the oil comparatively concentrated, and even if the oil was moved bodily to the west, its position would have been partially restored overnight by the reversal of the tidal stream (as was the case of the oil at the site of the second series of attacks). This possibility is strengthened by the lack of any submarine-like asdic contact in the position of the first attacks.[266]

Having moved back to the overnight datum south of Wolf Rock, Layard decided after lunch that "we must throw something more at it and had just said I was going to drop charges blindly at end of oil slick" when Stacey finally picked up the original contact.[267] He took *Saint John* over the contact and got an echo sounder trace, but could not attack immediately because the ship was moving at only 4 knots, too slowly to avoid the damaging effects of the depth charge explosions. To make matters more difficult, at this point the main asdic set displays broke down, and Stacey had to use centre bearings called out by Bradley or Haagenson to skilfully manoeuvre the ship round on to the target so that it could be picked up by the narrow beam of the echo-sounder – no mean feat in tidal Channel waters. Stacey, by consummate ship-handling, got *Saint John* over the target within 15 minutes, and at 1404 hours carried out another five-charge "tin-opening" attack.

"The result," Layard observed from his bridge, "was startling."[268] Stacey's asdic team heard "a heavy tearing explosion," followed by a further flow of oil and, at last, wreckage.[269] Both ships lowered boats to collect the evidence, which included "printed German books, letters, certificates and … clothing with German name marks."[270] By now the wind was getting up from the southwest, and it started raining, making it difficult to recover wreckage. Nevertheless, amongst the flotsam of broken panelling and woodwork recovered was an engine room certificate marked "*U-247*," other items marked with the U-boat's identity, and a German rubber life-raft.[271] Privately, Layard confessed in his diary, it "Really looks like a kill."[272] Layard signalled at 1445 hours to the C-in-C with the details of the attack and its spectacular result.[273] He may also have shared the success verbally with Prentice by TBS, for it was about this time that EG11 passed heading east towards the Lizard at the eastern extremity of their coastal patrol, where HMCS *Assiniboine* joined them, having sailed from Plymouth that morning.[274]

This latest attack had shattered *U-247*, for not only did wreckage come to the surface but the "peak" on the echo-sounder trace in the centre of the hull had disappeared and the overall extent of the target had been reduced. As a result the target was even more difficult to pick up.[275] In spite of several attempts, no further attacks could be carried out using the main asdic or the echo sounder because no contact persisted long enough to provide a fire control solution. Frustrated, Stacey decided to place one 13-charge attack by eye on the head of the oil slick, and this was followed by a similar 10-charge attack by *Swansea* at 1623 hours.[276]

While these attacks were in progress, Layard received new instructions from C-in-C, Plymouth, who was reacting to a sighting report from a Liberator of 547 Squadron. This aircraft had taken off from St Eval on a short air test in mid-morning of 1 September, crewed in part by Canadians. Its pilot, Squadron Leader R.B. Fleming, reported that he had sighted a periscope off the north coast of Cornwall in a position which confirmed the earlier report by Liberator C/53.[277] C-in-C, Plymouth, who had not yet seen Layard's signal of 1445 hours about *Saint John's* attack 40 minutes earlier that had brought up the large quantities of wreckage, therefore ordered Layard to take *Swansea*, *Saint John* and *Port Colborne* to join *Monnow* and her division as soon as possible off the north coast of Cornwall. At the same time, Prentice's EG11 was to take over EG9's contact. Prentice was also told to act on the assumption that both EG9's contacts to the east and south of Wolf Rock "were the same U-boat now damaged and retiring slowly to the south-west...."[278] Given that the C-in-C had not yet received Layard's signal concerning the wreckage brought to the surface by *Saint John's* latest attack, when the assessment was made, it made sense. The positions of the two contacts were about 7 miles apart and, if they were on the same U-boat, the enemy would have needed to move at some 2¼-3¼ knots over the ground, an entirely feasible speed given the initially westerly tide. In the early hours of the morning the tide would have begun to set to a more northerly direction, impeding the progress of the U-boat. At this point, or when EG9 approached, she probably bottomed, stemming the tide on a heading of about 285°, which is how Stacey found her.[279]

Having collected the items of wreckage, at 1545 hours Layard sent a signal by light to Stacey saying:

> I most heartily congratulate you on what appears to be a successful kill. It has been achieved entirely by your patience and skill and I shall have no hesitation in saying so.[280]

Stacey, in reply, flashed: "VMT. Would like to carry out further attack if at all possible to get contact."[281] Clearly C-in-C, Plymouth, shared Stacey's view, for, having read Layard's description of the wreckage brought up, he signalled: "Many congratulations on most promising results. Patience is its own reward." He added

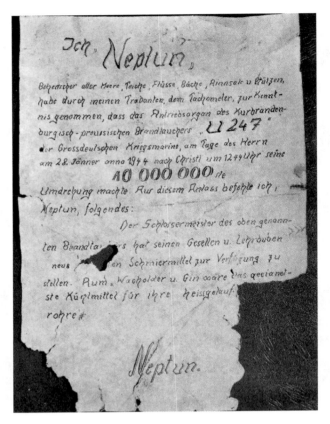

HMCS *Saint John*'s attack at 1404 produced a startling array of debris, including this Engine Room certificate shown here. Shore authorities insisted on clear evidence of destruction during the inshore campaign, because of the prevalence of wrecks and non-sub contacts in British coastal waters. The devastating attacks by HMCS *Saint John* produced all the evidence needed. ("Monthly Anti-Submarine Report, October 1944," DAUD, CB 04050/44(10), 15 November 1944, Naval Historical Branch (U.K.))

a rider, however, instructing "EG11 [to] continue attacking until destruction is absolutely certain."[282] Thus, when, at about 1730 hours, HMCS *Assiniboine* from EG11 arrived at the scene, Layard set off with *Swansea* and *Port Colborne* in haste to join *Monnow* off Trevose Head, leaving *Saint John* to assist *Assiniboine*.[283] Even with a fresh ship to help, *Saint John* found it impossible to get a firm contact on the battered *U-247*, in spite of its position being clearly marked by the oil which continued to well up. Around 2025 hours Stacey dropped a 10-charge pattern on the head of the oil slick, followed 10 minutes later by another 10-charge pattern from *Assiniboine*. The two attacks brought up a little more oil. Both ships then stood by the contact during the night, as the wind rapidly deteriorated to a full west to northwest gale with a high, steep sea. In these conditions further attacks were impossible.[284] At 1050 hours the following morning, 2 September, *Saint John* was ordered to rejoin EG9, and *Assiniboine* to rejoin EG11, now operating to the west of the Scillies.[285] At last the wreck of *U-247* was left in peace.

When it came to writing up his report of the action, Layard was as good as his word, for he rightly gave all the credit to *Saint John* and C-in-C, Plymouth, adding that but for Stacey's insistent enthusiasm and the C-in-C's instruction to remain with the contact, he would have abandoned the contact at an early stage and continued with his instructions to make for the north Cornish coast.[286] For this action, Stacey and Bradley were both awarded the DSC, while Haagenson and Royds both received the DSM. Four other member of *Saint John*'s crew were mentioned in despatches.[287] Some six weeks after the action, when the U-boat Assessment Committee had reviewed the results of the attacks, the Admiralty signalled:

> We congratulate HM Canadian Ships *Saint John* and *Swansea* on destruction of *U-247* on 1 September. The tenacity of *Saint John* in holding a difficult contact was most commendable.[288]

And the self-effacing Layard, referring to the Admiralty's signal, predictably added:

> We appreciate our mention although we feel we were really only onlookers. I am delighted that your very fine effort has been recognised in the last sentence of the signal.[289]

This remarkable echo sounder trace from HMCS *Saint John* shows oil rising from the bottomed *U-247*. Echo sounders proved very important sensors in the hunt for U-boats in shallow coastal waters. ("Echo Sounder Trace, [U-247]," n.d., National Archives (U.K.), Envelope ADU.1561/44, Enclosure Box 704, ADM 199/1462)

Endnotes

Abbreviations Used in Endnotes

BdU	*Befehlshaber der U-boote* (Commander in Chief U-boats)
DHH	Directorate of History and Heritage, Department of National Defence, Canada
KTB	*Kriegstagebuch* (War Diary)
LAC	Library and Archives Canada (formerly National Archives of Canada)
MG	Manuscript Group
MoD(UK)	Ministry of Defence (United Kingdom)
NAUK	National Archives of the U.K. (formerly Public Record Office)
NARA	National Archives and Records Administration (U.S.A.)
NHB	Naval Historical Branch, Ministry of Defence (U.K.)
RG	Record Group
RoP	Report of Proceedings
WD	War Diary

Introduction

1. This book is not intended to explore tactical theory directly. Those interested in this aspect are suggested to start with works such as Captain W.P. Hughes's USN (Ret) *Fleet Tactics: Theory and Practise* (Annapolis, 1987).
2. AAP 6, <http://www.nato.int/docu/logi-en/1997/defini.htm>.

Prelude: The Royal Navy at Quebec, 1759

1. Fred Anderson, *Crucible of War* (New York, 2001), 52-60.
2. Winston Churchill: Letter to General Marshall, 1944 in Robert Debs Heinl, Jr., *Dictionary of Military and Naval Quotations* (Annapolis, Maryland, 1966), 166.

Chapter 1: "A perfect good understanding between the Army and the Navy": British Seapower and the Siege of Quebec

1. "Heart of Oak," the traditional Royal Navy tune which was played, not sung, before ships of the sailing navy engaged in battle, dates from 1759, Britain's *annus mirabilis*, the year that saw three famous victories: Minden, Quebec and Quiberon. William Boyce wrote the tune and David Garrick supplied the words.
2. The historiography of the siege of Quebec is very lengthy and a listing of titles relating to it published between 1759 and 2006 will be found on pages 276 to 290 of C.P. Stacey, *Quebec, 1759: The Siege and the Battle,* edited and with new material by Donald E. Graves, which appeared in 2002 and was reissued in a new format in 2006. All page references below to Stacey are to the 2006 edition.

    There are, however, relatively few published works on the naval side of the operation. The most helpful remains William H. Wood, *The Logs of the Conquest of Canada* (Toronto, 1909) which contains excerpts from the logs of the warships in the British fleet. Although dated, Edward Salmon, *Life of Admiral Sir Charles Saunders, K.B.* (London, 1914) contains some useful correspondence. More recently, I wrote a short essay entitled "'Justice to the Admirals': The Royal Navy and the Siege of Quebec, 1759" which forms the basis of this present study. This essay appeared as an appendix in the revised edition of Stacey's book, *Quebec, 1759.*

    Interesting scientific information on navigational matters around Quebec will be found in Donald

W. Olson, William D. Liddle, Russell L. Doescher, Leah M. Behrends, Tammy D. Silakowski, and François-Jacques Saucier, "Perfect Tide, Ideal Moon: An Unappreciated Aspect of Wolfe's Generalship at Québec, 1759," *William and Mary Quarterly*, vol 59, No. 4 (October 2002), 957-974. This study, however, should be read in conjunction with my article, "The Anse au Foulon 1759: Some New Theories and Some New Evidence," *Northern Mariner*, vol 14, no. 4 (October 2005), 61-72. A very helpful article on the background of British combined operations in the 18th century is David Syrett, "The Methodology of British Amphibious Operations during the Seven Years and American Wars," *Mariner's Mirror*, vol 58, no 3 (August 1972), 267-280.

3. Martin Garrod, "Amphibious Warfare: Why?," *Journal of the Royal United Services Institute*, 113 (Winter 1988), 26.

4. The first published work on the subject of amphibious warfare – Thomas Molyneux's *Conjunct Operations; or, expeditions that have been carried on jointly by the fleet and navy*, which contained a commentary on "a littoral war," appeared only in 1759 and there is no evidence that it was widely read by naval officers. A second edition appeared in 1780 and, it is interesting to note, an American edition was brought out in 1778. Molyneaux was followed in 1763 by John MacIntire's *Military Treatise on the Discipline of the Marine Forces ... with short instructions for detachments sent to attack on shore*, a helpful practical guide by a serving officer of the Royal Marines which contained sound directions for junior officers of that corps. A similar, but not so comprehensive a work by another marine officer, Terence O'Loghlen, entitled *The Marine Volunteer*, was released in 1766 but was concerned mainly with sea duty. Finally, 1780 saw the publication of Wolfe's *Instruction to Young Officers ... with the orders and signals used in embarking and debarking an army*, a useful work with a supplement based on Wolfe's orders in the 1758 Louisbourg and 1759 Quebec operations. This latter title appears to have been the only period published work based on actual operational experience.

5. H.W. Richmond, *The Navy as an Instrument of Policy, 1558-1727* (Cambridge, 1953), 265.

6. Philip Woodfine, "Ideas of naval power and the conflict with Spain, 1737-1741" in Jeremy Black and Philip Woodfine, eds., *The British Navy and the Use of Naval Power in the Eighteenth Century*, (Leicester, 1988), 82.

7. Woodfine, "Ideas of naval power and the conflict with Spain," in Black and Woodfine, eds., *The British Navy and the Use of Naval Power*, 82-83.

8. Molyneaux, *Conjunct Operations*, 175

9. William L. Clowes, *The Royal Navy, A History*. vol. 3 (London, 1898); Woodfine, "Ideas of naval power and the conflict with Spain," in Black and Woodfine, eds, *The British Navy and the Use of Naval Power*, 83.

10. Earl of Stanhope, *Notes of Conversations with Lord Wellington* (London, 1889), 182.

11. Alfred Marini, "Parliament and the Marine Regiments, 1739," *Mariner's Mirror*, 62 (1976), 55-65.

12. Most critics of Hawke seem to rely on the evidence of Wolfe, who participated in the operation, see Beckles Willson, *The Life and Letters of James Wolfe* (London, 1909), 335, Wolfe to Walter Wolfe, 18 October, 1757 and 339, Wolfe to Rickson, 5 November, 1757. Hawke's biographer defends his subject, see Ruddock Mackay, *Admiral Hawke* (Oxford, 1965), 168-170.

13. Wolfe to Rickson, 5 November, 1757, Willson, ed., *Letters of Wolfe*, 339.

14. Robert Beatson, *Naval and Military Memoirs of Great Britain from 1727 to 1783* (6 vols., London, 1804; Boston, 1972), II, 167.

15. N.A.M. Rodger, *The Command of the Ocean. A Naval History of Britain, 1649-1815* (New York, 2004), 419.

16. Molyneaux, *Conjunct Operations*, 213.

17. Cunningham to Sackville, 30 May, 1758, in J.S. MacLennan, ed., *Louisbourg. From its Foundation to its fall, 1736-1758* (London, 1918; reprinted Sydney, 1969), 239. This concern about properly training the troops for the assault has been attributed solely to Wolfe but it is clear from the source that all the brigadier generals in Amherst's army were involved in it.

18. LAC, MG 18, L 21, vol. 1, p. 12, order of Boscawen, 9 June, 1758.

19. Boscawen to his wife, September 1756, quoted in Rodger, *Command of the Ocean*, 306.

20. On Minorca, see Rodger, *Command of the Ocean*, 265-266.

21. *Dictionary of Canadian Biography*, entries for David Kirke, William Phipps and Hovenden Walker.

The best account of the disastrous Walker expedition remains Gerald Graham, *The Walker Expedition to Quebec, 1711* (Toronto, 1953).

22. *Dictionary of Canadian Biography*, entries for Philip Durell, Charles Holmes and Charles Saunders.

23. Stephen Brumwell, *Paths of Glory: The Life and Death of General Wolfe* (London, 2006), 176.

24. Wolfe to Amherst, 29 December 1758, NAB, WO 34, vol 46b, (part 2), transcript in LAC, MG 13.

25. Wolfe's instructions in *Northcliffe Collection* (Ottawa, 1926), 131-132.

26. It should be noted that these 140 merchant vessels were only those that accompanied the army up the St. Lawrence in May and June 1759. After the Royal Navy had cleared the river, there was a constant traffic of merchant ships between Quebec, Louisbourg and Halifax, some chartered by the Crown and others entrepreneurial venturers wishing to sell to the forces besieging the town.

27. For the strength of the naval and military forces, see Richard Middleton, *The Bells of Victory; The Pitt–Newcastle Ministry and the Conduct of the Seven Years' War, 1757-1762* (Cambridge, 1985), 103-106; and Stacey, *Quebec 1759*, 228-237.

28. Pitt to Durell, 29 December 1758, in Gertrude Kimball, ed., *Correspondence of William Pitt ... with Colonial Governors and Military and Naval Commanders in America* (2 vols, New York, 1906), vol 1, 444.

29. Stacey, *Siege of Quebec*, 26.

30. Wolfe to Amherst, 1 May 59, in Willson, *Life and Letters*, 425; Saunders to Pitt, 1 May 59, Kimball, ed., *Correspondence of Pitt*, vol 2, 92. The source of Wolfe's antipathy toward Durell is not known; he had earlier complained directly to Pitt that Durell was "vastly unequal to the weight of the business," see Wolfe to Pitt, 24 December 1758, see Kimball, ed., *Correspondence of Pitt*, vol 1, 379.

31. Wolfe to Walter Wolfe, 19 May 59, in Willson, *Life and Letters*, 427.

32. Jonathan R. Dull, *The French Navy in the Seven Years' War* (Lincoln, 2005), 143.

33. E.A. Smillie, "The Achievement of Durell in 1759," *Transactions of the Royal Society of Canada*, 3rd Series, 1925. For a more recent appraisal of the man, see W.A.B. Douglas, "Nova Scotia and the Royal Navy, 1713-1766" (Unpublished Ph.D. Thesis, Queen's University, Kingston, 1973), chapter 9.

34. Durell to Admiralty, 19 March 1759, in C.H. Little, ed., *Despatches of Rear-Admiral Philip Durell 1758-1759 and Rear-Admiral Lord Colville 1759-1761* (Halifax, 1958).

35. Saunders to Admiralty, 2 May 1759, in Little, ed., *Despatches of Vice-Admiral Sir Charles Saunders, 1759-1760* (Halifax, 1958).

36. Wolfe to Pitt, 6 June 59, in Kimball, ed., *Corrspondence of Pitt*, vol. 2, 118.

37. *Dictionary of Canadian Biography*, entries for Philip Durell, Jacques Kanon, John Rous and Jean Vauquelin.

38. Saunders to Admiralty, 6 June 1759, in Little, ed., *Despatches of Saunders*.

39. Colville to Cleveland, 10 April 1761, in Little, ed., *Despatches of Durell and Colville*.

40. Sailing Orders and Instructions, By His Excellency Admiral Saunders, Louisbourg, 15 May 1759, contained in Wood, *Logs of the Conquest*, 99.

41. Wolfe, 30 April 1759, Halifax, in *Wolfe's Instructions*, 63. On the organization, embarkation and orders for the fleet, see LAC: MG 11, Co 5, embarkation return, 6 June 1759; 1759; MG 18 L 6, Thomas Bell, Journal; M, Northcliffe Collection, Series 1 and 2 and Series 3, vol. 2, Journal of Captain Alexander Schomberg, RN; N18, Siege of Quebec; N21, George Williamson Papers; MG 40 L1, "Directions for sailing from the Harbour of Louisbourgh to Quebec ... by Captain James Cook; A.G. Doughty, ed., *John Knox, An Historical Journal of the Campaigns in North America. 1769* (3 vols, Toronto, 1914-1916), vol 1, 350-370; and Wood, *Logs of the Conquest*, 87-97.

42. Orders, Louisbourg, 17 May 1759 in *Wolfe's Instructions*, 66.

43. David Syrett, "The Methodology of British Amphibious Operations during the Seven Years and American Wars, *Mariner's Mirror*, vol 58, no 3 (August 1972), 267-280; LAC, M, Northcliffe Collection, Series 1 and 2 and Series 3; A.G. Doughty, ed., *Knox Journal*, vol 1, 355-365.

44. "Journal de Bougainville," *Rapport de l'Archiviste de Québec* (123-1924), 310.

45. Gilles Proulx. *Between France and New France. Life Aboard the Tall Sailing Ships* (Toronto, 1984), 371-380.

46. Saunders to Delancey and Pownall, 10 March 1759, in Little, ed., *Despatches of Saunders*.

47. Colville to Admiralty, 22 December 1759, in Little, ed., *Despatches of Durell and Colville*, 13.

48. Stacey, *Quebec 1759*, 54-55.

49. Navigational information from *Sailing Directions. St. Lawrence River. Ile Verte to Québec* (Canadian Hydrographic Service, Ottawa, 1999); *St. Lawrence Pilot. With the Coast of Quebec … to Quebec* (Canadian Hydrographic Service, Ottawa, 1957); *The St. Lawrence River Pilot above Quebec* (Minister of the Naval Services of Canada, Ottawa, 1912); Proulx, *Between France and New France*, 77. The best account of the navigational difficulties of the St. Lawrence and the progress of the British fleet in 1759 is Wood, *Logs of the Conquest*, 110-136.

50. Wood, *Logs of the Conquest*, 137-138.

51. Montcalm to Bourlamaque, 18 June 1759, quoted in Stacey, *Quebec 1759*, 47.

52. Wood, *Logs of the Conquest*, 209-210, 263-264, Logs of HM Ships *Centurion* and *Pembroke*, 8-13 June 1759.

53. Wood, *Logs of the Conquest*, 136-139; Log of HMS *Neptune*, 7 June 1759.

54. Doughty, ed., *Knox Journal*, 370.

55. Doughty, ed., *Knox Journal*, vol 1, 371-372.

56. Wood, *Logs of the Conquest*, 134-136.

57. Information on the nautical characteristics of Quebec basin was derived from Canadian Hydrographic Service Charts: 1315, "St. Lawrence River, Quebec to Donnacona"; 1316, "St. Lawrence River, Port de Québec"; 1317, "Saint Lawrence River, Sault-au-Cochon to Québec"; and *Canadian Tide and Current Tables, 2001, Vol3: St. Lawrence and Saguenay Rivers* (Canadian Hydrographic Service, Ottawa, 2000); *Sailing Directions. St. Lawrence River, Cap-Rouge to Montreal* (Canadian Hydrographic Service, Ottawa, 1992); *Sailing Directions: St. Lawrence River, Ile Verte to Québec* (Canadian Hydrographic Service, Ottawa, 1999); *St. Lawrence Pilot. With the Coast of Quebec … to Quebec* (Canadian Hydrographic Service, Ottawa, 1957); *The St. Lawrence River Pilot above Quebec* (Minister of the Naval Services of Canada, Ottawa, 1912).

    Useful information on navigational conditions around Quebec will also be found in Olson *et al.*, "Perfect Tide, Ideal Moon: An Unappreciated Aspect of Wolfe's Generalship at Québec, 1759."

58. Wood, *Logs of the Conquest*, 203-316, Logs of HM Ships *Captain, Centurion, Diana, Dublin, Echo, Eurus, Hunter, Lowestoft, Medway, Neptune, Orford, Pembroke, Porcupine, Prince Fredrick, Princess Amelia, Racehorse, Richmond, Royal William, Scarborough, Seahorse, Squirrel, Stirling Castle* and *Sutherland*, 18 June-13 September 1759.

59. The description of the defences of the town is from Stacey, *Quebec 1759*, 33-42.

60. On Montcalm's strength and plans, see Stacey, *Quebec 1759*, 38-49 and 251-255.

61. This is almost certainly a transcription error as there was no 14-pdr. calibre gun in French service and what is probably meant is a 16-pdr. weapon.

62. "Batteries en barbette" are batteries where the guns are mounted to fire over the protective earth rampart, not through embrasures. This gave them a wider field of fire but their gun detachments had less protection.

63. "Journal of a French Officer," in A.G. Doughty and G.W. Parmelee, eds., *The Siege of Quebec and the Battle of the Plains of Abraham* (6 vols, Quebec, 1901), vol 4, 240-241.

64. For calculations of British military and naval strength, see Stacey, *Quebec 1759*, 228-237. On the population of New France, see W.J. Eccles, *The Canadian Frontier, 1534-1760* (Albuquerque, 1969), 174.

65. Wolfe to mother, 31 August 1759, in Robert Wright, *The Life of Major-General James Wolfe* (London, 1864), 553.

66. My summary of the events of the siege is drawn from Stacey, *Quebec 1759*, 57-107, which remains the most balanced account of the operation.

67. Gibson to Lawrence, August 1759, in Doughty and Parmalee, eds., *Siege of Quebec*, vol 5, 165.

68. Saunders to Pitt, 5 September 1759, in Little, ed., *Despatches of Saunders*.

69. LAC, MG 18, L6, Wolfe's Journal, 11 July 1759.

70. LAC, Micro A-652, Journal of an Anonymous officer. This journal was originally thought to be the work of Major Paulus Irving but recent research suggests that it was more likely written by Captain Matthew Leslie, Wolfe's assistant quartermaster-general, although this identification can not, at this point, be proven beyond a certainty. See Brumwell, *Paths of Glory*, 361-364, for an excellent discussion of the authorship.

Whoever was the author, a couple of points must be kept in mind about this source. Although it may have been based on a daily diary, the journal was written *post facto* in the winter of 1759-1760 and several errors of dates have been found in it. More important, the author was an extremely partisan and uncritical admirer of Wolfe and his comments are so blatantly biased his work must be treated with caution.

71. Wood, *Logs of the Conquest*, 144-145; Holmes to _____, 18 September 1759 in Doughty and Parmalee, eds. *Siege of Quebec*, vol 4, 296.

72. Wood, *Logs of the Conquest*, 210, Log of HMS *Centurion*, 29 June 1759. The attack actually took place on the night of 28/29 June but *Centurion* recorded it as having occurred on 29 June as in the Royal Navy until 1805, a new day began at noon, not midnight. Researchers not familiar with this fact should be very careful when dating events from ships' logs. Below I have indicated the correct (or rather commonly accepted) date when they differ from those given in the logs.

73. Wood, *Logs of the Conquest*, Log of HMS Pembroke, 29 June 1759 (actually 28/29 June 1759).

74. Wood, *Logs of the Conquest*, 219, Log of HMS *Dublin*, 28 June 1759, but the attack actually took place on the night of the 27/28 June.

75. Wood, *Logs of the Conquest*, 235, Log of HMS *Lowestoft*, 28 July 1759 (actually night of 27/28 July).

76. See Stacey, *Quebec 1759*, 65-66, for the text of these remarks.

77. Wood, *Logs of the Conquest*, 307, Log of HMS *Stirling Castle*, 13 July 1759 (actually 12 July 1759).

78. Wood, *Logs of the Conquest*, 277, Log of HMS *Racehorse*, 10 July 1759.

79. Wood, *Logs of the Conquest*, 278, Log of HMS *Racehorse*, 15 July 1759.

80. Wood, *Logs of the Conquest*, entries in logs of ships named in the text for 9-10 July 1759.

81. Wood, *Logs of the Conquest*, 209-210, Log of HMS *Centurion*, 16-17 July 1759 (actually 16 July).

82. Wood, *Logs of the Conquest*, 211, Log of HMS *Centurion*, 31 July and 1 August 1759 (actually 31 July).

83. Wolfe to Saunders, 30 August 1759, in Salmon, *Life of Saunders*, 460-462.

84. Stacey, *Quebec 1759*, 67.

85. Saunders to Pitt, no date (but probably 20 September 1759) in Salmon, *Life of Saunders*, 143.

86. See the daily entries in Wood, *Logs of the Conquest*, for examples of the variety of tasks carried out by the navy and marines during the siege.

87. Wood, *Logs of the Conquest*, 312, Log of HMS *Stirling Castle*, 31 August 1759.

88. Wood, *Logs of the Conquest*, 279, Log of HMS *Racehorse*, 12 September 1759.

89. Saunders to Pitt, 5 September 1759, in Little, ed., *Despatches of Saunders*.

90. Wood, *Logs of the Conquest*, 234-235, Log of HMS *Lowestoft*, 28 June (actually 27-28 June).

91. Wood, *Logs of the Conquest*, 292, Log of HMS *Scarborough*, 2 September (actually 1 September).

92. Wood, *Logs of the Conquest*, 292, Log of HMS *Scarborough*, 3 September (actually 2 September).

93. Wood, *Logs of the Conquest*, 212, Log of HMS *Diana*, 19-20 August (actually 18-19 August).

94. Saunders to Pitt, 5 September 1759, in Little, ed., *Despatches of Saunders*, 123.

95. Montcalm to Bougainville, 21 August 1759, quoted in Wood, *Logs of the Conquest*, 149.

96. Saunders to Pitt, 5 September 1759, in Little, ed., *Despatches of Saunders*. This movement of more than half of the specialized landing craft with the fleet above Quebec early in August is an important event that has been overlooked by other historians of the siege.

97. Saunders to Pitt, 5 September 1759, in Little, ed., *Despatches of Saunders*.

98. Wood, *Logs of the Conquest*, 237, Log of HMS *Lowestoft*, 28 August 1759 (actually 27 August).

99. Wood, *Logs of the Conquest*, 299, Log of HMS *Squirrel*, 5 September 1759; and 313-314, Log of HMS *Stirling Castle*, 5 September 1759.

100. Wood, *Logs of the Conquest*, 152.

101. Wolfe's Memorandum, contained in Stacey, *Quebec 1759*, 203-204.

102. Moncton, Murray and Townshend to Wolfe, 29 August 1759, in Stacey, *Quebec 1759*, 204-205.

103. Stacey, *Quebec 1759*, 116-118.

104. Wolfe to Pitt, 2 September 1759, in Stacey, *Quebec 1759*, 210.

105. Saunders to Pitt, 5 September 1759, in Little, ed., *Despatches of Saunders*.

106. Holmes to _____, 18 September 1759, Doughty and Parmalee, eds., *Siege of Quebec*, vol 4, 295.

107. Holmes to _____, 18 September 1759, Doughty and Parmalee, eds., *Siege of Quebec*, vol 4, 295.

108. This officer cannot be identified in any of the standard sources for RN officers in the 18th century and was almost certainly the master of one of the merchant vessels above the town – perhaps the saucy little schooner, *Terror of France*.

109. LAC, Micro A-652, Journal of an Anonymous officer.

110. Holmes to _____, 18 September 1759, Doughty and Parmalee, eds., *Siege of Quebec*, vol 4, 295.

111. LAC, Micro A-652, Journal of an Anonymous officer.

112. Wolfe to Burton, 10 September 1759, cited in Stacey, *Quebec 1759*, 123.

113. Orders, 11 September 1759, in *General Wolfe's Instructions*, 101.

114. LAC, Micro A-652, Journal of an Anonymous officer.

115. Wolfe to Monckton, 12 September 1759, contained in Stacey, *Quebec 1759*, 127.

116. Holmes to _____, 18 September 1759, Doughty and Parmalee, eds., *Siege of Quebec*, vol 4, 295.

117. Saunders to Wolfe, 25 August 1759, in Salmon, *Life of Saunders*, 131.

118. LAC, Micro A-652, Journal of an Anonymous officer.

119. Information on river conditions during the night of 12 September 1759 is contained in Olson *et al.*, "Perfect Tide, Ideal Moon: An Unappreciated Aspect of Wolfe's Generalship at Québec, 1759."

120. It has recently been suggested that, because the tidal and moonlight conditions on the night of 12/13 September were ideal for an landing operation downriver of Wolfe's position at St. Nicholas that he actually waited until that night to launch the Foulon landing. By doing so, he demonstrated "the perspicacity to use the best scientific data and analysis available in order to choose the most auspicious time and date for his enterprise," see Olson *et al.*, "Perfect Tide, Ideal Moon."

Unfortunately for this most interesting thesis, the choice of the night of 12/13 September was most fortuitous on Wolfe's part. He had actually planned to launch his operation upriver some days earlier but was delayed by bad weather, see Donald E. Graves, "The Anse au Foulon 1759: Some New Theories and Some New Evidence."

121. Orders, 11 September 1759, in *Wolfe's Instructions*, 101-103.

122. Orders, 11 September 1759, in *Wolfe's Instructions*, 102-103.

123. These three vessels, called variously "armed sloops" or "armed vessels" are not identified. They were civilian vessels and entries in the logs of HM Ships *Seahorse* and *Sutherland* indicate that they could have been either the *Good Intent*, *Prosperity* or *Resolution* which arrived with the main fleet in late June or the *Terror of France*, which arrived later – see Wood, *Logs of the Conquest*, 294-295, Log of HMS *Seahorse*, for September 1759 and, 299-301, Logs of HMS *Squirrel*, for September 1759.

124. Orders, 11 September 1759, *Wolfe's Instructions*, 103.

125. That is, the *Compagnies franches de la marine*, the colonial troops who garrisoned New France. The hail of the sentries at the Anse au Foulon and the British reply is a famous episode in Canadian history and has been the subject of much mythology. Stacey discusses it thoroughly in *Quebec 1759* (139-140, 226-227) and notes that the British officer who replied to the sentries' hail was a Captain Fraser of the 78th, who spoke fluent French. Many earlier accounts of this incident state that Fraser's response was "La Reine" or "De la Reine," (from "the *Régiment de la Reine*") but, as Stacey states, this unit was not at Quebec at that time and that the more apt response, "*Marine*," was later corrupted into "*De la Reine*."

126. Memoir of *Capitaine* de Vergor, probably 1761, contained in Stacey, *Quebec 1759*, 226-227. The translation is mine.

127. Journal of Lieutenant Gordon Skelly, HMS *Devonshire*, 13 September 1759. This document was offered for auction by Christie in 2003 but, unfortunately, the Library and Archives of Canada was outbid for the item by a private collector and its present whereabouts is unknown. Shortly before the sale, the author was asked to comment on its authenticity and, to do so, was provided with a typescript text of excerpts from the journal from which the quote in the text is taken. It is to be hoped that the entire document will one day be available to historians.

128. Holmes to _____, 18 September 1759, in Doughty and Parmalee, eds., *Siege of Quebec*, vol 4, 295.

129. Stacey, *Quebec 1759*, 147.

130. Saunders to Pitt, probably 20 September, in Salmon, *Life of Saunders*, 142.

131. Woods, *Logs of the Conquest*, 314-315, Log of HMS *Stirling Castle*, 13 September 1759 (actually 12/13 September 1759).

132. Holmes to _____, 18 September 1759, in Doughty and Parmalee, eds., *Siege of Quebec*, vol 4, 295.

133. Wood, *Logs of the Conquest*, 242, Log of HMS *Lowestoft*, 13 September 1759.

134. Holmes to _____, 18 September 1759, in Doughty and Parmalee, eds., *Siege of Quebec*, vol 4, 295.

135. Holmes to _____, 18 September 1759, in Doughty and Parmalee, eds., *Siege of Quebec*, vol 4, 295.

136. Saunders to Pitt, probably 20 September 1759, in Salmon, *Life of Saunders*, 144.

137. The most recent study of this sanguinary engagement is Ian McCulloch, "From April battles and Murray generals, good Lord deliver me!": The Battle of Sillery, 28 April 1760," in Donald E. Graves, ed., *More Fighting for Canada: Five Battles, 1760-1944* (Toronto, 2004), 17-70.

138. On the post-1759 operations at Quebec, see Stacey, *Siege of Quebec*, 183-187.

139. LAC, CO 5, vol. 51, Townsend to Pitt, 18 September 1759.

140. Saunders to Pitt, probably 20 September 1759, in Salmon, *Life of Saunders*, 143.

Chapter 2: "Old Ironsides'" Last Battle: USS *Constitution* versus HM Ships *Cyane* and *Levant*

1. Christopher McKee, *A Gentlemanly and Honorable Profession: The Creation of the U.S. Naval Officer Corps, 1794-1815* (Annapolis, Md., 1991), 473-484.

2. Philip Chadwick Foster Smith, *The Empress of China* (Philadelphia, Pa., 1984).

3. William S. Dudley, ed. *The Naval War of 1812: A Documentary History*, Vol. I (Washington, D.C. 1985), 16-17.

4. John F. Zimmerman, *Impressment of American Seamen* (New York, 1925), pp. 264-275; Dudley, *Naval War of 1812*, 61-68.

5. NARA, RG46, 12th Congress, Messages of the President (SEN12A-E2).

6. Ronald L. Hatzenbuehler and Robert L. Ivie, *Congress Declares War: Rhetoric, Leadership, and Partisanship* (Kent, Ohio, 1983; Reginald Horsman, *The Causes of the War of 1812* (New York, 1962).

7. Donald R. Hickey, *The War of 1812: A Forgotten Conflict* (Urbana and Chicago, Ill., 1989), 45-46.

8. Extract from Commodore Rodgers Journal, USS President, 23 June 1812, and Captain Richard Byron, R.N. to Vice Admiral Herbert Sawyer, 27 June 1812, in Dudley, *Naval War of 1812*, 153-160.

9. John Rodgers to Paul Hamilton, 3 June 1812, and Stephen Decatur to Paul Hamilton, 8 June 1812, in *Dudley, Naval War of 1812*, 117-124.

10. Geoffrey Footner, *USS Constellation: From Frigate to Sloop of War* (Annapolis, Md., 2003), 32-61.

11. Linda Maloney, *The Captain from Connecticut: The Life and Times of Isaac Hull* (Boston, Mass., 1986), 168-170.

12. George F.G. Stanley, *The War of 1812: Land Operations* (Ottawa, Ontario, 1983), 83-135.

13. Tyrone G. Martin, *A Most Fortunate Ship: A Narrative History of Old Ironsides* (Revised edition, Annapolis, Md., 1997), 171-179; Journal of Commodore William Bainbridge, 29 December 1812 and Lieutenant Henry D. Chads to Secretary of the Admiralty John W. Croker, 31 December 1812, in Dudley, *Naval War of 1812*, Vol. I, 639-648.

14. Berube and Rodgaard, *A Call to the Sea*, (Washington, D.C., 2005), 52-58.

15. Paul Hamilton to Charles Stewart, 22 June 1812, NA, RG45, Letters from the Secretary of the Navy to Captains, Ships of War.

16. Footner, *Constellation*, 75; Charles Stewart to Paul Hamilton, 12 November 1812 in *American State Papers*, VI, *Naval Affairs*, Vol. I (1789-1825), 278-279. Stewart sent this letter to Hamilton in response to his request in answering House of Representatives Naval Committee Chairman Burwell Bassett's queries concerning future naval construction. Captains Isaac Hull and Charles Morris concurred in Stewart's statement.

17. Lords Commissioners of the Admiralty to Admiral Sir John B. Warren, 26 December 1812, UkLPR, Adm. 2/1375 (Secret Orders and Letters), 337-338; see also Dudley, *Naval War of 1812*, I, 633-634.

18. Captain John Cassin to Secretary William Jones, 23 June 1813, in Dudley, *Naval War of 1812*, 359-360; Footner, *Constellation*, 93-98.

19. William Bainbridge to William Jones, 27 April 1813, in Dudley, *Naval War of 1812*, 429-430.

20. William Jones to Charles Stewart, 19 September 1813; Stewart to Jones, 5 December and 25 December 1813, in Dudley, *Naval War of 1812*, 292-293. On the blockade, see Wade G. Dudley, *Splintering the Wooden Wall: The British Blockade of the United States, 1812-1815* (Annapolis, Md., 2003), 79-109.

21. HM Schooner *Pictou*, 14 guns, was a recapture, originally the American privateer *Zebra* out of New

York. See Rif Winfield, *British Warships in the Age of Sail, 1793-1817: Design, Construction, Careers and Fates* (London, 2005), 369.

22. Martin, *Most Fortunate Ship*, 183-188.

23. Martin, *Most Fortunate Ship*, 188-190; see also end note 3, 377 comparing Stewart's cruise with that of John Rodgers in 1812 that lasted 81 days.

24. Winfield, *British Warships*, 122-123, 149. HMS *Newcastle*, dimensions: 176 feet length over all, 149 feet between perpendiculars, 44 feet beam, 15 feet draft; complement 450; ordnance upper deck 30 24-pounders, spar deck 28 42-pounder carronades, possibly 4 24-pounders in addition. HMS *Leander*, dimensions 174 feet overall, 145 feet between perpendiculars, 45 feet beam, 14 feet draft; complement 450; ordnance upper deck 30 24-pounders, spar deck 28 42-pounder carronades, forecastle 4 24-pounders. HMS *Acasta*, dimensions: 154 feet overall, 129 feet between perpendiculars, 40 feet beam, 14 feet draft; complement 320; ordnance upper deck 30 18-pounders, quarterdeck 8 9-pounders and 4 32-pounder carronades, forecastle 2 9-pounders, and 4 32-pounder carronades.

25. Gunades were a recent addition to *Constitution*'s armament, courtesy of the American privateer *Fox* that had captured them from the British brig *Stranger*. According to Tyrone Martin, they were "designed by Sir William Congreve in 1814 and each was 8' 6" long, but being of thinner barrel construction weighed only about 5000 pounds on carriage. The design was an attempt to combine the range of a long gun with the lighter weight of a carronade. The pair sat on carriages like the long guns, and it was expected that, since they were lighter, they could readily be shifted from side to side as combat required. To overcome the ship's weakness in firing straight ahead, he removed the officers' telephone booth-like "spice boxes" (johns) from their places forward on the gun deck so that 24-pounders could be fired dead ahead through the bridle ports."

26. Rear Admiral Hotham to J.W. Croker, Secretary of the Admiralty, 29 December 1814. NAUK, 1/508, 50.

27. Tyrone G. Martin, ed. *The USS Constitution's Finest Fight, 1815: The Journal of Acting Chaplain Assheton Humphreys, US Navy* (Mt. Pleasant, S.C., 2000), 10. Chaplain Humphreys describes *Lord Nelson* as "a perfect slop ship and grocery store, very opportunely sent to furnish a good rig [clothing] and bountiful cheer for Christmas …"

28. Martin, ed. *Finest Fight, 1815*, 66.

29. William James, *Naval Occurrences of The War of 1812: A Full and Correct Account of the Naval War Between Great Britain and the United States of America, 1812-1815* (London, 2004), Introduction by Andrew Lambert, 228-229.

30. Captain Gordon Falcon, Remarks on Board the *Cyane*, 20 February 1815. NAUK, ADM 1/5449.

31. HMS *Cyane*, 6th rate frigate; dimensions 118 feet overall, length between perpendiculars 98 feet, beam 32 feet, draft 10 feet. HMS Levant, 6th rate frigate, dimensions 116 feet overall, 98 feet between perpendiculars, 29 feet beam, 10 feet draft. Winfield, *British Warships*, 236, 238.

32. Spencer C. Tucker and Frank T. Reuter, *Injured Honor: The Chesapeake Leopard Affair, June 22 1807* (Annapolis, Md., 1996), 15. John Marshall, *Royal Navy Biography*, Supp. Part III (London, 1829), 161-162.

33. William James, *Naval Occurrences*, 228-236.

34. Stewart to Secretary of the Navy Benjamin W. Crowninshield, 15 May 1815, based on ship's log, states the ships were "about 300 yards distant" when the action started. Cessation of action with *Cyane* is given as 6:50 p.m. NARA, RG45, Captain's Letters to the Secretary of the Navy, 1815, Vol. 3, No. 93 [M125, Roll 44].

35. Gordon Falcon to George Douglas, 22 February 1815. NAUK, ADM 1/1740.

36. Douglas to Croker, 22 February 1815. NAUK, 1/ADM1740.

37. Theodore Roosevelt, *The Naval War of 1812 or the History of the United States Navy in the Last War with Great Britain* (New York, 1882), 418.

38. Roosevelt, *Naval War of 1812*, 421.

39. Martin, *A Most Fortunate Ship*, 196-197 and footnote 4, 378.

40. Charles Stewart to Benjamin W. Crowninshield, May [no date] 1815; with enclosure "Minuets [sic] of the action between the U.S. frigate *Constitution* and the HM ships *Cyane* and *Levant* on the 20th February 1815." NARA, RG45, Captain's Letters [CL], 1815, Vol. 3, No. 93. [M125, Roll 44].

41. Martin, ed. *The USS Constitution's Finest Fight*, 30-31.

42. Stewart suspected he could not depend on British respect for Portuguese neutrality. As he wrote in his report, "… from the little respect hitherto paid by the[m] to Neutral Waters, I deemed it most prudent to put to sea." This is a reference to the British attack on the American frigate *Essex* in Chilean territorial waters. See footnote no. 44 for source.

43. Commodore Collier's report on his search and the account of the Cape Verdes action is contained in Collier to Griffith, 12 March 1815. NAUK, ADM 1/509, 250-254.

44. Stewart to Crowninshield, May [no date] 1815, with enclosure "Minuets [sic] of the chace of the US frigate *Constitution* by an English squadron of 3 ships, from out of the harbor of Port [sic] Praya, island of St Iaga [sic]: Sunday, March 12th 1815. NARA. RG45, CL, 1815, Vol. 3, no. 93. [M125, Roll 44].

45. James, *Naval Occurrences*, 236.

46. Roosevelt, *Naval War of 1812*, 424. His reference is to James's *The Naval History of Great Britain : from the Declaration of War by France in 1793, to the Accession of George IV*, 6 vols., Vol. V, 558.

47. John Marshall, *Royal Navy Biography*, II, Vol. II (London, 1825), 536-538.

48. Collier to Griffiths, 12 March 1815. NAUK, ADM 1/509, 251.

49. *Niles Weekly Register*, 17 June 1815, p. 289. The source of this anecdote is not identified, though presumably a British officer, but its elements contain the ring of truth. The question of damage to the foretopsail spar remains to be corroborated. See also Abel Bowen, ed. *The Naval Monument, containing official accounts of all the battles fought between the United States and Great Britain during the late war and an account of the War with Algiers* (Boston, Mass., 1851 (1st edition, 1816)), 183-184.

50. Winfield, *British Warships in the Age of Sail*, 122.

51. Andrew Lambert, "Sir George Ralph Collier," in *The Oxford Dictionary of National Biography* (London, 2004), Vol. 12, 638-639.

52. Martin, *A Most Fortunate Ship*, 373.

53. The author wishes to acknowledge Tyrone Martin, Frederick Leiner and Donna Dudley for their editorial assistance and Andrew Lambert, Michael Crawford, Christine Hughes and Charles Brodine for their help in obtaining documentation for this chapter.

Chapter 3: Taking the President: HMS ***Endymion*** and the USS ***President***

1. I am indebted to my friends Michael Crawford, Bill Dudley, Don Hickey and Christopher McKee for their laudable efforts to improve my understanding of the U.S. Navy and of this action. Robert Gardiner, a friend of even longer standing, shared his expertise on the ships involved.

2. Adams, H. *History of the United States of America during the Second Administration of James Madison*, (Vol. III New York, 1890), 63. A thoughtful analysis, far removed from the partisan approach adopted by other 19th century American authors, who lionise Decatur. While Theodore Roosevelt *The Naval War of 1812*, (New York, 1882), 401-8 was critical of Decatur he was more concerned to score points off William James.

3. Marquardt, K.H. *The 44 Gun Frigate USS Constitution*.(London, 2005),23 and Gardiner, R. *Frigates of the Napoleonic wars*. (London, 2000), 43-4.

4. Canney, D.L. *Sailing Warships of the U S Navy*. (London, 2001), 23-41.

5. The issues were rehearsed many times by the Court on Inquiry: United States National Archive Record Group (RG) RG125: Vol. 6 no.202. McKee, C. *A Gentlemanly and Honourable Profession: The Creation of the U.S. Naval Officers Corps, 1794-1815*. (Annapolis, 1991) provides a brilliant study of the pre-1815 USN officer corps as a highly strung, self-absorbed society with a fatal penchant for duelling.

6. De Kay., J. T. *Chronicles of the Frigate Macedonian, 1809 -1922*. (New York, 1995) for the financial implications of equal combat, 95-6 and 102-8. McKee, 346-7, gives Decatur's total wartime prize money as $30,099 – his share of the *Macedonian*.

7. Broke to his Wife 8.1.1813: Padfield, P. *Broke and the Shannon* (London, 1968), 119.

8. James, W. *Naval Occurrences of the War of 1812* (London, 1817), 63. New edn, London 2004. See 62-3 for a detailed comparison of the 44s with the standard "Leda" class British frigates.

9. De Kay, 72-3. Significantly Adams cited James's work as an authority, rather than Theodore Roosevelt's work.

10. *Dictionary of American Naval Fighting Ships Volume V.* (Washington, 1970), 371.

11. Tucker, S. *Stephen Decatur: A Life Most Bold and Daring* (Annapolis, 2005), is the latest of many biographies. See 64-70 for the Tripoli incident.

12. Long, D.F. *Sailor-Diplomat: A Biography of Commodore James Biddle, 1783-1848* (Boston, 1983), 48.

13. Tucker, S. *Arming the Fleet: U.S. Navy Ordnance in the Muzzle-Loading Era* (Annapolis, 1989), 183.

14. Tucker, *Decatur*, 128-130.

15. Albion, R.G. with Pope, J.B. *The Rise of New York Port, 1815-1860* (New York, 1939), 16-27.

16. Albion, 8-9.

17. Secretary of Navy to Decatur 8.8.1814: *Calendar of the Correspondence of James Madison.* (Washington, 1894), 428.

18. Owsley, F.J. Jr. "William Jones" in Coeletta, P. ed. *American Secretaries of the Navy. Vol. I* (Annapolis, 1980), 100-112.

19. William Jones to Decatur 17.11.1814: RG45, CLS (Confidential Letters Sent) 1814.

20. Decatur–Jones 23.11.1814 RG45 Captains Letters 1814 (CL) vol.8 no.30.

21. Jones–Decatur 25.11.1814 RG45 CLS 1814.

22. Long, *Biddle*, 50.

23. Decatur–Jones 30.12.1814: RG45 CL 1818 vol. 8 no. 148.

24. Decatur–Jones 23.11.1814: RG45 Captains Letters 1814 (CL) vol. 8 no. 30.

25. Dudley, W. *Splintering the Wooden Wall: The British Blockade of the United States, 1812-1815* (Annapolis, 2003), 127.

26. Admiralty to Admirals, various 25.11.1814: ADM 2/1381 Secret orders. No. 44, 46.

27. Admiralty to Admirals, various 26.11.1814: ADM 2/1381 Secret orders. No. 46, 56.

28. Admiralty to Hotham 19.12.1814: & to Admiral Sir Byam Martin (Plymouth) 19.12.1814: ADM 2/1381, 80-4.

29. These urgent orders were sent in a "fast sailing vessel," directed not stop on any account, and Hotham was directed to send by return "the latest intelligence … of the enemy's preparations and movements." However, the timely arrival of orders and intelligence remained prey to wind and weather. The 150-ton Bermudan schooner HMS *Bramble* left Plymouth on the 23rd, endured a tempestuous voyage and ended up stuck on a rock off Bermuda on February 2nd. Log Book 1814-1815 HMS *Bramble*: ADM 51/2181.

30. Log Book HMS *Superb* December 1814–January 1815: ADM 51/2051.

31. These dispositions led to the final battle of the *Constitution*: see W.S. Dudley's essay in this collection.

32. Hotham to Admiralty 29.12.1814 *Superb* off Newfoundland rec. 2.2.1815: ADM 1/508 f107.

33. Log 28.12.1814 HMS *Superb*: ADM 51/2051. Unable to spare any more ships from the blockade Hotham sent this urgent intelligence on a Portuguese merchant ship. He took the precaution of sending it in "the telegraphic vocabulary," a plain language copy would follow on the next naval vessel. The copy reached London 16 days after the original.

34. Endorsement on the deciphered copy "JWC Feb 1.," that is *before* the message was registered as having arrived: ADM 1/508 f107. The reply was also sent before the letter was registered. Admiralty to Hotham 1.2.1815: ADM 2/1381 165.

35. Hotham to Cochrane 2.1.1815: Off New London Cochrane NLS MS 2327 fol.114-5.

36. Hotham to Admiralty 5.1.1815: ADM 1/508 ff.128-9.

37. Hotham to Admiralty 16.1.1815 rec. 17.2: No.2 ADM 1/508 f.143 enclosing intelligence report of the 7th.

38. Hotham to Admiralty 5.1.1815; rec. 17.2: ADM 1/508 No.2 f.128-9.

39. Lecky, H.S. *The King's Ships.* Vol. III (London, 1910), 1-10.

40. Gardiner, R. *Frigates*, 91-2.

41. Gardiner, *Frigates*, 145.

42. Garitee, J.R. *The Republic's Private Navy; The American Privateering Business as practised by Baltimore during the War of 1812* (Middletown, Conn., 1977), xii-xiii. This ship was taken soon afterwards by Sir George Collier's squadron.

43. Harland. J. *Seamanship in the Age of Sail* (London, 1984), 196-8. Hayes had club-hauled the anchored

74 HMS *Magnificent* off a lee shore in the Basque Roads in 1812.

44. Gardiner, 50-3 & 193 fn 28 for the origins of Hayes's soubriquet.

45. Log Book of HMS *Forth*: ADM 51/2397. While heading south *Forth* ran into the same storm that crippled *President* and *Endymion*.

46. Gardiner, 25-6. *Tenedos* was a sister ship of the *Shannon* and the *Macedonian*, built 1810-1812 at Chatham Dockyard. *Pomone* was built as the French *Astrée* at Genoa in 1808 and taken in 1810. These two ships were of almost identical form and size, although the structural plan was distinct.

47. Decatur to Crowninshield 14.1.1815: RG45 CL 1815 vol. 1 no.41.

48. Cooper, J. F. *History of the Navy of the United States of America*. 3rd ed (Philadelphia, 1841), 429-432. The same account, almost word for word, appears in Shaw, E. *Narrative of his 21 years' service in the American Navy* 3rd ed (Rochester, NY, 1845), 48. I must thank my good friend William S. Dudley for his help in obtaining a copy of this scarce title. The very close correlation between the Shaw text, first published in 1843, and James Fenimore Cooper's account in his *History of the U.S. Navy* first ed, 1839, demonstrates extensive textual borrowing. In correspondence Christopher McKee suggested that "Shaw" is one of a number of bogus sailor texts that appeared at this time, a judgement which I find compelling.

49. Cooper, 430.

50. Hayes to Hotham 17.1.1815: ADM 1/508 f.769.

51. Log Book HMS *Majestic* 15.1.1815: ADM 51/2543.

52. Cooper, 430.

53. Cooper, 430.

54. Hayes to Hotham 17.1.1815: ADM 1/508 f.769.

55. *Endymion* Log. & James, 213-4.

56. *Pomone* Log.

57. *Endymion* log.

58. Tucker *Arming*, 94. James considered such projectiles so important that those found on the *President* featured in his frontispiece! Such ammunition is still displayed on board the USS *Constitution*.

59. Douglas H. *A Treatise on Naval Gunnery* (London, 4th ed., 1855), 480.

60. Douglas, 483.

61. Cooper, 431 repeated word for word in Shaw, 47.

62. Cooper, 431 repeated in Shaw, 48.

63. Cooper, 431.

64. "Autobiography of Commodore Hollins CSA. *Maryland Historical Magazine 34* September 1939, 229-30. Although charming, the memories of a first voyage midshipman, recorded at the end of a long and eventful life are hardly authoritative. Decatur made his officers swear to this at the Court of Inquiry.

65. Hollins, 230, and Cooper, 431.

66. *Majestic* Log.

67. Tucker, *Decatur*, 144.

68. Harland, 12-3, 33.

69. Lieutenant Shubrick: Court of Inquiry Testimony: RG125 vol. 6, no. 202.

70. "Thessaly" in *The Naval Chronicle 1817 vol. 37* (London, 1817), 194.

71. The decisive moment of the action features in a fine study by Thomas Buttersworth, published as an engraving on 1 June 1815 in London, based on "particulars and the position of the ships by Lieut [Francis] Ormond, [2nd] of the *Endymion*." The picture shows *President's* spanker in the act of being brailed up, and her lanterns have been put out, which provides a precise time. See Smith, E.N. *American Naval Broadsides* (New York, 1974), 191.

72. James, 1817, 385.

73. *Endymion* Log.

74. This is quite clear in *Endymion's* Log, and the copy made by the master.

75. The Court of Inquiry testimony of both lieutenants makes this clear.

76. *Pomone* Log. *Pomone's* times are roughly two hours ahead of those kept by *Endymion*. I have used *Endymion's* record.

77. Decatur to Crowninshield 17.1.1815: James *Naval Occurrences*, 380-1.

78. Captured muster books were used to determine the "head-money" due to the captors for each enemy combatant taken or slain.

79. There is a copy of Hope's account of the engagement in the Master's Logs of the *Endymion*. ADM 52/3904. In both logs the material has been pasted in as a separate page. ADM 51/2324. Hayes tried to cram his Log entry into a regular page of *Majestic's* log, the result is almost illegible.

80. Roosevelt, 407-8: Adams, 68.

81. Lecky, *King's Ships*, 9.

82. Douglas, 423.

83. James, W., *Naval Occurrences between Great Britain and the United States* (London, 1817), 226.

84. Marshall, J *Naval Biography: Supplementary Volume I* London 1827 p.316.

85. Hope to Hayes, 15.1.1815: James, 1817, 379.

86. McKee, 398.

87. "Thessaly" letter dated Plymouth 15.1.1817: *The Naval Chronicle*, vol. 37 (London, 1917), 193-4.

88. Hayes to Hotham 17.1.1815: ADM 1/508 f.769.

89. Log 22.1.1815 HMS *Superb*. ADM 51/2051.

90. Hotham to Cochrane 23.1.1815: ADM 1/508 f767.

91. Hotham to Cochrane 12.2.1815 rec. 20.3. no. 17 ADM 1/.508 f.827.

92. Long, *Biddle*, 50-55.

93. Hotham to Admiralty 13.2.1815 rec. 20.3. no. 8 ADM 1/508 f831.

94. HCA 49/098 Bermuda captures 26.12.1814 -25.5.1815.

95. I am indebted to Dr Edward Harris for this information.

96. Decatur to Crowninshield 20.2.1815 (written on board HMS *Narcissus*) RG45 CL 1815 vol. 1 no.144.

97. James, *Naval Occurrences*, 222-3.

98. James, W., *The Naval History of Great Britain* vol. VI, 534.

99. O'Byrne, 329. James, 224.

100. James, *Occurrences*, 223-4.

101. Duffy, S.W.H. *Captain Blakeley and the Wasp* (Annapolis, 2001), 167, 279-80.

102. James 1817, 218.

103. Commodore Andrew Evans (at Bermuda) to Admiralty 19.2. & 23.1815: ADM 1/1771 f 26, 36.

104. James, *Naval History Vol. VI* (London, 2004), 370.

105. Lecky, H. S. *The King's Ships*. Vol. III, 1-10.

106. Cited by Marshall, 317.

107. *Endymion* Log. ADM 51/2324.

108. Adams, 70.

109. Leiner F. C. *The End of Barbary Terror: America's 1815 War against the Pirates of North Africa.* (New York, 2005), 101.

110. Decatur to Secretary of the Navy 18.1.1815: James, *Naval Occurrences*, 1817, 379-381.

111. O.H. Perry to Crowninshield 28.1.1815: RG45 CL 1815 vol. 1 no. 82.

112. Leiner, 40-41.

113. Log 22.2.1815 HMS *Superb*. ADM 51/2051.

114. Decatur to Crowninshield 20.2.1815 (HMS *Narcissus*) RG45 CL 1815 vol. 1 no.144.

115. Decatur to Crowninshield 6.3.1815: from NY RG45 CL microfilm 125 reel 43 no. 16.

116. Decatur to Crowninshield 6.3.1815: James, *Naval Occurrences*, 382.

117. Crowninshield to Decatur 14.3.1815 RG45 Sec. of Navy letters Micro. 149 p. 60.

118. Crowninshield to Decatur 14.3.1815: Leiner, 56. Peabody Essex Museum Crowninshield Collection.

119. Decatur to Crowninshield 20.3.1815: Leiner, 56. Peabody Essex Museum Crowninshield Collection.

120. James, *Occurrences*, 213-219.

121. Transcript and Report of the Court on Inquiry: United States National Archives RG125: vol. 6 no. 202.

122. Commodore Murray to Crowninshield 17.4.1815: James, *Naval Occurrences*, 382-4.

123. Comments by Commodore John Rodgers: Leiner, 59.

124. James, 383.

125. Albion, 26.

126. Leiner, 87-150.

127. Leiner, 177.

128. Lambert, A.D. ed. Intro to James, W. *The Naval History of Great Britain: Vol III.* (London, 2003), x-xi.

129. Leiner, 178.

130. Lambert, A.D. ed. Intro. to James Vol. III, xi.

131. James, 1817, 225.

132. Brenton, E.P. *Naval History of Great Britain. Vol. II* (London, 1937), 537-8.

133. The only other USS *President* was a 12-gun sloop operating on Lake Champlain, taken by the British in 1814, and renamed *Icicle. Dictionary of American Naval Fighting Ships Volume V.* (Washington, 1970), 371-2.

134. Graham (First Lord of the Admiralty) to Lord Stanley 18.9.1833: Graham MS 28 Microfilm.

135. Lecky, 10. In 1904 Captain King-Hall asked if the two Bermuda vases could be loaned to his ship "nothing would have a greater effect on the Esprit de Corps of the Ship's company." ADM 203/9.

136. Graham to Cockburn 24.1.1834: Graham MS 52. Morriss, R. *Cockburn and the British Navy in Transition: Admiral Sir George Cockburn 1772-1853* (Exeter, 1997), 210, 222.

137. Smith, E.N. *American Naval Broadsides* (New York, 1974), 191.

138. Muster Book HMS *Endymion*: ADM 37/5729. *The Naval Chronicle.* Vol. 33 1815, 262.

Interlude: The Renewed Wolfpack Attacks of September 1943

1. In the event only a few of the smaller new Type XXIII U-boats saw service in 1945, while none of the large Type XXI U-boats saw action before the war ended.

Chapter 4: Hollow Victory: Gruppe Leuthen's Attacks on Convoy ON202 and ONS18)

1. Many assisted with this paper, but the author owes particular gratitude to the following individuals: Mr. Michael Whitby, Senior Naval Historian, Directorate of History and Heritage, Department of National Defence, Ottawa, Canada, who not only suggested the topic but generously assisted my efforts at locating sources; Dr Malcolm Llewellyn-Jones, Naval Historical Branch at the UK Ministry of Defence, who provided sources I could not locate elsewhere; Dr Marc Milner, the University of New Brunswick, who generously shared his research on the Allied reaction to the German Type V torpedo; and Lieutenant-Colonel (Rtd) Robert L. Hills, who provided many helpful improvements to the text.

2. DHH *BdU KTB* 15-24 September 1943; Kenneth Wynn, *U-Boat Operations of the Second World War, Volume 2: Career histories, U511-UIT25*, (Dubai, 2003), 73. *BdU* ordered 21 U-boats to operate as part of group *Leuthen* on 15 September 1943. *U-603* was ordered to the southern most position in the patrol line (*BdU KTB* 15 September 1943), and departed Brest 9 September. However, *U-603*, was considerably south of its intended start point when operations began, and never reached the vicinity of the convoys. It is therefore not included in the U-boats engaged in the battle.

3. DHH, ADM 199/353, "RoP after Torpedoing of HMS LAGAN," 25 September 1943 by Lieutenant Commander A. Ayre, R.N.R.(Rtd). Only six of 29 bodies were ever found.

4. RNR is the acronym for Royal Navy Reserve. The rule in the Royal Navy of the period was that commanders reached retirement age at 50, lieutenants and lieutenant-commanders at 45, although "Retirement from the Active List is not to disqualify any Officer for employment at or under the Admiralty." Navy List 1944, 79-81, provided by Malcolm Llewellyn-Jones via e-mail correspondence 21 July 2006. In short, Ayre was an older officer still serving at sea because of the war.

5. Ayre's "RoP" sets out the actions of *Lagan* clearly and concisely.

6. Grand Admiral Karl Doenitz, *Memoirs: Ten Years and Twenty Days* (Annapolis, 1990), 340. The term frightful was used when describing the loss of 31 boats by the 22nd of the month. By the end of May 1943 41 U-boats had been lost to all causes: Michael Gannon, *Black May: The Epic Story of the Allies Defeat of the German U-Boats in May 1943*, (New York, 1998), 407.

7. Robert C. Stern, *Type VII U-boats*, (London, 1998), 18.

8. Stern, *Type VII U-boats*, 124-125

9. Research Report 27, "Anti-Submarine Operations by CVE Based Aircraft," ASWORG/87-32, 1 April 1944, pages 31-32. Posted at <www.uboatarchive.net/ASWORGReport27.htm>.

10. *Leuthen* was a famous victory of Frederick the Great, whose successful multi-front war became an important historical model for Hitler as prospects became ever grimmer for the Third Reich.

11. W.A.B. Douglas, *The Creation of a National Air Force: The Official History of the Royal Canadian Air Force, Volume* II (Toronto, 1986), 562.

12. W.J.R. Gardner, *Decoding History: The Battle of the Atlantic and Ultra* (Annapolis, 1999).

13. *BdU KTB* 14 September 1943.

14. NHB ADM 223/184, Signal Rodger Winn, DDIC to CinCWA, etc, 182125A, Serial H.2816, F.2288, 18 September 1943, courtesy of Dr. Malcolm Llewellyn-Jones.

15. Arnold Hague, *The Allied Convoy System 1939-1945: Its Organization, Defence and Operation* (Annapolis, 2000), 128.

16. General outline of EG 9 movements provided in "A Christmas Festival: EG 9's First and Last Operation," DHH, RCN Monthly Review, Number 24, December 1943, 51-52, by Lt A.F. Pickard, RCNR, CO of HMCS *Chambly*. Relevant messages from Admiralty and EG C-5 (advising of attack on HS-256) found with RoP of HMCS *Chambly*, 29 September 1943 in DHH ADM 199/383.

17. Jurgen Rohwer, *Axis Submarine Successes of World War II* (Annapolis, 1999), 171. Pickard, 52.

18. LAC RG 24 Series D-2, Vol 7185, Deck Log HMCS *Chambly*.

19. Course diversion in DHH, Admiralty Monthly Anti-Submarine Report, October, 1943 (published 15 November 1943) (hereafter Admty Monthly ASW Review Oct 43). Prior course of convoys not directly given in documents, but determined by plotting noon positions of ON202, as found in DHH ADM 199/583, Convoy Commodore's Report for Convoy ON202.

20. Details of the rush to produce *Zaunkoenig* are in *BdU KTB* 24 Sep 1943.

21. NHB MOD(UK), Admiralty message 200126A. Copy kindly provided by Dr Malcolm Llewellyn-Jones. F.H. Hinsley et al, *British Intelligence in the Second World War*, Volume 3 Part 1 (New York, 1984), 222, provides a general overview. Peter Coy, *The Echo of a Fighting Flower: The Story of HMS Narcissus & B3 Ocean Escort Group* (Upton upon Severn, 1997), 140-141 correctly details the decryption of details of the Type V torpedo on 24 September, and suggests that this is why no prior warning was provided to escorts of the impending use of the Type V. I differ from this analysis in that my research indicates that the OIC was aware of the impending deployment of a German acoustic homing torpedo, but assessed the use of this torpedo against escorts as the least likely prospect.

22. NAUK ADM 219/52 "Some Operational Implications of a Homing Torpedo," L. Solomon, Report 36/43, 1 June 1943. The greatest concern regarding the new torpedo was its possible use against independent merchant ships in the south Atlantic, with merchant ships in convoys being the next most serious concern. My thanks to Dr Malcolm Llewellyn-Jones for bringing this report to my attention, and providing me with a copy.

23. Rainer Busch and Hans-Joachim Roll, *German U-boat Commanders of World War II: A Biographical Dictionary* (London, 1999), 193; and Wynn, *U-boat Operations*, Vol I, 191.

24. Starshell were illumination rounds fired from the main gun, with fuses timed to (hopefully) ignite the round just after it passed the range of the target intended for illumination.

25. Admty Monthly ASW Review Oct 43.

26. DHH ADM 199/353, RoP for Commander M.B. Evans, B3 Escort Group, 29 September 1943, para 16 (hereafter Evans RoP).

27. This was Hepp's first patrol as CO of *U-238*, but he had been CO of *U-272* in the latter part of 1942, surviving the sinking of that submarine when it collided during training exercises in the Baltic.

28. Wynn, *U-Boat Operations*, Vol I, 171.

29. Rohwer, *Axis Submarine Successes*, 171; Admty Monthly ASW Review October 1943, 16. Hits on the port sides of these merchant ships indicate *U-238* fired from outside the convoy.

30. Admty Monthly ASW Review October 1943.

31. Sources are conflicting on the matter of survivors from these two ships. Evans RoP contains a message indicating that a total of 107 survivors were rescued from these two ships, which seems to

account for 70 from the *Douglass* and 37 from the *Weld*. The number on the *Douglass* was broken down into 40 crew, 29 armed guards and one female stowaway. The Admty Monthly ASW Review October 1943 details that 37 of the *Weld* crew were rescued, and that the "entire crew of 40" were picked up from the *Douglass*.

32. *BdU KTB* 13 September 1943.

33. B3 consisted of the destroyers HMS *Keppel* and *Escapade*, the frigate HMS *Towy*, the British corvettes HMS *Narcissus* and *Orchis*, the Free French corvettes FFS *Roselys*, *Lobelia* and *Renoncule* as well as HM Trawler *Northern Foam*.

34. A. Hague, *Allied Convoys*.

35. Syrett, *Defeat*, pp 147-148. The literature on Allied code-breaking in the Battle of the Atlantic is vast, that for German code-breaking less extensive. Some of the better works included David Kahn, *Seizing the Enigma: The German U-boat Codes 1939-1943* and *Hitler's Spies: German Military Intelligence in World War*; David Syrett, *Defeat and Battle of the Atlantic and Signals Intelligence* (excellent collection of documents), and F.H. Hinsley, et al *British Intelligence in the Second World War*, Vol 3, Part I New York: Cambridge University Press, 1984.

36. David Syrett, *The Battle of the Atlantic and Signals Intelligence: U-Boat Tracking Papers 1941-1947* (Burlington, VT, 2002), 238-9, 242.

37. DHH ADM 223/1, Admiralty Appreciation of the U-boat Situation, 1 August 1943; W.A.B. Douglas and J. Rohwer, "Canada and the Wolf Packs" in Douglas (ed), *The RCN in Transition, 1910-1985* (Vancouver 1988), 163.

38. Evans RoP.

39. DHH, CB 04050/43 (9) Admiralty "Monthly Anti-Submarine Report September 1943," 15 October 1943, "Explosion of Hedgehog Projectiles in HMS *Escapade.*"

40. Willem Hackmann, *Seek & Strike*, (London, 1984) 306; Marc Milner, *The U-boat Hunters: The Royal Canadian Navy and the Offensive against Germany's Submarines* (Toronto, 1994), 44.

41. The measures were apparently not completely effective. On 11 April 1945 HMCS *Strathadam* suffered a disaster similar to the one that befell *Escapade*. In this case 11 Hedgehog projectiles exploded above the ship's foc'sle, killing 6 and wounding 13. DHH NHS 8440 EG 25 RoP G 25 20 March to 12 April 1945.

42. Hackmann, *Seek & Strike*, 307-308.

43. DHH Royal Canadian Navy Monthly Review, December 1943, [hereafter RCNMR December 1943] "A Christmas Festival: EG 9's First and Last Operation," Lt A.F. Pickard, 53-54.

44. *BdU KTB* for 20 Sep 43, para IV a.

45. Admty Monthly ASW Review October 1943 tabulates the flights conducted and provides British perspective on the engagements. Douglas, *Creation of a National Air* Force, 564, provides short overview.

46. Admty Monthly ASW Review October 1943.

47. DHH ADM 223/16, "OIC Special Intelligence Summary," 30 September 1943.

48. Admiralty ASW Review October 43; and LAC RG 24 Series D-2 Vol 7262 Deck Log, HMCS *Drumheller* 20 September 1943. Times given in Deck Log and Admiralty ASW Review not completely consistent, and therefore approximations are used for most of this account. Syrett, *Defeat of the German U-boats*, indicates that *Drumheller* opened fire at only 1,400 yards (page 166) but this is far too short a range, especially given the large amount of ammunition (30 rounds) expended by *Drumheller* according to the Deck Log of that vessel.

49. Ibid.

50. *BdU KTB* for 20 September 43. The time of the signal is 1713, Central German Time, which is approx 1513 GMT.

51. Wynn, *U-boat* Operations, Vol I, 227, credits *Drumheller*. Rohwer and Douglas, "Canada and the Wolfpacks," 173, credit Liberator N/120. Axel Niestle, *German U-boat Losses During World War II* (Annapolis, 1998),226 indicates that original credit for *U-338*'s destruction was awarded to Liberator F of 120 Squadron, but that attack was conducted on *U-386* and that no known cause for the loss of *U-338* presently exists.

52. Admty Monthly ASW Review October 1943.

53. Evans RoP.

54. DHH U-386 *KTB*. This boat reported to *BdU* that it had been heavily depth charged at 1730 on 20 September 1943 while attacking a destroyer group, *BdU KTB* 20 September 1943. Allied perspectives from: Admty Monthly ASW Review for October 1943; and DHH 81/520/8280, "Post Convoy Conference St. John's Newfoundland, 28 September 1943."

55. Evans RoP.

56. Evans RoP, "General Conclusions.... the best way to defeat these new tactics is to sink or at least thoroughly incommode the U-boat and … press home our attacks with the utmost vigour."

57. DHH DEFE 3/721 Pt 5 ZTPGU 16961B, 0648 /20/9/43. Message is quoted in Rohwer and Douglas, "Canada and the Wolfpacks," 168.

58. LAC RG 24, 83/84 Accession, TMs 1000-97B Vol 8 *Signals,* CinC WA to All 201539Z September 43. (Notes on this file originally taken by Mr. Michael Whitby, and then graciously forwarded to me by Dr. Marc Milner).

59. A year and a half after the battle of ON202/ONS18, one sea going commander would note in his diary that "The thing I still find hardest of all to make up my mind about is the best and proper anti-Gnat procedure." M. Whitby, ed, *Commanding Canadians: The Wartime Diary of Commander A.F.C. Layard, DFC* (Vancouver, 2005) Diary entry for 10 March 1945, 297.

60. Admty Monthly ASW Review October 1943.

61. DHH ADM 223/16, "OIC Special Intelligence Summary," 30 September 1943.

62. Admty Monthly ASW Review October 1943 indicates two torpedoes struck *St. Croix* at 1956, but Rohwer, *Axis Submarine* Successes, 171, indicates only one Type V was used. Bercuson and Herwig, *Deadly Seas*, 263, square this by suggesting a depth charge, presumably jarred by the torpedo, exploded shortly after the torpedo hit.

63. Evans RoP.

64. Rohwer, *Axis Submarine Successes*, 171-172, details eight torpedo attacks by six different U-boats that most likely were involved in these engagements.

65. The distinction between support escort group and close escort group is greater than the apparently small difference in name suggests. Close escort groups were assigned to stay with convoys for the ocean crossing. Support escort groups were introduced to operate in areas of U-boat concentration, acting as aggressive hunters as opposed to shepherds. Support groups had only become routine in the North Atlantic in May 1943, and became increasingly important from that point in the war on.

66. Like many of the escort commanders, Dobson was an older man – although records are incomplete, at least 10 of the 21 warship commanding officers involved in the battle were Naval Reserves, either merchant mariners with part time naval careers, or professional naval officers past the normal age of retirement. A further three escort commanders were Voluntary Reserve officers, usually young but not very experienced sailors whose energy and enthusiasm compensated for their limited sea time. Three more were Free French officers, while only two were regular RN professionals – Commander Burnett, SO of EG C2 and commanding officer of HMCS *Gatineau*, and Lieutenant-Commander Dyer, RN, captain of HMS *Icarus*. Commander Evans was also a professional RN officer, but he did not command the ship he rode in. The precise age of most of the escort warship captains is unavailable, but the number of reserve officers suggests that the average age of the Allied commanders in this battle was significantly greater than for U-boat skippers, whose mean age was just over 27 years. German commanding officers were also more likely to be professional navy – 17 of the 20 involved in the battle entered the *Kriegsmarine* before the war. Only 5 of the U-boat commanders were on their first war patrol, while almost half – 9 out of 20 – had held command of at least one other U-boat before this operation. As a group they reflected the intense effort put forth by *BdU* to identify potential commanders, and then to train them and their crew to a high pitch before committing them to action. Statistical information on Allied officers from various sources, primarily RN Navy List for August and October 1943 provided by Dr Malcolm Llewellyn-Jones, NHB MoD (UK); Ken McPherson and John Burgess, *The Ships of Canada's Naval Forces 1910-1985* (Toronto, 1985), appendix 4. German statistics derived from Rainer Busch and Hans-Joachim Roll, *German U-Boat Commanders of World War II: A Biographical Dictionary* (Annapolis, 1999). See also Timothy Mulligan, *Neither Sharks nor Wolves: The Men of Germany's U-Boat Arm 1939-1945* (Annapolis, 1999).

67. Biographical details drawn primarily from Bercuson and Herwig, *Deadly Seas*, 134-138. DHH Personnel File LCdr A.H. Dobson.
68. *Deadly Seas*, 138-142 and
69. *Deadly Seas*, 159-160 and Niestle, 44.
70. *Deadly Seas*, 231-232, and Niestle *U-Boat Losses*, 41.
71. *Deadly Seas*, 291.
72. <www.naval-museum.mb.ca/battle_atlantic/st.croix/survivors-account.htm>. This web link provides a news story from the *Winnipeg Free Press* of 1 October 1943 detailing Able Seaman W.A. Fisher's experiences. This reference is valuable as Fisher would ultimately be the only survivor of St Croix.
73. DHH 81/520/8280, "Post Convoy Conference St. John's Newfoundland, 28 September 1943"; and Winnipeg Free Press article 1 October 1943.
74. *BdU KTB* 20 September 1943
75. *BdU KTB* 22 September 43: "Remote escort was still very strong even after the sinking of several destroyers. It is to be assumed that the prospect of losing large numbers of ships has caused the enemy to strengthen his naval escort by further chaser groups."
76. LAC RG 24, 83/84 Accession, TMs 1000-97B Vol 8 *Signals*, CinC WA to All 212332Z September 43 (Courtesy of Dr. Milner and Mr Whitby).
77. Whitby, *Commanding Canadians*, 22 September 1943, 36.
78. *BdU KTB* 20 September 1943, para IVa "The convoy was proceeding up to then on the Great Circle [route] westward and in the night of 20-21st turned slowly aside on the Great Circle to the S.W. The speed estimated was 9 knots which corresponds to that of the ONS. This must, therefore, have been ONS202....there were great variations in fixes from the reports received, which made it very difficult for the boats to find the convoy."
79. Evans RoP and Admty Monthly ASW Review October 1943.
80. Evans RoP
81. *BdU KTB* 21 September 1943.
82. Admty Monthly ASW Review October 1943.
83. *BdU KTB* 21 September 1943
84. Admty Monthly ASW Review October 1943; Evans RoP; DHH ADM 199/353, "*Voyage Report M.V. Empire McAlpine,*" 21 October 1943.
85. *BdU KTB* 21 September 1943
86. *BdU KTB* 21 September 1943, Para IV indicates that seven U-boats reported either contact with the convoy or sighting destroyers proceeding singly, which meant that the boats were in close proximity to the convoy. Escorts reported between five and seven U-boats engaged during the night. Although not all the U-boats were attempting to attack – *U-229* in particular was clearly shadowing, not attacking, when she was sunk – the encounters still indicate a rather aggressive effort on the part of the U-boats given the poor visibility and primitive sensors they were fitted with.
87. Evans RoP. The Admty Monthly ASW Review October 1943 indicates that *Roselys* had only one contact. After the FFS *Roselys* arrived in St. John's, her Commanding Officer, Capitaine de Corvette J. Boutron, was relieved of command. The records are not entirely clear, but it appears evident that Evans and the other commanding officers of escorts in EGB3 had lost confidence in Boutron's judgment, apparently as a result of incidents such as this one where reports of enemy contacts were assessed as exaggerated.
88. Evans RoP.
89. Douglas and Rohwer, *Wolfpacks*, 175.
90. Evans RoP.
91. Douglas and Rohwer, *Wolfpacks*, 175. Details of the attack from Rohwer, *Axis Submarine Successes*, 172.
92. *U-229* fired two Type V torpedoes earlier in the battle, during the night of 20/21 September 1943, although neither are assessed to have hit anything. Rohwer, *Axis Submarine Successes*, 171.
93. Evans RoP. Assuming a closing speed of 37 knots – 25 for *Keppel* and 12 for the unsuspecting U-boat – the 600 yards that needed to be covered from the time the U-boat first became aware of the threat from *Keppel* (when the destroyer was 800 yards from the boat) until the destroyer came in to

visibility range at about 200 yards would have been covered in just under thirty seconds. It would have been extremely difficult for *U-229* to have fired an effective Type V shot in that short period, given the surprise achieved by *Keppel*.

94. *BdU KTB* 21 September 1943.

95. Evans RoP.

96. *BdU KTB*, 22 September 1943, "Contact with the convoy was maintained by day, through hydrophone."

97. Evans RoP.

98. This may have been U-758. *BdU KTB* 22 September 1943: "…U 758 reported being pursued and heavily depth-charged."

99. Evans RoP. Evans wording "… *Roselys* … was apparently in radar contact with a submarine.…" again suggests he lacked confidence in reports from this ship.

100. Evans RoP.

101. Gap distance between convoys from Evans RoP. Identification of which convoy trailed from RC-NMR December 1943.

102. Relative position of U-boats engaged with convoy estimated by using report of attacks on U-boats in DHH ADM 199/353 File "Convoys ONS 18 & ON 202: Reports of Proceedings 19-25 September 1943." Estimated convoy positions from Deck Log of HMCS *Chambly*. All positions at this point, after some 36 hours in the fog, were DR positions and therefore unlikely to be very accurate, but the relative positions seem reasonable.

103. DHH *KTB* U-270; Douglas, *Creation of a National Air Force*, 564; and DHH ADM 199/353 "Convoys ONS 18 & ON 202; Reports of Proceedings 19-25 September 1943."

104. Clay Blair, *Hitler's U-boat War: The Hunters, 1939-1942* (New York, 1996), 478; Montgomery C. Meigs, *Slide rules and Submarines* (Washington, 1990), 130-131.

105. "temperamental weapon" from RCAF Douglas, *Creation of a National Air Force*, 565.

106. DHH 85/77, *KTB* U-377, 22 September 1943.

107. Wynn *U-Boat Operations Vol I*, 250; Douglas, *Creation of a National Air Force*, 564.

108. Evans RoP.

109. DHH ADM 199/353, "*Voyage Report M.V. Empire McAlpine*," 21 October 1943; Evans RoP; Admty Monthly ASW Review October 1943.

110. Douglas, *Creation of a National Air Force*, 565 identifies U-boat as *U-275*. Wynn, *U-Boat Operations, Vol I*, 193, indicates U-275 took no active part in the ON202 ONS18 operation.

111. *BdU KTB* for 22 September 1943 indicated that seven U-boats reported attacks by aircraft on this day. Allied records identify five occasions when aircraft knowingly attacked. The difference may be the result of U-boats that sighted aircraft and reported this but were not directly attacked.

112. Evans RoP.

113. *BdU KTB* 22 September 1943.

114. Evans RoP; Appendix I to Evans RoP prepared by Commander Burnett, SO C2 EG, investigating the circumstances surrounding the loss of HMS *Itchen*; Admty Monthly ASW Review October 1943; Coy, *Echo of a Fighting Flower*, 134-135; Rohwer, *Axis Submarine Successes*, 172.

115. The details of this incident have been difficult to reconstruct. Commander Evans identified the merchant ship that rescued the sailors as the SS *Wisla*, both in his RoP and at the Escort Commanders conference following the battle in St. John's, Newfoundland. The Admiralty Monthly ASW Review for October 1943 identifies the rescuing vessel as the SS *James Smith*, pennant 31, in convoy ON202. Commander Evans noted that SS *James Smith* only reported that survivors were in the water. The Commodore of convoy ON202 reported that SS *James Smith* slipped rafts into the water when it sighted survivors, but not that the vessel picked up any survivors. Complicating matters further, Commander Evans repeatedly wrote *Wilha* when referring to *Wisla* and his transcribed comments from St John's render this as *Wahela*. Several works of history now refer to the *Wahela*, which this author can only find in the one historical document, and assesses is a transcription error. SS *James Smith* was directly astern of the location where *Itchen* exploded, making her well placed to sight survivors, but she was also the lead ship of two convoys, making a stop risky not only for exposing her to torpedo attack, but to inadvertent collision with merchant ships astern of her. SS *Wisla* was

either pennant 21 or pennant 54 – sources differ. The vessel picking up the survivors was sighted by HMS *Keppel* astern of both convoys, a long way astern of station for SS *James Smith* but not as far for *Wisla*. The most likely explanation is that SS *Wisla* did indeed recover the survivors, but the records are not conclusive enough to state this definitively.

Other sources include: Survivor story from *Winnipeg Free Press* 1 October 1943; Peter Coy, *Echo of a Fighting Flower*, 135; Marc Milner, *The U-boat Hunters*, 70; <www.warsailors.com/convoys/on202. html> .This web site contains comments from Peter Coy, author of *Echo of a Fighting Flower*, and witness to the events.

116. Evans RoP: "Unfortunately at this moment[0215 on 23 September], he [*Icarus*] was shown the latest Admiralty advice on combating the new German weapon and this dissuaded him from closing the range until the submarine dived some 12 miles away from the convoy."

117. Rohwer, *Axis Submarine Successes*, 172.

118. DHH ADM 199/583, Commodore's Report of ON 202.

119. The most detailed accounts of the survivors of these three merchant ships can be found at <www. warsailors.com>.

120. Evans RoP; Rohwer, *Axis Submarine* Successes, 172; RCNMR December 1943, 60-61.

121. DHH 81/520/8280, "Post Convoy Conference St. John's Newfoundland, 28 September 1943." A number of other escort Commanding Officers also agreed with Commander Evans that the crew of *Steel Voyager* had behaved badly, whereas the crews of other torpedoed ships had displayed great valour.

122. RCNMR December 1943, 61.

123. During the forenoon of 23 September Commander Evans used the superior speed of ON202 to form it on the starboard beam of ONS18, increasing the speed of ON202 to 9 knots and emergency turning the formation 45 degrees to starboard, displacing the faster convoy from ahead of ONS18 onto its beam. Once in station he had ON202 emergency turn 45 degrees to port back onto the main course, slowing to 7.5 knots to match ONS18. LAC RG 24 Series D-2 Vol 7262 Deck Log, HMCS *Drumheller*, 23 September 1943.

124. *BdU KTB* 23 Sep 43. Deck Logs for escort ships indicated cloud and broken cloud with visibility of 6 to 7 miles for most of 23 September, so there is a clear discrepancy between *BdU*'s beliefs and reality around on the convoys for this day. LAC RG 24 Series D-2 Vol 7347, Deck Log HMCS *Gatineau*, 23 September 1943.

125. Evans RoP, 29 September 1943: For the Germans, the battle "… can hardly have been a very encouraging result." About six weeks later the Admiralty ASW Division agreed, stating in their Monthly ASW Review for October 1943, published 15 Nov 1943, that "The results of this renewed effort must have been disappointing."

126. Evans RoP.

127. Rohwer, *Axis Submarine* Successes, 171-172.

128. *BdU* provided only general direction regarding U-boats operating in groups. In its operational orders to *Leuthen* boats, sent out 13 September 1943, *BdU* directed that, "When hauling ahead on the surface, never form groups of more than 2 boats. Boats must aim at equal distributions round the convoy in order to split up the enemy defences." No discussion of decoy tactics or red lights is provided, although U-boat commanding officers may have developed these tactics on their own.

129. Unless otherwise indicates, all discussions of *BdU* Final Summary are from *BdU KTB* 24 September 1943.

130. Doenitz, *Ten years*, 91-92 and Appendix 3.

131. *BdU KTB* 24 September 1943.

132. HMCS *Esquimalt*, sunk by a Type V torpedo on 16 April 1945 within a few miles of Halifax, Nova Scotia, just three weeks before the end of the war, is a good example of the enduring threat posed by the new weapon. This small ship would have been a difficult target for a traditional straight running torpedo.

133. Research Report 27, "Anti-Submarine Operations by CVE Based Aircraft," ASWORG/87-32, 1 April 1944, pages 31-32. Posted at <www.uboatarchive.net/ASWORGReport27.htm>.

134. This was another of Frederick the Great's famous victories.

135. Blair, *Hitler's U-boat* War, Vol 2, 426-427.

136. Hessler, Section 381.

137. Blair, *Hitler's U-boat* War, Vol 2, 427.

138. This was the first flight of a Blohm & Voss 222 flying boat in support of the U-boats. There were only 13 of these huge six-engine aircraft ever built. Hessler Section 382.

139. *U-336, U-279* and *U-389*. Blair, *Hitler's U-boat* War, Vol 2, 427-428. Niestle, *German U-boat Losses*, 55, indicates the correct date for the sinking of *U-336* is 5 October 1943, but in all other respects his information is the same as Blair.

140. *U-419, U-643* and *U-610*. Blair, *Hitler's U-boat* War, Vol 2, 430.

141. Blair, *Hitler's U-boat* War, Vol 2, 428-429.

Interlude: Clearing the English Channel

1. The surface forces discussed here are limited to the warships of approximately destroyer size. Smaller craft such as E-boats and R-craft are outside the scope of this study.

Chapter 5: "Shoot, Shoot, Shoot!": Destroyer Night Fighting and the Battle of Île de Batz

1. The author would like to acknowledge the important contribution of Vice Admiral Harry DeWolf for his patience and cooperation in dozens of meetings, Captain Gil Lauzon, USN (Ret'd) for his insights into destroyer tactics and operations, Commodore Jan Drent (Ret'd) whose translation and analysis of *Kriegsmarine* documents illuminated the German side of the sorry, and to Goetz, Susanne and Franz-George Freiheir von Bechtolsheim for the warm hospitality and insights into the career of Theodor Freiheir von Bechtolsheim.

2. Viscount Cunningham, *A Sailor's Odyssey* (London, 1951), 336.

3. W. Hughes, *Naval Tactics* (Annapolis, 1986), 25. The emphasis is Hughes.

4. An Admiralty memo dated 7 July 1943 decreed that Radio Direction Finding (RDF) equipment was in future to be known as Radar. See NHB narrative, *The Naval Staff, 1943-1945, The Navy Transformed and Renewed Air Attack*, n 1.

5. S.W. Roskill, *The War at Sea*, Vol I (London, 1954), 47-48.

6. Interview with Franz-George Freiheir von Bechtolsheim, Landsuch, Germany, 2 December 2006. His father was CO of the *Karl Galster*.

7. M.J. Whitley, *German Destroyers of World War Two* (Annapolis, 1991), 83, 86-89. Whitley remains the essential source on *Kriegsmarine* destroyer operations.

8. Admiral Sir Reginald Bacon, *The Concise Story of the Dover Patrol* (London, 1932), 120. Bacon commanded the Dover Patrol for most of the Great War. For the *Kriegsmarine's* strategy see K. Bird, *Erich Raeder: Admiral of the Third Reich* (Annapolis, 2006), 55.

9. R. Hough, *Bless Our Ship: Mountbatten and the Kelly* (London, 1992), 104.

10. All times in this study have been converted to local Plymouth or "B" time.

11. D. Howse, *Radar at Sea: The Royal Navy in World War II* (Annapolis, 1993), 58. The first sea trials of Type 286 took place in July 1940, and only 32 ships had been fitted by the end of 1940.

12. Captain Lord Louis Mountbatten, "Report of Action Between 5th Destroyer Flotilla and German Destroyers on Night of 28/29th November 1940," 5 December 1940, NAUK, ADM 199/430.

13. A.F. Pugsley, *Destroyer Man* (London, 1957), 58-9.

14. Pugsley, *Destroyer Man*, 59. Pugsley's public criticism is remarkable considering that Mountbatten was not only a notable – and sensitive – Royal but was also First Sea Lord at the time of publication.

15. Mountbatten, Report of Action.

16. Pugsley, Report of Action, 2 December 1940. NAUK, ADM 199/430.

17. Whitley, *German Destroyers*, 113-4; C. Langtree, *The Kelly's: British J, K and N Class Destroyers of World War II* (London, 2002), 82-84; Hough, *Bless Our Ship*, 113.

18. Biographies of German naval officers are from W. Lohmann and H.H. Hildebrand, *Die Deutsche Kriegsmarine, 1939-1945, Band III, Verzeichnis aller Marineoffiziere* (Podzun Verlag, 1956).

19. *KTB Karl Galster*, 29 November 1940. NHB, Reel No. 824 (P); Whitley, *German Destroyers*, 111-114.

20. Captain C. Harcourt (DOD (H)) minute, 2 December 1940; Captain A.J. Power (ACNS (H)) minute, 7 January 1941, NAUK, ADM 199/430. Emphasis is Power's. This criticism evidently stung Mountbatten sharply. In something rarely seen, and perhaps belying his special status as a member of the Royal Family, he wrote a supplementary and then amplifying report to the Admiralty defending his actions, and blaming the delay in opening fire on the circumstances.

21. C-in-C Western Approaches, "Report of Action Between 5th Destroyer Flotilla and German Destroyers on Night of 28/29th November 1940," 5 December 1940; DTSD minute, ud, January 1941; DCNS minute, 21 January 1941, NAUK, ADM 199/430.

22. J. Campbell, *Naval Weapons of World War Two* (Annapolis, 1985), 225-6; Pritchard, *The Radar War: Germany's Pioneering Achievement, 1904-1945* (Wellingborough, 1989), 190-203. N. Friedman, *Naval Radar* (Annapolis, 1981), 205-6. D.E Graves ms, "German Naval Radar Detectors, Radar Foxing and IFF Equipment 1939-1945: An Overview and Descriptive Catalogue," DHH, 2000/5.

23. See Admiralty, Signals Division, "R.D.F. Bulletin," No. 3, 29 May 1943. NAUK, ADM 220/204.

24. Howse, *Radar at Sea*, 181.

25. The names and classification of British naval radars changed throughout the war. I have used those current at the time, listed in *Confidential Admiralty Fleet Order* (CAFO) 477/1944.

26. Friedman, *Naval Radar*, 195.

27. B. Lavery, *Churchill's Navy: The Ships Men and Organization, 1939-1945* (London, 2006), 96-98.

28. Admiralty, Signals Division, "RDF Bulletin No. 6," 3 March 1944, NAUK, ADM 220/204; A. Mitchell, "The Development of Radar in the Royal Navy (1935-45)," Part II in *Warship*, April 1990 No. 14, 123 and 129-30. In November 1943, Admiral Leatham wrote, "the radar of our destroyers is masked by the bridge superstructure from R.27 to G.27," C-in-C Plymouth minute, 13 November 1943. See also Capt D Plymouth minute, 29 November 1943, both NAUK, ADM 199/1038.

29. A. Hezlet, *The Electron and Seapower* (London, 1975), 264.

30. R. Hill, *Destroyer Captain* (London, 1975), 135.

31. M.J. Whitley, *Destroyers of World War Two: An International Encyclopedia* (Annapolis, 1988), 143.

32. For the design of the Hunts see N. Friedman, *British Destroyers and Frigates: The Second World War and After* (London, 2006), 69-80.

33. Type 1 and 2 Hunts had three twin 4-inch mountings but no torpedo armament, while the Type 3s that were used in the western Channel had two 21-inch torpedo tubes but only two twin 4-inch mounts.

34. Lieutenant-Commander R. Hill, "Night Action With Enemy Destroyers," 22 October 1943, NAUK, ADM 199/1038.

35. There is no evidence to suggest that the 3/4 October action was a factor in the change of command.

36. Phipps diary, cited in P. Smith, *Hold the Narrow Sea: Naval Warfare in the English Channel 1939-1945* (London, 1984), 190; DTSD minute, 27 November 1943, NAUK, ADM 199/1038.

37. *KTB 4. Torpedobootflotille*, 23 October 1943; ACNS(H) minute, 26 November 1943, NAUK, ADM 199/1038; Whitley, *German Destroyers*, 147.

38. C-in-C Plymouth, "Action with German Ships on the Night of October 22nd/23rd October, 1943," 13 November 1943, NAUK, ADM 199/1038.

39. "Remarks by *Führer der Zerstörer* on the *KTB* of the *4. Torpedobootflotille* for the period 16-23 October 1943."

40. ACNS(H) minute, 26 January 1943, NAUK, ADM 199/1038.

41. The Home Fleet War Diary complains loudly and consistently about how the lack of Fleet destroyers hampered operations. See for example, vol. III (1943), 5, 10, NAUK, ADM 199/1426.

42. Home Fleet War Diary, III, 5.

43. Friedman, *British Destroyers and Frigates*, 24-29. See also E.J. March, *British Destroyers, 1892-1953* (London, 1966), and M. Brice, *The Tribals: Biography of a Destroyer Class* (London, 1971).

44. C-in-C Home Fleet to Admiralty, 3 January 1944, LAC, RG 24, vol. 11751, CS151-1-1; Home Fleet War Diary, III, 210.

45. For the Canadian acquisition of the Tribals see W. Douglas, R. Sarty and M. Whitby, *No Higher Purpose: The Official Operational History of the Royal Canadian Navy in the Second World War, 1939-43*

(St. Catharines, 2003), 76-8; and M. Whitby, "Instruments of Security: The RCN's Procurement of the Tribal Class Destroyers, 1938-43," *The Northern Mariner* (July 1992), 1-15.

46. Admiral Sir Frederick Dreyer to Admiralty, 13 January 1940, NAUK, ADM 1/10608.

47. M. Hadley and R. Sarty, *Tin-Pots and Pirate Ships: Canadian Naval Forces and German Sea Raiders, 1880-1918* (Montreal, 1991), 268-69.

48. The Tribals stood as the RCN's most powerful warships until the cruiser HMCS *Uganda* commissioned on 21 October 1944.

49. Captain H.G. DeWolf, "Employment of Tribal Destroyers," 7 December 1942, LAC, RG 24, vol. 6797, NSS8375-355.

50. For the early Tribal operations see W. Douglas, R. Sarty and M. Whitby, *A Blue Water Navy: The Official Operational History of the Royal Canadian Navy in the Second World War, 1943-45* (St. Catharines, 2007). For the significance of the Home Fleet see Lavery, *Churchill's Navy*, and J. Levi, *The Royal Navy's Home Fleet in World War II* (New York, 2003).

51. *Tartar* and *Ashanti* received the lattice masts in scheduled refits while *Athabaskan* got hers as a result of battle damage received when she was struck by a glider bomb in August 1943.

52. Admiralty, "A Brief Description of the GEE System," NAUK, ADM 1/13418; J. Watkins, "Destroyer Action, Île de Batz, 9 June 1944," *The Mariner's Mirror* (August, 1992), 308.

53. Friedman, *British Destroyers and Frigates*, 107; Author's interview with Vice Admiral H.G. DeWolf, 20 August 1987. F.H. Hinsley, *British Intelligence in the Second World War, II* (London, 1981), 194-5.

54. Friedman, *British Destroyers and Frigates*, 106; Admiralty, Director of Gunnery Division, *Gunnery Review*, July 1945, 80-2, DHH, 89/235; and Admiralty, Tactical and Staff Duties Division, "*Guard Book of Fighting Experience*," December 1942, 13, DHH, 91/79.

55. DeWolf, Report of Proceedings, 20 January 1944, LAC, RG 24, vol. 11730, CS151-11-9.

56. CO *Black Prince*, Action Report, 2 May 1944, 8. NAUK, ADM 199/263; HMS *Bellona*, Deck Log, February 1944 and March 1944. NAUK, ADM 53/118941 and ADM 53/118942.

57. Much has been written on the battles in the South Pacific but the best recent works are V. O'Hara, *US Navy Against the Axis: Surface Combat, 1941-1945* (Annapolis, 2007); and T. Hone, "'Give Them Hell!': The US Navy's Night Combat Doctrine and the Campaign for Guadalcanal," *War in History* 2006 Vol 13 No 2, 171-199.

58. Captain W. Norris, Report of Proceedings, 27 February 1944, NAUK, ADM 199/532.

59. Commander St J. Tyrwhitt, Report of Proceedings, 2 March 1944, NAUK, ADM 199/532; *KTB*, Operations Division, German Naval Staff, 2 March 1944. DHH, SGL II/261.

60. Tyrwhitt, who ultimately rose to flag rank, was from a famous naval family and his father was Admiral of the Fleet Sir Reginald Tyrwhitt.

61. C-in-C Plymouth, "10th Destroyer Flotilla – Report on Tunnel, Night 1st/2nd March," 25 March 1944, NAUK, ADM 199/532.

62. Author's interview with Vice Admiral H.G. DeWolf, 20 August 1987.

63. Captain C.P. Nixon to author, 12 March 2001. See also M. Whitby, "Vice-Admiral Harry G. DeWolf: Pragmatic Navalist" in M. Whitby, R. Gimblett and P. Haydon (eds), *The Admirals: Canada's Senior Naval Leadership in the Twentieth Century* (Toronto, 2006) 213-246.

64. *KTB 4. Torpedobootflotille*, 16-23 October 1943.

65. F.H. Hinsley, *British Intelligence in the Second World War*, vol III, pt 1 (London, 1984), 280.

66. C-in-C Plymouth to SO Force 26, 25 April 1944, DHH, 81/520/1650-239/13; Hinsley, *British Intelligence in the Second World War*, III Pt 1, 287; *KTB* Group West, 25 April 1944.

67. C-in-C Plymouth to SO Force 26, 25 April 1944, DHH, 81/520/1650-239/13.

68. *KTB 4. Torpedobootflotille*, "Evaluation of Actions Fought by the 4th Torpedo Boat Flotilla," DHH, 81/520 HMCS *Athabaskan* 8000.

69. *KTB T-27*, 26 April 1944.

70. Plymouth *Daily Sketch* interview with Lieutenant M. Heslam, 29 April 1944, 3. The article is in the DeWolf Papers, LAC, MG 30, E509, vol 2.

71. Lieutenant-Commander J.S. Stubbs, Report of Proceedings, LAC, RG 24 (acc 83-84/167), Box 695, 1926-355/1.

72. LT Gasde recollection in *FdZ* Report 21 September 1944. PG 74796.

73. Stuart Kettles, "A Wartime Log: A Personal Account of Life in HMCS *Athabaskan* and as a Prisoner of War," DHH, 74/458. Portions of the diary are also at <www.webhome.idirect.com/~kettles>.

74. *KTB T-27*, 28 April 1944; *KTB 4.Torpedobootflotille*, "Evaluations of Actions Fought by the 4th Torpedo Boat Flotilla," 3-4. DHH, 81/520 *Athabaskan* 8000.

75. The German Fleet commander, Admiral Schniewind, initially wanted to court martial Gotzmann for "beaching his ship before fully exhausting his means of defence." Upon investigation he discovered the problem was "one of incompetence and deficient lack of responsibilities and duties rather than a lack of daring." *KTB*, Fleet Command, 1-15 May 1944.

76. There has been recent controversy over *Athabaskan*'s sinking with claims that the second explosion was the result of a torpedo fired by a British MTB. Research by Canadian and British official historians has refuted this myth. See P.A Dixon, "'I Will Never Forget the Sound of Those Engines Going Away': A Re-examination into the Sinking of HMCS *Athabaskan*, 29 April 1944," *Canadian Military History* (Spring, 1996) 16-25; M. Whitby, "The Case of the Phantom MTB and the Loss of HMCS *Athabaskan*," *Canadian Military History*, (Autumn 2002) 5-14. A popular study of *Athabaskan*'s last moments can be found in B. Gough, *HMCS Haida: Battle Ensign Flying* (St. Catharines, 2001), 75-84.

77. Some of them found graves on French soil, and they are still remembered by French people of the region. For a poignant memoir of *Athabaskan* and those who sailed in her see L. Burrow and E. Beaudoin, *Unlucky Lady: The Life and Death of HMCS Athabaskan* (Toronto, 1987).

78. DTSD minute, 8 June 1944, NAUK, ADM 199/263.

79. Admiralty to C-in-C Plymouth, 21 September 1944; DTSD memorandum, 28 August 1944, both NAUK, ADM 199/22.

80. Jones described his career in a family memoir *And So to Battle* (Battle, 1976). Jones retired to a cottage in the town of Battle on the site of the Battle of Hastings. Author interview with Michael Jones, 1988.

81. Commander B. Jones to Admiralty, 4 November 1944, NAUK, ADM 199/22. Emphasis is Jones's. For more on the debate over torpedoes see C-in-C Plymouth to Admiralty, "Torpedoes Fired in Action," 6 December 1944, and CO HMS *Defiance* to Admiralty, 26 October 1944, both NAUK, ADM 199/22. *Defiance* was the torpedo school at Plymouth.

82. Basil Jones, "A Matter of Length and Breadth," *The Naval Review*, (May 1950), 139.

83. Information on Namiesniowski and Gorazdowski from <www.polishnavy.pl>; and Namiesniowski Papers, LAC, MG30 E293. By the 18 November 1939 Anglo-Polish Naval Agreement, ships of the Polish navy were commanded and manned by Poles, but came under RN operational control.

84. Flotillas normally comprised eight destroyers, which were in turn divided into divisions of four and subdivisions of two. For Sinclair see "Rear-Admiral Erroll Sinclair," *The Telegraph*, 18 November 1993, and for Lewis see <www.unithistories.com>.

85. R. Hill, *Lewin of Greenwich: The Authorized Biography of Admiral of the Fleet Lord Lewin* (London, 2000), 58. Lewin rose to become First Sea Lord and Chief of the Defence Staff.

86. With Grant's contribution, from December 1943 to June 1944, ships commanded by Canadians – *Enterprise*, *Haida*, *Huron* and *Athabaskan* – had a hand in the destruction of seven German destroyers. The addition of the torpedo boat *Grief*, sunk in the Baie de la Seine in May 1944 by a Fairey Albacore from No 415 Squadron RCAF, makes for an impressive total.

87. Whitley, *German Destroyers of World War Two*, 201-3.

88. Whitley, *German Destroyers of World War Two*, 166.

89. *KTB T-24*, 29 April 1944.

90. *KTB Befehlshaber Sicherung West* (Commander Security West), 3-4 May 1944. The radar performance data revealed to the Germans was extremely accurate, to the extent that it virtually matches that discussed in RN radar manuals and current secondary sources cited in this study.

91. Interview with Franz-George Freiheir von Bechtolsheim.

92. Admiralty to ANCXF, 0435B/6 June 1944, NAUK, ADM 223/195.

93. Admiralty to C-in-C Plymouth, 1700B/6 June 1944 laid down the precise courses and speeds von Bechtolsheim's destroyers were to use from Royen to Brest. NAUK, ADM 223/195.

94. Michael Simpson (ed), *The Cunningham Papers: Volume II, The Triumph of Allied Sea Power, 1942-1946* (Ashgate, 2006), 202.

95. See B. Greenhous, S. Harris, W. Johnston and W. Rawling, *The Crucible of War: The Official His-*

tory of the RCAF, Vol III (Toronto, 1994), 456-459; C. Goulter, *A Forgotten Offensive: RAF Coastal Command's Anti-shipping Campaign, 1940-45* (London, 1995), 208-230; and Air Historical Branch, *The RAF in Maritime War: Vol V, The Atlantic and Home Waters, The Victorious Phase, June 1944-May 1945*, 8-9, DHH, 79/599.

96. ORB No 404 RCAF, 7 June 1944 cited in Greenhous, *The Crucible of War*, 465. The No 404 ORB is at LAC, RG 24, vol 22647, Spine No 39-11.

97. *KTB 8. Zerstörerflotille, Z-32, Z-24* and *ZH-1*, 7-8 June 1944, DHH, SGR II 340 vol 72 PG 74731.

98. Plymouth Command War Diary, 1-15 June, 1944, NAUK, ADM 199/1393.

99. Admiralty to C-in-C Plymouth, 1230B, 1405B, 1911B, 2310B 8 June 1944, and 0005B 9 June 1944, NAUK, ADM 223/195 and ADM 223/196. Expecting an action to take place off the North Brittany coast, the Germans warned their U-boats to avoid the area while the Allies cleared out their escort groups.

100. Commander B. Jones, "Report of Night Action – 9th June 1944," 14 June 1944, 1, NAUK, ADM 199/1644. F.H. Hinsley, *British Intelligence in the Second World War, III pt 2*, (London, 1988), 161-2. C. Barnett, *Engage the Enemy More Closely*, (London, 1991), 268. *Seekriegsleitung, FdZ- Nachricht Nr. 4, Gefecht 8.Zerstörerflotille Am 9.6.44*, DHH, SGR II 340, Vol. 51. Pg 74796. This last entry is the appreciation of the 9 June action written by von Bechtolsheim and commented upon by the *FdZ*.

101. Lieutenant P.A. Hazelton, "Enemy R/T Intercepted on 9th June 1944," 12 June 1944, 2, NAUK, ADM 199/1644. Hazelton was a Signals Officer on the staff at Plymouth.

102. Jones, *And So To Battle*, 15; Leatham, "Destroyer Action – Night of 8th/9th June, 1944," 18 June 1944, 2; Jones, "Report of Night Action – 9th June, 1944," 14 June 1944, 1-2, NAUK, ADM 199/1644.

103. Commander C.F. Namiesniowski, "Report of Action," ud, 1-2, NAUK, ADM 199/1644: Leatham, "Destroyer Action – Night of 8th/9th June 1944," 1-2.

104. Turning towards to avoid torpedoes was a key lesson absorbed after the Battle of Jutland. See J. Sumida, "The Best Laid Plans: The Development of British Battle-Fleet tactics, 1919-1942," *International History Review*, November 1992, 681-700.

105. Von Bechtolsheim, "*Gefecht 8.Zerstörerflotille am 9.6.1944.*" 4.

106. Meiklem, *Tartar Memoirs* (Glasgow, 1948), 83.

107. HMS *Tartar*, Award recommendations, 16 June 1944, NAUK, ADM 1/29685.

108. Jones, "Report of Night Action – 9th June 1944," 2.

109. *Tartar* Award Recommendations; HMCS *Huron*, Award recommendations, 9 June 1944, NAUK, ADM 1/29685. Stone's initiative was also mentioned in Commander Jones' action report.

110. Whitley, *German Destroyers of World War II*, 160; Watkins, "Destroyer Action, Île de Batz, 9 June 1944," 316; Hazelton, "Enemy R/T Intercepted on 9th June 1944," 2; Lieutenant Commander J.R. Barnes, "Report of Action," 9 June 1944, NAUK, ADM 199/1644.

111. Lieutenant John Watkins, RNVR, *Ashanti*'s navigator and torpedo officer, was awarded the DSC for torpedoing *ZH-1*. Reflecting on the Admiralty's criticism of the flotilla's poor torpedo performance Watkins later noted that "he rather than the gunnery officer, was decorated only because British destroyers rarely fired their torpedoes and rarely hit anything with them when they did." "Obituary: Professor John Watkins," *The Independent*, 5 August 1999.

112. *KTB ZH-1*, 8-9 June 1944. DHH, SGR II 340 vol 72 PG 74771B; Plymouth Command War Diary, 1-15 June.

113. Commander H.G. DeWolf, "Report of Action," 9 June 1944, 1, DHH, 81/520 HMCS Haida 8000.

114. *KTB Z-24*, 9 June 1944. "*TB Abschnitt Maschine Des Zerstörer Z-24*," DHH, SGR II 340, Vol. 72, PG 74646. The second item cited above is the Engineer's *KTB*.

115. *KTB T-24*, 9 June 1944. Lieutenant Commander H.S. Rayner, "Report of Action,"9 June 1944, 1-2, DHH, 81/520 HMCS Huron (Prior 1950) 8000.

116. DeWolf, "Report of Action," 4.

117. For Operation MAPLE see Admiralty Historical Section, *British Mining Operations, 1939-1945*, Vol 1 (London, 1973), NAUK, ADM 234/560.

118. Deputy Director of Operations Division (Mines) Minute, 28 September 1944, NAUK, ADM 199/1644.

119. DeWolf, "Report of Action," 2.

120. *KTB Z-24*, 1-15 June 1944. Von Bechtolsheim minute on *KTB T-24*, DHH, SGR II 340 vol 72 PG 74646.

121. Von Bechtolsheim, "*Gefecht 8.Zerstörerflotille*," 6; Jones, "Report of Night Action," 2-3.

122. DeWolf, "Report of Action," 2.

123. Von Bechtolsheim, "*Gefecht 8.Zerstörerflotille*," 6. *Kriegsmarine* action reports were written in the historic present.

124. Von Bechtolsheim, "*Gefecht 8.Zerstörerflotille*," 6-7.

125. DeWolf, "Report on Action," 3. Author's interview with Vice Admiral H.G. DeWolf, 2 November 1992.

126. Admiralty to C-in-C Plymouth, 0647B/9 June 1944, NAUK, ADM 223/196. This message was a summary of von Bechtolsheim's signals later in the action.

127. Von Bechtolsheim, "*Gefecht 8.Zerstörerflotille*," 7; *KTB Z-32*, 9 June 1944.

128. Von Bechtolsheim, "*Gefecht 8.Zerstörerflotille*," 7-8; DeWolf, "Report of Action," 3; Rayner, "Report of Action," 2-3.

129. Admiral Kreisch, "*Stellungnahme des FdZ*," 10-11, DHH, SGR II 340, Vol. 51, PG 74796.

130. Captain J. Cowie minute, NAUK, ADM 199/1644

131. For the 10 DF's operations throughout the summer of 1944 see Douglas, Sarty and Whitby, *A Blue Water Navy;* and M. Whitby, "*RCN Tribal Class Destroyer Operations in the Bay of Biscay July-October 1944*." (February 1989), DHH, 2000/5.

Chapter 6: On Britain's Doorstep: The Hunt for *U-247*

1. The author must thank Michael Whitby, Senior Naval Historian, DHH, Ottawa, for first suggesting the destruction of *U-247* as an interesting topic for research.

2. Signal, DDIC, Admiralty, to C-in-C, HF, C-in-C, WA, AOC-in-C, CC, Info. ACOS, etc., H.5082, 081245B, 8 June 1944, ADM 223/195. The U-boat *U-247*, Lieutenant (Junior Grade) Gerhard Matschulat had received these instructions barely 6 hours before the Admiralty were reading the same information! "Kriegstagebuch, *U-247*, 18 May to 28 July 1944," NID PG/30225/NID, NHB, p. 9.

3. Signal, DDIC, Admiralty, to C-in-C, WA, AOC-in-C, CC, ACOS, etc., H.5129, 091715B, 9 June 1944, ADM 223/196.

4. "U-boat Situation, Week Ending 12 June 1944," Captain Rodger Winn, RNVR, OIC 8S, OIC/SI.976, [12 June 1944], ADM 223/172.

5. "Diary, 26 June 1944," in "War Diary," 16-30 June 1944, C-in-C, Western Approaches, WA.2267/0745/91, TSD.5208/44, 29 July 1944, ADM 199/1392; "Report of Attack on U-boat, Form S.1203," HMS *Bulldog*, 26 June 1944, ADM 199/472; "Report of Proceedings, Operation "CW," 20 June 1944 to 1 July 1944, A/Lieutenant Commander L.P. Denny, RCN, Commanding Officer, HMCS *St. Thomas*, 1 July 1944, ADM 199/472.

6. "Western Approaches Command, General Survey of Events, 1-15 July 1944," in "War Diary," 1-15 July 1944, C-in-C, Western Approaches, WA.2447/0745/92, TSD.5209/44, 14 August 1944, ADM 199/1392; "[Report of Proceedings, 1-14 July 1944]," Captain H.T.T. Bayliss, HMS *Vindex*, Vindex 2100/0133, 15 July 1944, ADM 199/497, p. 1.

7. 825 was a composite A/S Squadron equipped with 12 Swordfish Mk. III (with ASV Mk. XI and RATOG, 6 Hurricanes Mk. IIC (fitted with RP), and 2 Hurricanes IIC (air reserve). Appendix I to "[Report of Proceedings, 1-14 July 1944]," Captain H.T.T. Bayliss, HMS *Vindex*, Vindex 2100/0133, 15 July 1944, ADM 199/497.

8. For a more detailed description of these factors, see M. Llewellyn-Jones, *The Royal Navy and Anti-Submarine Warfare, 1917-49* (London: Routledge, 2006), Chapter 3 *passim*.

9. "U-boat Situation, Week Ending 28 August 1944," Captain Rodger Winn, RNVR, OIC/SI.1062, n.d., ADM 223/172.

10. By mid-1944, about half of U-boat captains had less than 3 months experience in command. Minute, "Amount of Experience of U-boat Commanding Officers now on Operations," S.R. Fiske, OIC/SI.1009, 12 July 1944, ADM 223/172. See also [ACNS(UT)] to Churchill, [11 March 1944], ADM 223/320. See, for example, the report of a U-boat which entered harbour in order to charge her batteries: "Monthly Anti-Submarine Report, June 1944," DAUD, CB 04050/44(6), 15 July 1944, NHB, p. 6.

11. "Detection of U-boats in the English Channel and Approaches (Rough analysis of the period D to D+10)," E.J. Williams and L. Solomon, Report No. 48/44, 19 June 1944, ADM 219/131; "Note on the Value of "Snort" to U-boats," L. Solomon and E.J. Williams, DNOR, Report No. 62/44, 19 August 1944, ADM 219/144.

12. Ralph Erskine, "British High Frequency Direction Finding in the Battle of the Atlantic," paper presented to the 14th Naval History Symposium, US Naval Academy, Annapolis, 23-25 September 1999, DHH, p. 14 and fn. 153. A copy of this paper was kindly supplied by Michael Whitby, DHH.

13. There is some evidence to suggest that the British HF/DF stations and plotting system was less efficient that that operated by the Canadians. See: "Naval HF/DF Organisation, 1939-1945," Commander P. Kemp, RN, Historical Memorandum No. 43, n.d., HW 3/147.

14. "U-boat Situation. Week Ending 12 June 1944," in David Syrett (ed.), *The Battle of the Atlantic and Signal Intelligence: U-boat Situations and Trends, 1941-1945* (Aldershot: Ashgate, 1998); "26 August 1944," in "Translation of PG/30353, *BdU*'s War Log, 16-31 August 1944," Godt, Chief of Operations Department, *BdU*, NHB; Samuel Eliot Morison, *History of United States Naval Operations in World War II*, Vol. X, *The Atlantic Battle Won, May 1943-May 1945* (Boston: Little, Brown and Co., 1956), p. 325.

15. "Naval Headlines, No. 1071," Naval Section, 9 June 1944, HW 1/2919.

16. The lack of radio reports from sea also left *BdU* largely uncertain of actual progress made by the U-boats, see: Karl Doenitz, *Memoirs: Ten Years and Twenty Days*, tr. R. H. Stevens with introduction by Jurgen Rohwer (London: Greenhill Books, 1990), 421.

17. This policy was a significant change over that implied up to mid-1943, as described in: Patrick Beesly, *Very Special Intelligence: The Story of the Admiralty's Operational Intelligence Centre, 1939-1945* (London: Hamish Hamilton, 1977), 189.

18. "Operations (General)," Part II, "Coastal Command Manual of Anti-U-Boat Warfare," May 1944, AIR 15/294, Article 21.

19. Patrick Beesly, *Very Special Intelligence: The Story of the Admiralty's Operational Intelligence Centre, 1939-1945* (London: Hamish Hamilton, 1977), 242 and 260.

20. "Report of Proceedings on Recent U-boat Action off Cape Wrath," Vice Admiral H. Harwood, Admiral Commanding Orkney and Shetland, 912/OS.0253, 1 August 1944, ADM 199/497, pp. 1-2.

21. "Report of Proceedings on Recent U-boat Action off Cape Wrath," Vice Admiral H. Harwood, Admiral Commanding Orkney and Shetland, 912/OS.0253, 1 August 1944, ADM 199/497, p. 1; "U-boat Situation, Week Ending 19 June 1944," Captain Rodger Winn, RNVR, OIC 8S, OIC/SI.978, [19 June 1944], ADM 223/172.

22. "Report of Proceedings on Recent U-boat Action off Cape Wrath," Vice Admiral H. Harwood, Admiral Commanding Orkney and Shetland, 912/OS.0253, 1 August 1944, ADM 199/497, p. 1.

23. "July, 1944," in "15 Group, Headquarters, Operations Record Book, Form 540," 1944, AIR 25/255. However, 15 Group Narrative does not include any mention of this sortie, though there is one, as will be seen for the following night (6/7 July), see: "Wednesday, 5 July 1944, Headquarters No. 15 Group, Narrative," 15 Group Narrative Office, n.d, AIR 25/296, and "Thursday, 6 July 1944, Headquarters No. 15 Group, Narrative," 15 Group Narrative Office, n.d, AIR 25/296.

24. Signal, ACOS to NOIC Aultbea, RNO Stornoway (R) C-in-C, HF, C-in-C, WA, Capt D3, *Vindex*, *Huntsville*, 052247B, 5 July 1944, ADM 199/497. The "Striking Force" organization had been set up before the war, see: "Requirements of Auxiliary Anti-Submarine Vessels at various Empire Ports in the Event of War in Europe," Admiralty, M.05818/38, 25 October 1938, in "Naval War Memorandum (Germany)," Admiralty, 1937-1939, Case 00244, Vol. II, Naval Historical Branch.

25. Signal, C-in-C, HF, to ACOS (R) *Milne, Marne, Verulam*, Capt (D) Scapa, 052231B, 5 July 1944, ADM 199/497.

26. "Western Approaches Command, General Survey of Events, 1-15 July 1944," in "War Diary," 1-15 July 1944, C-in-C, Western Approaches, WA.2447/0745/92, TSD.5209/44, 14 August 1944, ADM 199/1392; "[Report of Proceedings, 1-14 July 1944]," Captain H.T.T. Bayliss, HMS *Vindex*, Vindex 2100/0133, 15 July 1944, ADM 199/497, pp. 1-2. This disposition was a long-standing tactical concept, see, for example: "Progress in Tactics, 1931," Tactical Division, CB 3016/31 (BR 1876/31), August 1932, Admiralty Library, p. 37.

27. "Summary of Air Operations," Appendix 4 to "[Report of Proceedings, 1-14 July 1944]," Captain H.T.T. Bayliss, HMS *Vindex*, Vindex 2100/0133, 15 July 1944, ADM 199/497.

28. "Report on the Serviceability of ASV Mk. XI in 825 Naval Air Squadron embarked in HMS *Vindex*, during operations from 1 to 14 July 1944," Appendix 3 to "[Report of Proceedings, 1-14 July 1944]," Captain H.T.T. Bayliss, HMS *Vindex*, Vindex 2100/0133, 15 July 1944, ADM 199/497.

29. The technical difficulties of producing a receiver fitted to the schnorkel and capable of detecting centimetric radar transmission defeated the Germans during the war the British believed. However, it appears that the enemy had developed the "Tunis" consisting of the "Fleige" and "Mücke" systems capable of detecting 9-cm and 3-cm radars. These were later made pressure-tight. "Monthly Anti-Submarine Report, April and May 1945," ADM 199/2062; Eberhard Rössler, *The U-boat: The Evolution and Technical History of German Submarines*, tr. Harold Erenberg (London: Arms and Armour Press, 1981), 196. See also "Operation of Tunis installation ("Fleige" and "Mücke")," Current Order No. 27. June 1944 edition, in "Translation of PG/30349, *BdU*'s War Log, 16-30 June 1944," Naval Historical Branch, 389.

30. "[Report of Proceedings, 1-14 July 1944]," Captain H.T.T. Bayliss, HMS *Vindex*, Vindex 2100/0133, 15 July 1944, ADM 199/497, p. 5.

31. D. Howse, *Radar at Sea. The Royal Navy in World War 2* (London: Macmillan for The Naval Radar Trust, 1993), 311; F.A. Kingsley (ed.), *The Development of Radar Equipments for the Royal Navy, 1935-1945* (London: Macmillan Press, for the Naval Radar Trust, 1995), 357; F.A. Kingsley (ed.), *The Applications of Radar and Other Electronic Systems in the Royal Navy in World War 2* (London: Macmillan Press, for the Naval Radar Trust, 1995), 184.

32. "[Report of Proceedings, 1-14 July 1944]," Captain H.T.T. Bayliss, HMS *Vindex*, Vindex 2100/0133, 15 July 1944, ADM 199/497, p. 1.

33. "[Report of Proceedings, 1-14 July 1944]," Captain H.T.T. Bayliss, HMS *Vindex*, Vindex 2100/0133, 15 July 1944, ADM 199/497, p. 2.

34. "The Captain (D), Third Destroyer Flotilla, No. 01, 11 July 1944," Appendix VI to, "Report of Proceedings on Recent U-boat Action off Cape Wrath," Vice Admiral H. Harwood, Admiral Commanding Orkney and Shetland, 912/OS.0253, 1 August 1944, ADM 199/497.

35. "The Commanding Officer, HMS *Bulldog*'s Report, 5 July 1944," Appendix V to, "Report of Proceedings on Recent U-boat Action off Cape Wrath," Vice Admiral H. Harwood, Admiral Commanding Orkney and Shetland, 912/OS.0253, 1 August 1944, ADM 199/497.

36. "A U-boat Hunt off Cape Wrath, 5-17 July," in "Monthly Anti-Submarine Report, August 1944," DAUD, CB 04050/44(8), 15 September 1944, NHB, p. 19; "Report of Proceedings on Recent U-boat Action off Cape Wrath," Vice Admiral H. Harwood, Admiral Commanding Orkney and Shetland, 912/OS.0253, 1 August 1944, ADM 199/497, p. 2.

37. Signal, ACOS to *Huntsville*, D3, *Vindex*, etc., 061044B, 6 July 1944, ADM 199/497.

38. Signal, C-in-C, HF, to D3 (R) *Marne, Verulam, Duke of York, Fencer*, C-in-C, WA, *Vindex, Khedive*, Capt (D) Scapa, 061039B, 6 July 1944, ADM 199/497.

39. Signal, ACOS to *Vindex* (info) *Bulldog*, 061224B, 6 July 1944, ADM 199/497.

40. "Report of Proceedings on Recent U-boat Action off Cape Wrath," Vice Admiral H. Harwood, Admiral Commanding Orkney and Shetland, 912/OS.0253, 1 August 1944, ADM 199/497, p. 4.

41. "No. 18 Group Headquarters Narrative," 6 July 1944, AIR 25/441.

42. "Report of Proceedings on Recent U-boat Action off Cape Wrath," Vice Admiral H. Harwood, Admiral Commanding Orkney and Shetland, 912/OS.0253, 1 August 1944, ADM 199/497, p. 4.

43. "Report of Proceedings on Recent U-boat Action off Cape Wrath," Vice Admiral H. Harwood, Admiral Commanding Orkney and Shetland, 912/OS.0253, 1 August 1944, ADM 199/497, p. 7.

44. Signal, ACOS to *Huntsville, Vindex* (R) C-in-C, Rosyth, for 18 Group, 061917B, 6 July 1944, ADM 199/497.

45. "No. 18 Group Headquarters Narrative," 6 July 1944, AIR 25/441; "No. 162 Squadron, RCAF, Operations Record Book," January-December 1944, AIR 271074.

46. "Report of Proceedings," Lieutenant Commander P.S. Evans, HMS *Inman*, Ref. 1511/1, 20 July 1944, ADM 199/497; "[Report of Proceedings, 1-14 July 1944]," Captain H.T.T. Bayliss, HMS *Vindex*, Vindex 2100/0133, 15 July 1944, ADM 199/497, p. 2.

47. "Thursday, 6 July 1944, Headquarters No. 15 Group, Narrative," 15 Group Narrative Office, n.d,

AIR 25/296; "No. 120 Squadron Operations Record Book," January to December 1944, AIR 27/912, p. 57. There is some confusion in the records over the side-letter of this aircraft, with some stating A/120, but most have B/120 and that has been adopted here.

48. "Report of Proceedings on Recent U-boat Action off Cape Wrath," Vice Admiral H. Harwood, Admiral Commanding Orkney and Shetland, 912/OS.0253, 1 August 1944, ADM 199/497, p. 5; Signal, ACOS to *Vindex* (R) *Manners, Burgess*, 070021B, 7 July 1944, ADM 199/497.

49. "18 Group, Headquarters, Operations Record Book, Form 540," 1944, AIR 25/381.

50. "No. 18 Group Headquarters Narrative," 7 July 1944, AIR 25/441.

51. "Report of Proceedings on Recent U-boat Action off Cape Wrath," Vice Admiral H. Harwood, Admiral Commanding Orkney and Shetland, 912/OS.0253, 1 August 1944, ADM 199/497, p. 6.

52. "Report of Proceedings on Recent U-boat Action off Cape Wrath," Vice Admiral H. Harwood, Admiral Commanding Orkney and Shetland, 912/OS.0253, 1 August 1944, ADM 199/497, p. 6.

53. Between these two positions the U-boat would have had to make good about 3¼ knots. "Report of Proceedings on Recent U-boat Action off Cape Wrath," Vice Admiral H. Harwood, Admiral Commanding Orkney and Shetland, 912/OS.0253, 1 August 1944, ADM 199/497, p. 6.

54. "Report of Proceedings on Recent U-boat Action off Cape Wrath," Vice Admiral H. Harwood, Admiral Commanding Orkney and Shetland, 912/OS.0253, 1 August 1944, ADM 199/497, p. 6.

55. "Report of Proceedings on Recent U-boat Action off Cape Wrath," Vice Admiral H. Harwood, Admiral Commanding Orkney and Shetland, 912/OS.0253, 1 August 1944, ADM 199/497, p. 6.

56. Signal, ACOS to *Vindex*, Captain D3, *Bulldog* (R) C-in-C, Rosyth, for 18 Group, 071109B, 7 July 1944, ADM 199/497.

57. Signal, DDIC, Admiralty, to C-in-C, Plymouth, AOC-in-C, Coastal Command, C-in-C, WA, etc., 071356B/July, 7 July 1944, ADM 223/198.

58. Signal, ACOS to Admiralty (R) C-in-C, Rosyth, C-in-C, WA, C-in-C, HF, 071915B, 7 July 1944, ADM 199/497.

59. Minute, G.E. Colpoys, DNI, NID/8, 2 November 1944, ADM 199/497.

60. Signal, ACOS to *Vindex, Bulldog, Huntsville* (R) C-in-C, Rosyth, for 18 Group, 072331B, 7 July 1944, ADM 199/497.

61. "Report of Proceedings on Recent U-boat Action off Cape Wrath," Vice Admiral H. Harwood, Admiral Commanding Orkney and Shetland, 912/OS.0253, 1 August 1944, ADM 199/497, p. 8.

62. Signal, ACOS to C-in-C, WA, (R) Admiralty, C-in-C, HF, 081001B, 8 July 1944, ADM 199/497; Signal, C-in-C, WA, to ACOS (R) Admiralty, C-in-C, HF, C-in-C, Rosyth, 081110B, 8 July 1944, ADM 199/497; "July, 1944," in "15 Group, Headquarters, Operations Record Book, Form 540," 1944, AIR 25/255.

63. For Horton's anticipatory orders, see: "Standing Operation Orders for U-boat Hunts off Northern Ireland (Short Title: Operation "CW")," C-in-C, Western Approaches, WA.00770/104, M.056090/44, 24 February 1944, ADM 199/468.

64. Signal, DDIC, Admiralty, to C-in-C, WA, AOC-in-C, CC, ACOS, etc., 121350B, Serial No. H.5175, 12 June 1944, ADM 223/196. The OIC had, however, assumed that *U-247* was the candidate for the sinking of *Noreen Mary*. Signal, DDIC, Admiralty, to C-in-C, Plymouth, AOC-in-C, Coastal Command, C-in-C, WA, etc., 061426B/July, 6 July 1944, ADM 223/198.

65. Signal, DDIC, Admiralty, to C-in-C, Plymouth, AOC-in-C, Coastal Command, C-in-C, WA, etc., 081331B/July, 8 July 1944, ADM 223/198.

66. "Saturday, 8 July 1944," in "War Diary (Naval), 1-15 July 1944," Vol. 167, NHB, pp. 184-185.

67. "Saturday, 8 July 1944, Headquarters No. 15 Group, Narrative," 15 Group Narrative Office, n.d, AIR 25/296.

68. "[Report of Proceedings, 1-14 July 1944]," Captain H.T.T. Bayliss, HMS *Vindex*, Vindex 2100/0133, 15 July 1944, ADM 199/497, p. 3.

69. "[Report of Proceedings, 1-14 July 1944]," Captain H.T.T. Bayliss, HMS *Vindex*, Vindex 2100/0133, 15 July 1944, ADM 199/497, p. 4.

70. Signal, DDIC, Admiralty, to C-in-C, Plymouth, AOC-in-C, Coastal Command, C-in-C, WA, etc., 091331B/July, 9 July 1944, ADM 223/198.

71. The problems of U-boat navigation during the inshore campaign is discussed in [Günter Hessler],

The U-boat War in the Atlantic, 1939-1945, Vol. III, *June 1943–May 1945* (London: HMSO, 1989), p. 74. For the original Bletchley Park signals, see: ZIP/ZTPGU/27743, TOI 2311Z, 9 July 1944, DEFE 3/732; ZIP/ZTPGU/27744, TOI 2335Z, 9 July 1944, DEFE 3/732.

72. "Kriegstagebuch, *U-247*, 18 May to 28 July 1944," NID PG/30225/NID, NHB, pp. 27-28.

73. "Monday, 10 July 1944, Headquarters No. 15 Group, Narrative," 15 Group Narrative Office, n.d, AIR 25/296.

74. "Monday, 10 July 1944," in "War Diary (Naval), 1-15 July 1944," Vol. 167, NHB; "[Report of Proceedings, 1-14 July 1944]," Captain H.T.T. Bayliss, HMS *Vindex*, Vindex 2100/0133, 15 July 1944, ADM 199/497, p. 4.

75. "[Report of Proceedings, 1-14 July 1944]," Captain H.T.T. Bayliss, HMS *Vindex*, Vindex 2100/0133, 15 July 1944, ADM 199/497, p. 5.

76. "[Report of Proceedings, 1-14 July 1944]," Captain H.T.T. Bayliss, HMS *Vindex*, Vindex 2100/0133, 15 July 1944, ADM 199/497, p. 4; "Report of Proceedings," Lieutenant Commander P.S. Evans, HMS *Inman*, Ref. 1511/1, 20 July 1944, ADM 199/497.

77. Peter Nash, "The Contribution of Mobile Logistic Support to Anglo-American Naval Policy, 1945-1953," (PhD Thesis, King's College, London, 2006), p. 90.

78. ZIP/ZTPGU/27743, TOI 2311, 9 July 1944, DEFE 3/732; ZIP/ZTPGU/27744, TOI 2335, 9 July 1944, DEFE 3/732; Signal, DDIC, Admiralty, to C-in-C, WA, ACOS, C-in-C, Rosyth, AOC-in-C, Coastal Command, etc., 110505B/July, 11 July 1944, ADM 223/198.

79. Signal, DDIC, Admiralty, to C-in-C, WA, ACOS, C-in-C, Rosyth, AOC-in-C, Coastal Command, etc., 110505B/July, 11 July 1944, ADM 223/198.

80. "Naval Headlines, No. 1103," Naval Section, 11 July 1944, HW 1/3056.

81. "[Report of Proceedings, 1-14 July 1944]," Captain H.T.T. Bayliss, HMS *Vindex*, Vindex 2100/0133, 15 July 1944, ADM 199/497, p. 4; "Monday, 10 July 1944, Headquarters No. 15 Group, Narrative," 15 Group Narrative Office, n.d, AIR 25/296.

82. "HQCC Narrative No. 995 for 24 hours ending 2359/11/7/44," in "Headquarters, Coastal Command, Operations Record Book, Appendices," AHB IIM/A3/2A-R, July 1944, AIR 24/416; "No. 120 Squadron Operations Record Book," January to December 1944, AIR 27/912, p. 59; "Précis of Attack by Liberator Aircraft "A" of 120 Squadron, 11 July 1944," U-boat Assessment Committee, Serial No. 86, AUD.1240/44, 14 August 1944, from, "Proceedings of U-Boat Assessment Committee, July–September 1944," Vol. 16, NHB, p. 98.

83. "No. 120 Squadron Operations Record Book," January to December 1944, AIR 27/912, pp. 58-59. The standard tactical reaction is described in: "Operations (General)," Part II, "Coastal Command Manual of Anti-U-Boat Warfare," May 1944, AIR 15/294.

84. "Report of Proceedings – 1 July to 15 July 1944," A/Lieutenant Commander Denny, RCNR, HMS *St. Thomas*, St.T. 0-5-1, 20 July 1944, ADM 199/497.

85. "Search "A,"" in "Report of Proceedings – 1 July to 15 July 1944," A/Lieutenant Commander Denny, RCNR, HMS *St. Thomas*, St.T. 0-5-1, 20 July 1944, ADM 199/497; "No. 120 Squadron Operations Record Book," January to December 1944, AIR 27/912, pp. 58-59.

86. "[Report of Proceedings, 1-14 July 1944]," Captain H.T.T. Bayliss, HMS *Vindex*, Vindex 2100/0133, 15 July 1944, ADM 199/497, p. 4.

87. Signal, *Vindex* to *St. Thomas*, 111628, 11 July 1944, ADM 199/497.

88. In his report, Carne confuses the position of the attack by A/120 with that of the contact made by *Vindex*'s Swordfish. "Narrative," Appendix 1 to, "Report of Proceedings for the Period 10 July 1944 to 19 July 1944," Captain W.P. Carne, RN, HMS *Striker*, 2542/01/4, 19 July 1944, ADM 199/497, p. 1.

89. "[Report of Proceedings, 1-14 July 1944]," Captain H.T.T. Bayliss, HMS *Vindex*, Vindex 2100/0133, 15 July 1944, ADM 199/497, pp. 4-5; "Summary of Air Operations," Appendix 4 to "[Report of Proceedings, 1-14 July 1944]," Captain H.T.T. Bayliss, HMS *Vindex*, Vindex 2100/0133, 15 July 1944, ADM 199/497.

90. "Narrative," Appendix 1 to, "Report of Proceedings for the Period 10 July 1944 to 19 July 1944," Captain W.P. Carne, RN, HMS *Striker*, 2542/01/4, 19 July 1944, ADM 199/497, p. 1.

91. Signal, DDIC, Admiralty, to C-in-C, Plymouth, C-in-C, WA, AOC-in-C, Coastal Command, etc., 111251B/July, 11 July 1944, ADM 223/198.

92. "The Schnorkel Smoke Myth," Appendix V, in "The RAF in Maritime War, Vol. V: The Atlantic and Home Waters, The Victorious Phase, June 1944–May 1945," Air Historical Branch (1), Air Ministry, n.d., AIR 41/74, p. 1; "Précis of Attack by Liberator Aircraft "A" of 120 Squadron, 11 July 1944," U-boat Assessment Committee, Serial No. 86, AUD.1240/44, 14 August 1944, from, "Proceedings of U-Boat Assessment Committee, July–September 1944," Vol. 16, NHB, p. 98.

93. "Wednesday, 12 July 1944," in "War Diary (Naval), 1-15 July 1944," Vol. 167, NHB, pp. 273-274; "Wednesday, 12 July 1944, Headquarters No. 15 Group, Narrative," 15 Group Narrative Office, n.d, AIR 25/296.

94. That is, 30° to the right of the aircraft's nose.

95. The depth charges were held in the Sunderland's hull and in order to be dropped had to pass through hatches on runners which ran under the wings on either side. The weapons were moved by simple electric motors. The author is grateful to Air Vice Marshal George Chesworth for this information. Telephone conversation with Air Vice Marshal George Chesworth, 11 March 2003.

96. "No. 423 Squadron ORB," DHH 77/1887, n.d., DHH. The author is grateful for a transcription of this ORB supplied by Michael Whitby, Chief, Naval History Team, DHH, 22 October 2002.

97. This account is based on "Précis of Attack by Sunderland Aircraft "J" of 423 Squadron [RCAF], 12 July 1944," U-boat Assessment Committee, Serial No. 85, AUD.1255/44, 14 August 1944, from, "Proceedings of U-Boat Assessment Committee, July–September 1944," Vol. 16, NHB, p. 103; "No. 423 Squadron ORB," DHH 77/1887, n.d., Directorate of History and Heritage, Department of National Defence, Ottawa, Canada. The author is grateful for a transcription of this ORB which was supplied by Michael Whitby, Chief, Naval History Team, DHH, via e-mail Whitby.MJ@forces.gc.ca, 22/10/02 20:10:30 GMT. Also: "Kriegstagebuch, U-247, 18 May to 28 July 1944" NID PG/30225/ NID, Naval Historical Branch.

98. The construction of the searches was based on the underwater performance deduced from trials with U-570 captured in 1941 (and re-commissioned as HMS Graph). "Coastal Command Manual of Anti-U-Boat Warfare," May 1944, AIR 15/294, Article 28.

99. "Operations (General)," Part II, "Coastal Command Manual of Anti-U-Boat Warfare," May 1944, AIR 15/294, Article 29; "No. 423 Squadron ORB," DHH 77/1887, n.d., Directorate of History and Heritage, Department of National Defence, Ottawa, Canada.

100. "Précis of Attack by Sunderland Aircraft "J" of 423 Squadron [RCAF], 12 July 1944," U-boat Assessment Committee, Serial No. 85, AUD.1255/44, 14 August 1944, from, "Proceedings of U-Boat Assessment Committee, July–September 1944," Vol. 16, NHB, p. 103.

101. This is helped by the survival of U-247's log; see "Kriegstagebuch, U-247, 18 May to 28 July 1944," NID PG/30225/NID, NHB.

102. "Western Approaches Monthly News Bulletin, November 1944," 18 December 1944, Records of the Office of the Chief of Naval Operations: Registered Publications Section, Foreign Navy and Related Foreign Military Publications, 1913-1960, Box 396, RG 38, NARA2.

103. "Kriegstagebuch, U-247, 18 May to 28 July 1944," NID PG/30225/NID, NHB, pp. 25-26.

104. Signal, DDIC, Admiralty, to C-in-C, WA, AOC-in-C, CC, H.4944, F.3225, 311712B, 31 May 1944, ADM 223/195; Signal, DDIC, Admiralty, to C-in-C, WA, AOC-in-C, CC, C-in-C, Rosyth, etc., H.5001, F.3237, 060910B, 6 June 1944, ADM 223/195. In this habit, the C-in-C, Western Approaches, thought, the Germans continued into September 1944. "Western Approaches Monthly Air Summary, No. 9, September/October 1944," Admiral Max Horton, Commander-in-Chief, Western Approaches, 20 November 1944, Records of the Office of the Chief of Naval Operations: Registered Publications Section, Foreign Navy and Related Foreign Military Publications, 1913-1960, Box 396, RG 38, NARA2.

105. "No. 423 Squadron ORB," DHH 77/1887, n.d., DHH.

106. "Narrative," Appendix 1 to, "Report of Proceedings for the Period 10 July 1944 to 19 July 1944," Captain W.P. Carne, RN, HMS Striker, 2542/01/4, 19 July 1944, ADM 199/497, p. 1.

107. "Search "C,"" in "Report of Proceedings – 1 July to 15 July 1944," A/Lieutenant Commander Denny, RCNR, HMS St. Thomas, St.T. 0-5-1, 20 July 1944, ADM 199/497.

108. Signal, C-in-C, WA, to Force 34 (R) 15 Group, BN.822, 121148, 12 July 1944, ADM 199/497; "Search 'E,'" in "Report of Proceedings – 1 July to 15 July 1944," A/Lieutenant Commander Denny, RCNR, HMS St. Thomas, St.T. 0-5-1, 20 July 1944, ADM 199/497.

109. "English Channel U-boat Estimate No. 37," Signal, DDIC, Admiralty, to C-in-C, Plymouth, C-in-C, WA, AOC-in-C, Coastal Command, etc., 131414B/July, 13 July 1944, ADM 223/199.

110. "Estimate No. 38 for English Channel U-boat," Signal, DDIC, Admiralty, to C-in-C, Plymouth, C-in-C, WA, AOC-in-C, Coastal Command, etc., 141346B/July, 14 July 1944, ADM 223/199.

111. "Thirty Ninth Estimate for English Channel U-boat," Signal, DDIC, Admiralty, to C-in-C, Plymouth, C-in-C, WA, AOC-in-C, Coastal Command, etc., 151358B/July, 15 July 1944, ADM 223/199.

112. "Narrative," Appendix 1 to, "Report of Proceedings for the Period 10 July 1944 to 19 July 1944," Captain W.P. Carne, RN, HMS *Striker*, 2542/01/4, 19 July 1944, ADM 199/497, p. 4.

113. "Report of Proceedings of Third Escort Group, 12 to 19 July 1944, Commander R.G. Mills, DSO, DSC*, RN, Senior Officer Third Escort Group, HMS *Duckworth*, No. 1244/46, 19 July 1944, ADM 199/497.

114. "July, 1944," in "15 Group, Headquarters, Operations Record Book, Form 540," 1944, AIR 25/255; "Thursday, 13 July 1944, Headquarters No. 15 Group, Narrative," 15 Group Narrative Office, n.d, AIR 25/296; "Friday, 14 July 1944, Headquarters No. 15 Group, Narrative," 15 Group Narrative Office, n.d, AIR 25/296.

115. Signal, C-in-C, WA, to Force 34, 151225B, 15 July 1944, ADM 199/497.

116. "Thirty Ninth Estimate for English Channel U-boat," Signal, DDIC, Admiralty, to C-in-C, Plymouth, C-in-C, WA, AOC-in-C, Coastal Command, etc., 151358B/July, 15 July 1944, ADM 223/199.

117. "Saturday, 15 July 1944, Headquarters No. 15 Group, Narrative," 15 Group Narrative Office, n.d, AIR 25/296; "Sunday, 16 July 1944, Headquarters No. 15 Group, Narrative," 15 Group Narrative Office, n.d, AIR 25/296.

118. Signal, C-in-C, WA, to Force 34, 151225B, 15 July 1944, ADM 199/497.

119. "Convoy HXM 298," in "HX" File, NHB.

120. Signal, Admiralty, to AIG 331, CominCh, 111135B, 11 July 1944, ADM 199/308.

121. "16 July 1944, in "Convoy Positions, 1 July 1944 to 31 August 1944," Vol. 16, NHB; Signal, DDIC, Admiralty, to C-in-C, Plymouth, AOC-in-C, Coastal Command, etc., 161247B/July, 16 July 1944, ADM 223/199.

122. "Channel U-boats Estimate No. 42," Signal, DDIC, Admiralty, to C-in-C, Plymouth, C-in-C, WA, AOC-in-C, Coastal Command, etc., 181346B/July, 18 July 1944, ADM 223/199.

123. "Tuesday, 18 July 1944, Headquarters No. 15 Group, Narrative," 15 Group Narrative Office, n.d, AIR 25/296; "17 July 1944," in "Convoy Positions, 1 July 1944 to 31 August 1944," Vol. 16, NHB.

124. "Thursday, 20 July 1944," in "War Diary (Naval), 16-30 July 1944," Vol. 168, NHB; Signal, DDIC, Admiralty, to C-in-C, Plymouth, C-in-C, WA, AOC-in-C, Coastal Command, etc., 191336B/July, 19 July 1944, ADM 223/199.

125. "Narrative," Appendix 1 to, "Report of Proceedings for the Period 10 July 1944 to 19 July 1944," Captain W.P. Carne, RN, HMS *Striker*, 2542/01/4, 19 July 1944, ADM 199/497, p. 5; Signal, C-in-C, WA, to Force 34 (R) Admiralty, C-in-C, Plymouth, FOCT, 15 Group CC, FOIC Greenock, 171303B, 17 July 1944, ADM 199/497; "Western Approaches Monthly News Bulletin, July 1944," 19 August 1944, NHB.

126. "July, 1944," in "15 Group, Headquarters, Operations Record Book, Form 540," 1944, AIR 25/255; "Wednesday, 19 July 1944, Headquarters No. 15 Group, Narrative," 15 Group Narrative Office, n.d, AIR 25/296.

127. "Thursday, 20 July 1944, Headquarters No. 15 Group, Narrative," 15 Group Narrative Office, n.d, AIR 25/296.

128. "No. 224 Squadron, Operations Record Book," January 1944–June 1945, AIR 27/1389, p. 131; "Tasking Signal," 1236, 13 July 1944, AIR 25/496.

129. "No. 224 Squadron, Operations Record Book," January 1944–June 1945, AIR 27/1389, p. 131.

130. "Night Actions and the Use of Radar," Part IV, "Coastal Command Manual of Anti-U-Boat Warfare," May 1944, AIR 15/294, Article 87.

131. "No. 224 Squadron, Operations Record Book," January 1944–June 1945, AIR 27/1389, p. 131.

132. "Kriegstagebuch, *U-953*, 18 June to 22 July 1944," NID PG/30788/NID, NHB, p. 20; Signal, DDIC,

Admiralty, to C-in-C, Plymouth, C-in-C, WA, AOC-in-C, Coastal Command, etc., 201413B/July, 20 July 1944, ADM 223/199; Signal, DDIC, Admiralty, to C-in-C, Plymouth, C-in-C, WA, AOC-in-C, Coastal Command, etc., 201656B/July, 20 July 1944, ADM 223/199.

133. "War and Post War Service," n.d., File 6, Captain M.J. Evans, RN, Papers, IWM 65/25/1; Letter, Admiral Max Horton to Martin Evans, 8 May 1944, File 4, Captain M.J. Evans, RN, Papers, IWM 65/25/1; Letter, Commander B. Biskupski to Commander Evans, 30 May 1944, File 5, Captain M.J. Evans, RN, Papers, IWM 65/25/1; "Legion of Merit, Degree of Legionnaire, awarded to Commander Martin James Evans, OBE, Royal Navy," Secretary of the Navy, Washington, n.d., File 14, Captain M.J. Evans, RN, Papers, IWM 65/25/1.

134. He had first been indoctrinated as Staff Officer (Operations) in Malta earlier in the war. Letter, Martin Evans to Gretton, 10 September 1981, "Battle of the Atlantic," Gretton, 23 Part 1, Box 1, Corresp. E-M, MS93/008, NMM(G).

135. This Group, bore the same name as the Group under Commander C.E. Bridgeman, RNR, described in the Chapter on the "Hollow Victory: *Gruppe Leuthen*'s attacks on Convoy ON 202 and ONS18," by Doug McLean. After that episode, however, the 9th Escort Group had been reconstituted at the end of 1943, when Layard took command. Michael Whitby (ed.), *Commanding Canadians: The Second World War Diaries of A.F.C. Layard* (Vancouver: UBC Press, 2005), pp. 72-73.

136. Michael Whitby (ed.), *Commanding Canadians: The Second World War Diaries of A.F.C. Layard* (Vancouver: UBC Press, 2005), pp. 24-25.

137. Michael Whitby (ed.), *Commanding Canadians: The Second World War Diaries of A.F.C. Layard* (Vancouver: UBC Press, 2005), pp. 178.

138. "[Report of Proceedings]," Commander A.F.C. Layard, Senior Officer, 9th Escort Group, 25 July 1944, ADM 199/1644.

139. "Thursday, 20 July 1944," Commander A.F.C. Layard, Diary, RNM.

140. "Thursday, 20 July 1944," Commander A.F.C. Layard, Diary, RNM.

141. "Report of Proceedings, 14th Escort Group, Period 15–26 July 1944," [Commander R.A. Currie, RN, Senior Officer, EG14, 27 July 1944], ADM 199/497.

142. "Report of Proceedings, 14th Escort Group, Period 15–26 July 1944," [Commander R.A. Currie, RN, Senior Officer, EG14, 27 July 1944], ADM 199/497; "Kriegstagebuch, *U-953*, 18 June to 22 July 1944," NID PG/30788/NID, NHB, p. 21.

143. Signal, DDIC, Admiralty, to C-in-C, Plymouth, C-in-C, WA, AOC-in-C, Coastal Command, etc., 201352B/July, 20 July 1944, ADM 223/199; Signal, DDIC, Admiralty, to C-in-C, Plymouth, C-in-C, WA, AOC-in-C, Coastal Command, etc., 211349B/July, 21 July 1944, ADM 223/199; Signal, DDIC, Admiralty, to C-in-C, Plymouth, C-in-C, WA, AOC-in-C, Coastal Command, etc., 211707B/July, 21 July 1944, ADM 223/199.

144. "Report of Proceedings, 14th Escort Group, Period 15–26 July 1944," [Commander R.A. Currie, RN, Senior Officer, EG14, 27 July 1944], ADM 199/497.

145. "21 July 1944," in "Translation of PG/30351, *BdU*'s War Log, 16-31 July 1944," NHB; "Kriegstagebuch, *U-953*, 18 June to 22 July 1944," NID PG/30788/NID, NHB, p. 21.

146. Marbach had completed 7 war patrols over the previous year, the last in mid-Channel where he had sunk a merchant ship and unsuccessfully attacked three RCN destroyers. Peter Sharpe, *U-boat Fact File, Detailed Service Histories of the Submarines Operated by the Kriegsmarine, 1935-1945* (Leicester: Midland Publishing, 1998), pp. 153-154.

147. M. Llewellyn-Jones, "Trials with HM Submarine *Seraph* and British Preparations to Defeat the Type XXI U-Boat, September–October, 1944," *The Mariner's Mirror*, Vol. 86, No. 4 (November 2000), pp. 434-451.

148. "Plymouth Command Diary of Events, 0001-2400, 22 July 1944," ADM 199/1394.

149. Signal, DDIC, Admiralty, to C-in-C, Plymouth, C-in-C, WA, AOC-in-C, Coastal Command, etc., 221335B/July, 22 July 1944, ADM 223/199.

150. Signal, DDIC, Admiralty, to C-in-C, Plymouth, C-in-C, WA, AOC-in-C, Coastal Command, etc., 261359B/July, 26 July 1944, ADM 223/200; Signal, DDIC, Admiralty, to C-in-C, Plymouth, C-in-C, WA, AOC-in-C, Coastal Command, etc., 261732B/July, 26 July 1944, ADM 223/200.

151. This was Captain Johnnie Walker's famous Support Group. Walker had, however, died barely three

weeks before. "Tasking Signal," 1545, 26 July 1944, AIR 25/496; "[Report of Proceedings]," Commander N.W. Duck, RNR, Senior Officer, Second Support Group, No. 048, 1 August 1944, ADM 199/1460.

152. "Plymouth Command Diary of Events, 0001-2400, 27 July 1944," ADM 199/1394; "Kriegstagebuch, *U-247*, 18 May to 28 July 1944," NID PG/30225/NID, NHB, p. 36; Signal, DDIC, Admiralty, to C-in-C, Plymouth, C-in-C, WA, AOC-in-C, Coastal Command, etc., 271509B/July, 27 July 1944, ADM 223/200.

153. "Friday, 28 July 1944," Commander A.F.C. Layard, Diary, RNM; "Plymouth Command Diary of Events, 0001-2400, 28 July 1944," ADM 199/1394.

154. Michael Whitby (ed.), *Commanding Canadians: The Second World War Diaries of A.F.C. Layard* (Vancouver: UBC Press, 2005), p. 184.

155. Richard Overy, *Russia's War* (London: Allen Lane, 1998), p. 245.

156. Chester Wilmot, *The Struggle for Europe* (London: The Reprint Society, 1954), pp. 428, 430 and 433.

157. "U-boat Trend, Period 21 to 28 August," J.W. Clayton, DDIC, 28 August 1944, ADM 223/20.

158. "U-boat Situation, Week ending 12 June 1944," Captain Rodger Winn, RNVR, OIC.IS/976, 12 June 1944, ADM 223/21.

159. "U-boats in Bay Ports," Lieutenant Commander P. Beesly, RNVR, OIC/SI.1030, 24 August 1944, ADM 223/172; "Anti-U-Boat Results of Neptune," Captain Rodger Winn, RNVR, OIC/SI.1064, 28 August 1944, ADM 223/172.

160. "U-boat Trend, Period 7 to 16 August," J.W. Clayton, DDIC, 16 August 1944, ADM 223/20; "U-boat Trend, Period 21 to 28 August," J.W. Clayton, DDIC, 28 August 1944, ADM 223/20.

161. "Western Approaches Monthly News Bulletin, August 1944," 20 September 1944, NHB.

162. There is not room to describe these operations in detail, but they may be deduced from a study of "War Diaries, Plymouth, July-December 1944," War History, Case 8794, 1944, ADM 199/1394 and "19 Group, Headquarters, Operations Record Book, Form 540," AHB, August 1944, AIR 25/497, in conjunction with, "H Series, 5901-6100," July-August 1944, ADM 223/200 and "H Series, 6501-6700," August 1944, ADM 223/203.

163. Stephen Roskill, *The War at Sea 1939-1945*, Vol. III, Part II, *The Offensive 1 June 1944-14 August 1945* (London: HMSO, 1961), p. 130.

164. F.H. Hinsley, E.E. Thomas, C.A.G. Simkins & C.F.G. Ransom, *British Intelligence in the Second World War: Its Influence on Strategy and Operations*, Vol. III, Part 2 (London: HMSO, 1988), p. 464.

165. "The Development of British Naval Aviation, 1919-1945," Naval Staff History, Second World War, Vol. II, Historical Section, Admiralty, HS.20/54, CB 3307(2) [BR 1736(53)(2)], 5 December 1956, NHB, p. 162.

166. Michael Whitby (ed.), *Commanding Canadians: The Second World War Diaries of A.F.C. Layard* (Vancouver: UBC Press, 2005), p. 196.

167. "[Report of Proceedings of EG9, 17-28 August and 31 August to 10 September 1944]," Commander Layard, RN, Senior Officer EG9, 10 September 1944, ADM 199/1462; Signal, C-in-C, Plymouth to EG9 (R) Force 26, Force 28, 19 Group, 241049B, 24 August 1944, ADM 199/1462; Signal, C-in-C, Plymouth to EG9 (R)...Force 26, 28, 241908B, 24 August 1944, ADM 199/1462.

168. Roy Conyers Nesbit, *The Strike Wings: Special Anti-Shipping Squadrons, 1942-45* (London: HMSO, 1995), pp. 175-180; Stephen Roskill, *The War at Sea 1939-1945*, Vol. III, Part II, *The Offensive 1 June 1944-14 August 1945* (London: HMSO, 1961), p. 131; "[Report of Proceedings of EG9, 17-28 August and 31 August to 10 September 1944]," Commander Layard, RN, Senior Officer EG9, 10 September 1944, ADM 199/1462.

169. "Meeting of Force 28 with HM Ships *Port Colborne* and *Meon*," Commanding Officer, HMS *Diadem*, No. 191, 14 October 1944, ADM 199/1462" Minute, Captain C.D. Howard-Johnston, DAUD, AU Division, File No. 8001/9, AUD.1561/44, 19 September 1944, ADM 199/1462.

170. "U-boat Trend, Period 3-12 July," J.W. Clayton, DDIC, 12 July 1944, ADM 223/20.

171. Karl Doenitz, *Memoirs: Ten Years and Twenty Days*, tr. R.H. Stevens with introduction by Jurgen Rohwer (London: Greenhill Books, 1990), p. 421; "U-boat Situation, Week Ending 28 August 1944," Captain Rodger Winn, RNVR, OIC/SI.1062, n.d., ADM 223/172.

172. "U-boat Trend, Period 14 to 23 August," J.W. Clayton, DDIC, 23 August 1944, ADM 223/20; "U-

boat Situation, Week Ending 28 August 1944," Captain Rodger Winn, RNVR, OIC/SI.1062, n.d., ADM 223/172.

173. "U-boat Trend, Period 31 July to 9 August," J.W. Clayton, DDIC, 9 August 1944, ADM 223/20.

174. "U-boat Trend, Period 21 to 28 August," J.W. Clayton, DDIC, 28 August 1944, ADM 223/20; Signal, DDIC, Admiralty, to C-in-C, Plymouth, C-in-C, WA, AOC-in-C, Coastal Command, etc., 271405B/August, 27 August 1944, ADM 223/203; "State of U-boats in Biscay Area on 26 August 1944," Captain Rodger Winn, RNVR, OIC/SI.1055, 26 August 1944, ADM 223/172.

175. "Bombing of U-boat Biscay Bases," DNI, NID.0998, 15 March 1943, ADM 205/30; Letter, Air Marshal A.T. Harris, Commanding-in-Chief, Bomber Command, to the Under Secretary of State, Air Ministry, BC/S.23746/C-in-C, 27 January 1943, AIR 2/8694; "U-boat Trend, Period 7 August 1944 to 14 August 1944," in David Syrett (ed.), *The Battle of the Atlantic and Signal Intelligence: U-boat Situations and Trends, 1941-1945* (Aldershot: Ashgate, 1998), pp. 436-437.

176. "U-boats in Biscay Ports – Bombing Targets," J.W. Clayton, DDIC, OIC/SI.998, 8 July 1944, ADM 223/172.

177. "Report on Bombing of U-boat Shelters at Brest," Captain E. Terrell (Admiralty), Dr H.M. Jenkins (USSTAF), Lieutenant Colonel D.L. Sutherland (HQ 8th AF), Wing Commander G.H. Everitt (Air Ministry), Lieutenant N.G.H. Thomas (Admiralty), 12 October 1944, ADM 199/240; "No. 617 Squadron Operations Record Book," April 1943 to May 1945, AIR 27/2128; [Günter Hessler], *The U-boat War in the Atlantic, 1939-1945*, Vol. III, *June 1943–May 1945* (London: HMSO, 1989), p. 77.

178. "Bombing of Major Naval Targets, No. 44, (Excluding attacks by Coastal Command and the Tactical Air Force), Week ending 15 August 1944," H. Clanchy, DDNI(H), NID UC Report No. 523, 16 August 1944, in "UC Reports, Nos. 501-657," SL.4213, NID, 1944-1945, NHB.

179. "Plymouth Command, General Survey of Events, 16-31 August 1944," ADM 199/1394; "Plymouth Command Diary of Events, 0001-2400, 25 August 1944," ADM 199/1394.

180. "War Diary, British Assault Area," C-in-C, Portsmouth, No. 7252/0/514/17, TSD.5259/44, 29 October 1944, ADM 199/1396.

181. The author is grateful to Michael Meech for the information on these an other bombing missions against Brest. Michael Meech, e-mail, 1434, 16 April 2006.

182. "27 August 1944," in "Translation of PG/30353, *BdU's* War Log, 16-31 August 1944," Godt, Chief of Operations Department, *BdU*, NHB.

183. The various sources used in the description of the action that follows do not agree in all respects. Where conflicts arise, the author has therefore taken the evidence of the those most directly involved in a particular aspect.

184. Signal, DDIC, Admiralty, to C-in-C, Plymouth, C-in-C, WA, AOC-in-C, Coastal Command, etc., 251324B/August, 25 August 1944, ADM 223/203.

185. Signal, DDIC, Admiralty, to C-in-C, Plymouth, C-in-C, WA, AOC-in-C, Coastal Command, 251812B/August, 25 August 1944, ADM 223/203.

186. Signal, DDIC, Admiralty, to C-in-C, Plymouth, C-in-C, WA, AOC-in-C, Coastal Command, 251812B/August, 25 August 1944, ADM 223/203. The list of the OIC's two-letter identifiers for this period has not yet been located in the archive record. Those for those up to the end of 1942, see: "Intelligence Papers," 1941-1942, ADM 223/3, pp. 509-510 and 544-547.

187. "'Neptune' Plan 3," in L.J. Pitcairn-Jones, *Operation "Neptune," Landings in Normandy, June 1944* (London: HMSO, 1994); Stephen Roskill, *The War at Sea 1939-1945*, Vol. III, Part II, *The Offensive 1 June 1944-14 August 1945* (London: HMSO, 1961), pp. 179-180.

188. "26 August 1944," in "Translation of PG/30353, *BdU's* War Log, 16-31 August 1944," Godt, Chief of Operations Department, *BdU*, NHB; "Landwirte," n.d., HW 18/405, pp. 122-123.

189. Signal, DDIC, Admiralty, to C-in-C, Plymouth, C-in-C, WA, AOC-in-C, Coastal Command, etc., 261349B/August, 26 August 1944, ADM 223/203; "State of U-boats in Biscay Area on 26 August 1944," Captain Rodger Winn, RNVR, OIC/SI.1055, 26 August 1944, ADM 223/172.

190. Signal, DDIC, Admiralty, to C-in-C, Plymouth, C-in-C, WA, AOC-in-C, Coastal Command, etc., *271402*B/August, 27 August 1944, ADM 223/203.

191. Signal, DDIC, Admiralty, to C-in-C, Plymouth, C-in-C, WA, AOC-in-C, Coastal Command, etc., 271405B/August, 27 August 1944, ADM 223/203.

192. "Anti-U-Boat Results of Neptune," Captain Rodger Winn, RNVR, OIC/SI.1064, 28 August 1944, ADM 223/172.

193. Signal, DDIC, Admiralty, to C-in-C, Plymouth, C-in-C, WA, AOC-in-C, Coastal Command, etc., 281029B/August, 28 August 1944, ADM 223/203.

194. Signal, DDIC, Admiralty, to C-in-C, Plymouth, C-in-C, WA, AOC-in-C, Coastal Command, etc., 281350B/August, 28 August 1944, ADM 223/203.

195. This assertion is deduced by, for example, the comparison of "War Diary, 16-31 August 1944," C-in-C, Plymouth, No. 3651/Ply.1505, TSD.4900/44, 3 October 1944, ADM 199/1394 with "19 Group, Headquarters, Operations Record Book, Form 540," AHB, August 1944, AIR 25/497, against the SI material in "H Series, 6501-6700," August 1944, ADM 223/203.

196. "Tasking Signal," 19 Group, Headquarters, 28 August 1944, AIR 25/497.

197. Signal, C-in-C, Plymouth to EG1, EG15, 272322B, 27 August 1944, ADM 199/1462; Signal, C-in-C, Plymouth to EG15, EG1, 281105B, 28 August 1944, ADM 199/1462.

198. "[Report of Proceedings, 15th Escort Group, 20 August–31 August 1944]," Commander L.A.B. Majendie, RN, Senior Officer, 15th Escort Group, HMS Louis, RP.4/44, 31 August 1944, ADM 199/1462.

199. Signal, C-in-C, Plymouth to EG15, 291108B, 29 August 1944, ADM 199/1462.

200. "No. 224 Squadron, Operations Record Book," January 1944–June 1945, AIR 27/1389, p. 163.

201. "Wednesday, 30 August 1944," in "War Diary (Naval), 17-31 August 1944," Vol. 170, NHB.

202. Signal, C-in-C, Plymouth to EG15, 291108B, 29 August 1944, ADM 199/1462.

203. Signal, C-in-C, Plymouth to EG15, 291922B, 29 August 1944, ADM 199/1462.

204. "Narrative," 19 Group, Headquarters, 29 August 1944, AIR 25/497.

205. "Plymouth Command Diary of Events, 0001-2400, 28 August 1944," ADM 199/1394; "Report of Proceedings from 24 August–4 September," Commander C. Gwinner, RN, Senior Officer, 1st Escort Group, RP/10, 4 September 1944, ADM 199/498.

206. See also "Naval Headlines, No. 1150," Naval Section, 27 August 1944, HW 1/3191, which predicted an area somewhat close to Land's End.

207. "Narrative," 19 Group, Headquarters, 29 August 1944, AIR 25/497; Signal, C-in-C, Plymouth to EG1, EG15, 290128B, 29 August 1944, ADM 199/498.

208. "Operations Record Book, 179 Squadron, August 1944," n.d., AIR 27/1128; "Narrative," 19 Group, Headquarters, 29 August 1944, AIR 25/497.

209. "Report of Proceedings from 24 August–4 September," Commander C. Gwinner, RN, Senior Officer, 1st Escort Group, RP/10, 4 September 1944, ADM 199/498.

210. "Report of Proceedings from 24 August–4 September," Commander C. Gwinner, RN, Senior Officer, 1st Escort Group, RP/10, 4 September 1944, ADM 199/498.

211. "Report of Proceedings from 24 August–4 September," Commander C. Gwinner, RN, Senior Officer, 1st Escort Group, RP/10, 4 September 1944, ADM 199/498; Signal, C-in-C, Plymouth to EG1, 300821B, 30 August 1944, ADM 199/498.

212. "History of Patrol Bombing Squadron One Hundred Five," Enclosure to Lieutenant Commander L.E. Harmon, USN, Commander Patrol Bombing Squadron One Hundred Five, FVPB-105/A12, Serial 015, 5 February 1945, NHC(OA), p. 11. The subsequent US Fleet analysis, however, only gave a "Probably Damaged" assessment. "Anti-Submarine Operations," in "United States Fleet Anti-Submarine Bulletin, February 1945," NHB. It is of interest to note that all other reports of this engagement refer to the use of "bombs."

213. "Narrative," 19 Group, Headquarters, 30 August 1944, AIR 25/497.

214. "Report of Proceedings from 24 August–4 September," Commander C. Gwinner, RN, Senior Officer, 1st Escort Group, RP/10, 4 September 1944, ADM 199/498.

215. "Report of Proceedings from 24 August–4 September," Commander C. Gwinner, RN, Senior Officer, 1st Escort Group, RP/10, 4 September 1944, ADM 199/498.

216. "Report of Proceedings from 24 August–4 September," Commander C. Gwinner, RN, Senior Officer, 1st Escort Group, RP/10, 4 September 1944, ADM 199/498.

217. "Report of Proceedings from 24 August–4 September," Commander C. Gwinner, RN, Senior Officer, 1st Escort Group, RP/10, 4 September 1944, ADM 199/498.

218. "Report of Proceedings from 24 August–4 September," Commander C. Gwinner, RN, Senior Officer, 1st Escort Group, RP/10, 4 September 1944, ADM 199/498. Subsequently, none of Gwinner's attacks were analysed by the U-boat Assessment Committee, and 19 Group noted that Edwards' attack was "disallowed," though the US Fleet analysis was more generous and gave it an assessment "Probably Damaged." "Narrative," 19 Group, Headquarters, 30 August 1944, AIR 25/497; "Anti-Submarine Operations," in "United States Fleet Anti-Submarine Bulletin, February 1945," NHB.

219. Signal, DDIC, Admiralty, to C-in-C, Plymouth, C-in-C, WA, AOC-in-C, Coastal Command, etc., 301350B/August, 30 August 1944, ADM 223/203.

220. "Submarine Warfare in the Channel," Commander J.D. Prentice, RCN, Senior Officer EG11, HMCS *Ottawa*, to Commodore (D), Western Approaches, 17 July 1944, Folder CNA 7-6-1, Vol. 11022, RG 24, LAC.

221. Minute, Commodore G.W.G. Simpson, RN, Commodore (D), Western Approaches, No.DW.40/603.OP, 26 July 1944, File D 01-18-0, Vol. 11575, RG 24, LAC.

222. "Choice of Weapons for "Opening-Up" U-boats," Section V, "The Anti-Submarine Report, September, October, November and December 1945," DTASW, CB04050/45(7), 19 December 1945, NHB.

223. "U-Boat Tactics," Secretary, Navy Board, Melbourne, 22 September 1944, NAA(M): MP1185/8, 1932/3/45.

224. "Experience Gained during Anti-U-boat Inshore Operations," Captain C.D. Howard-Johnston, DAUD, Ref: D.218, 11 September 1944, ADM 223/20, p. 2. [emphasis supplied]

225. See, for example: Marc Milner, "The Dawn of Modern Anti-Submarine Warfare: Allied responses to the U-boats, 1944-45," *RUSI Journal* (Spring 1989), pp. 61-68; and, W.A.B. Douglas, *The Creation of a National Air Force: The Official History of the Royal Canadian Air Force*, Vol. II (Toronto: University of Toronto Press, 1986).

226. Signal, DDIC, Admiralty, to C-in-C, Plymouth, etc., (information C-in-C, WA, AOC-in-C, Coastal Command, etc., 301216B/August, 30 August 1944, ADM 223/203; "State of U-boats in Biscay Area on 31 August 1944," Captain Rodger Winn, RNVR, OIC/SI.1068, 31 August 1944, ADM 223/172.

227. "Narrative," 19 Group, Headquarters, 30 August 1944, AIR 25/497.

228. "[Report of Proceedings, 15th Escort Group, 20 August–31 August 1944]," Commander L.A.B. Majendie, RN, Senior Officer, 15th Escort Group, HMS *Louis*, RP.4/44, 31 August 1944, ADM 199/1462; "Diary of Events, 11th Escort Group," Commander J.D. Prentice, RCN (Temp), Senior Officer 11th Escort Group, File: EG11-0-5, 8 September 1944, ADM 199/1462.

229. "Plymouth Command Diary of Events, 0001-2400, 29 August 1944," ADM 199/1394; Doug McLean, e-mail, 0415, 20 November 2006. For Prentice's determination and experience see: Minute, C.D. Howard-Johnston, DAUD, 12 October 1944, ADM 199/1462; "Channel Hunts," [A/Commander J. Plomer, RCNVR], Tactical Unit, Training Commander, Captain (D), Halifax, Staff Minute Sheet, [August 1944], A/S Warfare, File D 01-18-0, Vol. 11575, RG 24, LAC. Prentice was also one of the few escort commanders to commit to paper his experiences of shallow water operations. See: "Submarine Warfare in the Channel," Commander J.D. Prentice, RCN, Senior Officer EG11, HMCS *Ottawa*, to Commodore (D), Western Approaches, 17 July 1944, Conduct of A/S Warfare Operations, Folder CNA 7-6-1, Vol. 11022, RG 24, LAC. These papers are also in "Hints on Escort Work," 1943-1944, ADM 1/13749.

230. Michael Whitby (ed.), *Commanding Canadians: The Second World War Diaries of A.F.C. Layard* (Vancouver: UBC Press, 2005), p. 200.

231. Commander A.F.C. Layard, Diary, RNM, *passim*.

232. Marc Milner, *The U-boat Hunters: The Royal Canadian Navy and the Offensive against Germany's Submarines* (Annapolis, Maryland: Naval Institute Press, 1994), p. 144.

233. Michael Whitby (ed.), *Commanding Canadians: The Second World War Diaries of A.F.C. Layard* (Vancouver: UBC Press, 2005), p. 200.

234. The initial tasking signal is missing from the file, but see: Michael Whitby (ed.), *Commanding Canadians: The Second World War Diaries of A.F.C. Layard* (Vancouver: UBC Press, 2005), p. 200.

235. Signal, DDIC, Admiralty, to C-in-C, Plymouth, C-in-C, WA, AOC-in-C, Coastal Command, etc.,

271405B/August, 27 August 1944, ADM 223/203; Signal, DDIC, Admiralty, to C-in-C, Plymouth, C-in-C, WA, AOC-in-C, Coastal Command, etc., 311335B/August, 31 August 1944, ADM 223/204.

236. "Thursday, 31 August 1944," in "War Diary (Naval), 17-31 August 1944," Vol. 170, NHB, p. 734.

237. "Narrative," 19 Group, Headquarters, 30 August 1944, AIR 25/497; "No. 53 Squadron Operations Record Book," January to December 1944, AIR 27/506, p. 138.

238. Stephen Roskill, *The War at Sea 1939-1945*, Vol. III, Part II, *The Offensive 1 June 1944-14 August 1945* (London: HMSO, 1961), pp. 179-180; Signal, DDIC, Admiralty, to C-in-C, Plymouth, C-in-C, WA, AOC-in-C, Coastal Command, etc., 281029B/August, 28 August 1944, ADM 223/203.

239. Signal, C-in-C, Plymouth to EG9, 311111B, 31 August 1944, ADM 199/1462; "Plymouth Command Diary of Events, 0001-2400, 31 August 1944," ADM 199/1394.

240. Signal, DDIC, Admiralty, to C-in-C, Plymouth, etc., (information C-in-C, WA, AOC-in-C, Coastal Command, etc., 301216B/August, 30 August 1944, ADM 223/203.

241. "[Report of Proceedings of EG9, 17-28 August and 31 August to 10 September 1944]," Commander Layard, RN, Senior Officer EG9, 10 September 1944, ADM 199/1462.

242. "Work-up Programme – HMCS *Saint John*," Commander R.F. Harris, RCNR and A/Captain W.L. Puxley, RN, Captain (D), Halifax, File D.21-13-2, 5 June 1944, File CS/148-36-4, Vol. 11726, RG24, DHH; "Recommendation for Award, Lieutenant James Richard Bradley, RCNVR, HMCS *Saint John*," Lieutenant Commander W.R. Stacey, 1 September 1944, ADM 1/30162; "Recommendation for Award, AB HSD Lloyd Palmer Haagenson, RCNVR, HMCS *Saint John*," Lieutenant Commander W.R. Stacey, 1 September 1944, ADM 1/30162.

243. Michael Whitby (ed.), *Commanding Canadians: The Second World War Diaries of A.F.C. Layard* (Vancouver: UBC Press, 2005), pp. 185-200, *passim*.

244. Michael Whitby (ed.), *Commanding Canadians: The Second World War Diaries of A.F.C. Layard* (Vancouver: UBC Press, 2005), p. 200; "[Report of Proceedings of EG9, 17-28 August and 31 August to 10 September 1944]," Commander Layard, RN, Senior Officer EG9, 10 September 1944, ADM 199/1462.

245. Michael Whitby (ed.), *Commanding Canadians: The Second World War Diaries of A.F.C. Layard* (Vancouver: UBC Press, 2005), pp. 185-200 *passim*.

246. "Wreck System – Vessel Details Print, Radius 7 Nautical Miles of Wolf Rock," Peter Chew, 2 October 2002, Hydrographic Office.

247. "[Report of Proceedings of EG9, 17-28 August and 31 August to 10 September 1944]," Commander Layard, RN, Senior Officer EG9, 10 September 1944, ADM 199/1462; Michael Whitby (ed.), *Commanding Canadians: The Second World War Diaries of A.F.C. Layard* (Vancouver: UBC Press, 2005), p. 200.

248. "Plymouth Command Diary of Events, 0001-2400, 31 August 1944," ADM 199/1394; "Plymouth Command Diary of Events, 0001-2400, 1 September 1944," ADM 199/1394.

249. QH, otherwise known as GEE, was a radio aid that could be used to fix a ship to within ½ mile at 100 miles from the reference stations. "Navigational Aspect of the Passage and Assault," Director of Radio Equipment (DRE) to Captain Superintendent, Admiralty Signal Establishment, Haslemere, dated 6 September 1944, ADM 1/16664; "Wrecks and other Non-Sub Contacts in British Inshore Waters." E.J. Williams and W.F.F., DNOR, Report No. 86/44, 4 December 1944, ADM 219/166.

250. Michael Whitby (ed.), *Commanding Canadians: The Second World War Diaries of A.F.C. Layard* (Vancouver: UBC Press, 2005), p. 200.

251. The RoP mistakenly states the course as 080°. [Report of Proceedings of EG9, 17-28 August and 31 August to 10 September 1944], Commander Layard, RN, Senior Officer EG9, 10 September 1944, ADM 199/1462. "EI" turns were where ships in line abreast turn in succession in this manoeuvre, commencing with the ship furthest away from centre of the turn. The second ship turns as the first passes approximately astern of it. When all ships have turned, the order of ships in line abreast will have been reversed. All ships will then be proceeding in the new direction at the same interval as before. Most importantly, all of the area in the vicinity of the change of direction will be swept equally, which would not be the case if the ships performed an extended wheel in line abreast formation to alter direction. This explanation is taken from, Doug McLean, "Muddling Through: Canadian Anti-submarine Doctrine and Practice, 1942-45," in M.L. Hadley, *et al*, (eds), *A Nation's Navy: In quest of a Canadian Naval Identity* (Montreal & Kingston, 1996), p. 388.

252. These courses and speeds, are calculated by the author from the various reports. "Report of Attack on U-boat," A/Lieutenant Commander W.R. Stacey, RCNR, Commanding Officer, HMCS *St John*, 1 September 1944, ADM 199/1462.

253. Michael Whitby (ed.), *Commanding Canadians: The Second World War Diaries of A.F.C. Layard* (Vancouver: UBC Press, 2005), p. 166; "Monthly Anti-Submarine Report, July 1944," DAUD, CB 04050/44(7), 15 August 1944, NHB, p. 22; "Summary of Attacks and Evidence Recovered from Submarine believed to be *U-247* ... on 1 September 1944," A/Lieutenant Commander W.R. Stacey, RCNR, HMCS *St John*, 3 September 1944, ADM 199/1462.

254. Signal, C-in-C, Plymouth to EG9, 010119B, 1 September 1944, ADM 199/1462.

255. "Summary of Attacks and Evidence Recovered from Submarine believed to be *U-247*...on 1 September 1944," A/Lieutenant Commander W.R. Stacey, RCNR, HMCS *St John*, 3 September 1944, ADM 199/1462.

256. "Summary of Attacks and Evidence Recovered from Submarine believed to be *U-247* ... on 1 September 1944," A/Lieutenant Commander W.R. Stacey, RCNR, HMCS *St John*, 3 September 1944, ADM 199/1462; "[Track Chart of Attacks on *U-247*]," n.d., Envelope ADU.1561/44, Enclosure Box 704, ADM 199/1462.

257. Michael Whitby (ed.), *Commanding Canadians: The Second World War Diaries of A.F.C. Layard* (Vancouver: UBC Press, 2005), p. 201.

258. "Summary of Attacks and Evidence Recovered from Submarine believed to be *U-247* ... on 1 September 1944," A/Lieutenant Commander W.R. Stacey, RCNR, HMCS *Saint John*, 3 September 1944, ADM 199/1462.

259. "Sinking of *U-247* by *St. John* and *Swansea* 1 September 1944," Anon, n.d., File 81/520 *St. John* 8000, DHH; Michael Whitby (ed.), *Commanding Canadians: The Second World War Diaries of A.F.C. Layard* (Vancouver: UBC Press, 2005), p. 201.

260. "Summary of Attacks and Evidence Recovered from Submarine believed to be *U-247*...on 1 September 1944," A/Lieutenant Commander W.R. Stacey, RCNR, HMCS *Saint John*, 3 September 1944, ADM 199/1462; "Report of Attack on U-boat," A/Lieutenant Commander W.R. Stacey, RCNR, Commanding Officer, HMCS *Saint John*, 1 September 1944, ADM 199/1462.

261. "Summary of Attacks and Evidence Recovered from Submarine believed to be *U-247*...on 1 September 1944," A/Lieutenant Commander W.R. Stacey, RCNR, HMCS *Saint John*, 3 September 1944, ADM 199/1462.

262. Michael Whitby (ed.), *Commanding Canadians: The Second World War Diaries of A.F.C. Layard* (Vancouver: UBC Press, 2005), p. 201; "Obtaining Evidence of a Kill," Section V, "The Anti-Submarine Report, September, October, November and December 1945," Torpedo, Anti-Submarine and Mine Warfare Division, CB 04050/45(7), 19 December 1945, NHB.

263. "Friday, 1 September 1944," in "War Diary (Naval), 1-14 September 1944," Vol. 171, NHB, p. 16.

264. Signal, C-in-C, Plymouth to EG9 and EG11 (R) Monnow, 011300B, 1 September 1944, ADM 199/1462; "Plymouth Command Diary of Events, 0001-2400, 1 September 1944," ADM 199/1394.

265. Michael Whitby (ed.), *Commanding Canadians: The Second World War Diaries of A.F.C. Layard* (Vancouver: UBC Press, 2005), p. 201.

266. [Report of Proceedings of EG9, 17-28 August and 31 August to 10 September 1944], Commander Layard, RN, Senior Officer EG9, 10 September 1944, ADM 199/1462; "English Channel," Chart 2675, 28 May 1982, NHB.

267. Michael Whitby (ed.), *Commanding Canadians: The Second World War Diaries of A.F.C. Layard* (Vancouver: UBC Press, 2005), p. 201; "Précis of Attack by HMC Ships *St John* and *Swansea*, 1 September 1944," U-boat Assessment Committee, Serial No. 8, AUD.1561/44, 9 October 1944, from, "Proceedings of U-Boat Assessment Committee, October 1944–April 1945," Vol. 17, NHB.

268. [Report of Proceedings of EG9, 17-28 August and 31 August to 10 September 1944], Commander Layard, RN, Senior Officer EG9, 10 September 1944, ADM 199/1462; Michael Whitby (ed.), *Commanding Canadians: The Second World War Diaries of A.F.C. Layard* (Vancouver: UBC Press, 2005), p. 201.

269. "Summary of Attacks and Evidence Recovered from Submarine believed to be *U-247*...on 1 September 1944," A/Lieutenant Commander W.R. Stacey, RCNR, HMCS *Saint John*, 3 September 1944, ADM 199/1462.

270. Michael Whitby (ed.), *Commanding Canadians: The Second World War Diaries of A.F.C. Layard* (Vancouver: UBC Press, 2005), p. 201.

271. "Summary of Attacks and Evidence Recovered from Submarine believed to be *U-247*...on 1 September 1944," A/Lieutenant Commander W.R. Stacey, RCNR, HMCS *Saint John*, 3 September 1944, ADM 199/1462. The certificate is reproduced in "Monthly Anti-Submarine Report, October 1944," DAUD, CB 04050/44(10), 15 November 1944, NHB. (See also: ADM 199/2061).

272. [Report of Proceedings of EG9, 17-28 August and 31 August to 10 September 1944], Commander Layard, RN, Senior Officer EG9, 10 September 1944, ADM 199/1462; Michael Whitby (ed.), *Commanding Canadians: The Second World War Diaries of A.F.C. Layard* (Vancouver: UBC Press, 2005), p. 201.

273. "Friday, 1 September 1944," in "War Diary (Naval), 1-14 September 1944," Vol. 171, NHB, p. 16.

274. "Diary of Events, 11th Escort Group," Commander J.D. Prentice, RCN (Temp), Senior Officer 11th Escort Group, File: EG11-0-5, 8 September 1944, ADM 199/1462; "Plymouth Command Diary of Events, 0001-2400, 1 September 1944," ADM 199/1394.

275. Compare the E/S trace in "Echo Sounder Trace, [*U-247*]," n.d., Envelope ADU.1561/44, Enclosure Box 704, ADM 199/1462, with that in "Monthly Anti-Submarine Report, October 1944," DAUD, CB 04050/44(10), 15 November 1944, NHB.

276. "Summary of Attacks and Evidence Recovered from Submarine believed to be *U-247*...on 1 September 1944," A/Lieutenant Commander W.R. Stacey, RCNR, HMCS *Saint John*, 3 September 1944, ADM 199/1462.

277. "No. 547 Squadron Operations Record Book," August 1944 to July 1945, AIR 27/2034, p. 30; "No. 53 Squadron Operations Record Book," January to December 1944, AIR 27/506, p. 138.

278. Signal, C-in-C, Plymouth to EG9 and EG11 (R) *Monnow*, 011300B, 1 September 1944, ADM 199/1462.

279. "English Channel," Chart 2675, 28 May 1982, NHB.

280. Signal, EG9 to *Saint John*, 1545, 1 September 1944, Stacey Papers. The author is most grateful to Lieutenant Commander Ray Stacey, CD (Ret), the son of *Saint John*'s captain, for access to copies of signals relating to this action.

281. Signal, *Saint John* to EG9, 1555, 1 September 1944, Stacey Papers.

282. Signal, C-in-C, Plymouth, to SO EG9, *Saint John* (R) EG11, C-in-C, WA, 011816B, 1 September 1944, Stacey Papers.

283. "9th Escort Group, 17 August to 14 September 1944, Summary of Report of Proceedings," [DAUD], n.d., ADM 199/1462; [Report of Proceedings of EG9, 17-28 August and 31 August to 10 September 1944], Commander Layard, RN, Senior Officer EG9, 10 September 1944, ADM 199/1462.

284. "Plymouth Command Diary of Events, 0001-2400, 1 September 1944," ADM 199/1394.

285. "Summary of Attacks and Evidence Recovered from Submarine believed to be *U-247*...on 1 September 1944," A/Lieutenant Commander W.R. Stacey, RCNR, HMCS *Saint John*, 3 September 1944, ADM 199/1462.

286. Michael Whitby (ed.), *Commanding Canadians: The Second World War Diaries of A.F.C. Layard* (Vancouver: UBC Press, 2005), p. 201.

287. Minute, Chairman, Honours and Awards Committee, 1 March 1945, ADM 1/30162.

288. Signal, Admiralty to *Saint John*, *Swansea* (R) *Verity*, C-in-C, N/A, C-in-C, WA, C-in-C, Plymouth, ANCXF, 201022A, 20 October 1944, Stacey Papers. This signal probably came from Howard-Johnston, Director of the Anti-U-Boat Division and a member of the U-boat Assessment Committee.

289. Signal, *Swansea* to *Saint John*, 241615, 24 October 1944, Stacey Papers.

Index

The contributors to *Fighting at Sea*

Donald E. Graves is one of Canada's best known military historians. A member of the naval history team of the Department of National Defence, Canada, he is the author or editor of 15 books, including *In Peril on the Sea: The Royal Canadian Navy and the Battle of the Atlantic* and *Another Place, Another Time: A U-Boat Officer's Wartime Album*.

A retired officer of the United States Navy, **William S. Dudley** taught history at Southern Methodist University before joining the Naval Historical Center, eventually becoming director of the Center and Director of Naval History for the Chief of Naval Operations. Among William Dudley's published works are *Going South: U.S. Navy Resignations and Dismissals on the Eve of the Civil War* and he also edited a new edition of James Fenimore Cooper's classic sailing navy memoir, *Ned Myers: or, A Life Before the Mast*. William Dudley also served as the editor of the magisterial multi-volume works *Naval Documents of the American Revolution* and *The Naval War of 1812: A Documentary History*.

Professor of Naval History at King's College, London, **Andrew Lambert** has been described by one reviewer as "the outstanding British naval historian of his generation." Among his major books are *Trincomalee: The Last of Nelson's Frigates, The Foundations of Naval History* and a well-received biography of Nelson, *Nelson: Britannia's God of War* (2005). Most recently Andrew Lambert has prepared for re-publication a new edition of William James's landmark work, *Naval Occurrences of the War of 1812*.

Lieutenant-Commander **Douglas M. McLean**, Royal Canadian Navy (Retd.), held a number of operational and staff postings involving anti-submarine warfare during his career. He is the author of numerous articles on the subject, published in *Naval War College Review, Northern Mariner* and *Canadian Military History*, and also contributed to the 1996 anthology *A Nation's Navy*. Lieutenant-Commander McLean is both the editor of and a contributor to *Fighting at Sea*.

Michael Whitby is the Senior Naval Historian of the Department of National Defence, Canada. He is the prime author of *No Higher Purpose*, the first volume of the official history of the Canadian navy in the Second World War, supervised the completion of the second volume, *A Blue Water Navy*, and is currently overseeing the postwar naval history volume. Whitby is also the author of several articles on naval topics published in *Mariner's Mirror, Journal of Military History* (2002 Moncado Prize winner) and *Northern Mariner*. His most recent work is the award-winning *Commanding Canadians: The Second World War Diaries of A.F.C. Layard* (UBC Press, 2005).

Dr. Malcolm Llewellyn-Jones, MBE, is a retired officer of the Royal Navy's Fleet Air Arm currently serving in the Naval Historical Branch, Ministry of Defence, London. He is the author of a number of articles and books: *The Royal Navy and Anti-Submarine Warfare, 1917-1949; The Royal Navy and the Mediterranean Convoys;* and *British Naval Aviation during the Interwar Period, 1919-1939*. His most recent work is *The Royal Navy and the Arctic Convoys*.

Christopher Johnson, whose work has graced more than 30 books of military and naval history, is the cartographer/artist for *Fighting at Sea*. His work can be seen in numerous other Robin Brass Studio publications.